Despite its remote and seemingly rigorous environment, the Antarctic is the world's most important habitat for seals, currently supporting more seals than all other parts of the world combined. As various national Antarctic programmes were established to study these animals, the need to standardize techniques became apparent. This book, arising from work by the Scientific Committee on Antarctic Research (Group of Specialists on Seals), gives a detailed account of well-tried and, where possible, agreed methodologies, techniques, procedures and rationale for the collection and initial analysis of data on the biology and population ecology of Antarctic seals. This volume will not only help facilitate comparisons between different regions of Antarctica, but will also provide a guide for those studying seals in other parts of the world and those carrying out research on other large mammal species.

ANTARCTIC SEALS

ANTARCTIC SEALS

research methods and techniques

Edited by

R.M. LAWS
St. Edmund's College, Cambridge

CAMBRIDGE UNIVERSITY PRESS
Cambridge, New York, Melbourne, Madrid, Cape Town, Singapore, São Paulo, Delhi

Cambridge University Press
The Edinburgh Building, Cambridge CB2 8RU, UK

Published in the United States of America by Cambridge University Press, New York

www.cambridge.org
Information on this title: www.cambridge.org/9780521111768

First published 1993
This digitally printed version 2009

A catalogue record for this publication is available from the British Library

Library of Congress Cataloguing in Publication data
Antarctic seal research methods and techniques/Scientific Committee
on Antarctic Research, Group of Specialists on Seals; edited by
R.M. Laws.
    p. cm.
Includes index.
ISBN 0 521 44302 4 (hardback)
1. Seals (Animals) – Antarctic regions.   I. Laws, Richard M.
II. International Council of Scientific Unions.   Scientific
Committee on Antarctic Research.   Group of Specialists on Seals.
QL737.P64A57   1993
599.74'5 – dc20   92–47467   CIP

ISBN 978-0-521-44302-9 hardback
ISBN 978-0-521-11176-8 paperback

Contents

Contributors

S. S. Anderson
Sea Mammal Research Unit, c/o British Antarctic Survey, Madingley Road, Cambridge CB3 OET, UK.

J. L. Bengtson
NMFS National Marine Mammal Laboratory, 7600 Sand Point Way N. E., Seattle, Washington 98115, USA.

M. N. Bester
Mammal Research Institute, University of Pretoria 0002 Pretoria, South Africa.

W. N. Bonner
SCAR, c/o Scott Polar Research Institute, Lensfield Road, Cambridge CB2 1ER, UK.

W. M. Brown
Division of Biological Sciences and Museum of Zoology, University of Michigan, Ann Arbor, Michigan 48109, USA.

D. Costa
Long Marine Laboratory, Institute of Marine Sciences University of California, Santa Cruz California 95064, USA.

J. P. Croxall
British Antarctic Survey, Madingley Road, Cambridge CB3 OET, UK.

T. E. Dowling
Department of Zoology, Arizona State University, Tempe, Arizona 85287-1501, USA.

A. W. Erickson
School of Fisheries, University of Washington Seattle, Washington 98115, USA.

M. A. Fedak
Sea Mammal Research Unit, c/o British Antarctic Survey, Madingley Road, Cambridge CB3 OET, UK.

R. L. Gentry
NMFS National Marine Mammal Laboratory, 7600 Sand Point Way N. E., Seattle, Washington 98115, USA.

J. Harwood
Sea Mammal Research Unit, c/o British Antarctic Survey, Madingley Road, Cambridge CB3 OET, UK.

R. J. Hofman
Marine Mammal Commission, 1825 Connecticut Avenue, N. W., Washington DC 2009, USA.

R. M. Laws
St. Edmund's College, Cambridge CB3 OBN, UK.

T. S. McCann
c/o British Antarctic Survey, Madingley Road, Cambridge CB3 OET, UK.

P. D. Shaughnessy
CSIRO, Division of Wildlife Ecology PO Box 84, Lyneham ACT 2602, Australia.

A. A. Sinha
Veterans Hospital, 54th & 48th Avenue, South, Minneapolis, Minnesota 55417, USA.

D. B. Siniff
Department of Ecology and Behavioral Biology 109 Zoology Building, University of Minnesota Minneapolis, Minnesota 55455, USA.

I. Stirling
Canadian Wildlife Service, 5320 122 St., Edmonton Alberta, Canada.

Preface

The need for a handbook to standardize techniques for studying Antarctic seals was recognized during the initial meetings of the SCAR Group of Specialists on Seals. Even during various discussions between experts about the results from various national Antarctic programmes, it became clear that a lack of standardization sometimes made it difficult to make direct comparisons between studies. For example, even apparently small variations, such as the use of different types of tags placed in flippers for long-term identification of individuals had the potential to cause differences in estimates of population parameters. The problems of lack of standardization in data collection were even more critical when less experienced persons were involved. In remote areas such as the Antarctic where the number of personnel on a base or a ship may be limited, scientific and non-scientific staff with no training in pinniped research are sometimes required to record various types of observations on seals. Similarly, from time to time, non-experts may have unique opportunities to record very important observations on seals. Without a standardized format that could be readily available to everyone; it was clear that valuable opportunities and data would be lost and the full scientific value of some studies might not be realised. Consequently, the SCAR Group of Specialists on Seals undertook to produce a handbook of standardized techniques for the study of Antarctic seals, the benefits of which would also carry over to the study of Arctic seals and facilitate more bi-polar comparisons.

Completion of this volume has taken longer than was originally envisaged. Meetings of the Group of Specialists were often dominated by current issues and needs which, at the time, seemed more pressing. The limited resources that were available for this task also meant that everyone had to make sacrifices of their own time and funds to facilitate completion of the handbook. Similarly, as the volume was being developed, rapid advances were being made in several areas such as immobilization techniques, population estimation and genetics; to mention just a few. Consequently, the chapters have all undergone several revisions to make them as complete, practical and relevant as possible.

The resulting volume is a unique and comprehensive work that covers the techniques, procedures and rationale for the collection of data on the population ecology and biology of Antarctic seals. It will be of value for years to come as new and old seal scientists alike design and execute their research. It will be essential for less formally trained observers who, nonetheless, are required, or simply have the opportunity, to record important observations.

Recently, Dr Richard Laws stepped down as Convenor of the SCAR Group of Specialists on Seals. His contributions to the study of Antarctic biology as a whole, and the study of seals in particular, have been enormous. His critical eye and extensive experience that guided the production of this volume have significantly contributed to the quality of the final product. Recently, I became Convenor of the Group of Specialists and in that capacity I take great pleasure in congratulating Dr Laws and the other authors on the production of this valuable book.

Convenor   *Donald B. Siniff*
SCAR Group of Specialists on Seals

Introduction

R.M. LAWS

The Antarctic Treaty System – the Antarctic Treaty and related international conventions – provides a legal and diplomatic regime for the
Antarctic region. Its non-governmental counterpart, which is responsible
for promoting scientific co-operation, is the Scientific Committee on
Antarctic Research (SCAR) of the International Council of Scientific
Unions (ICSU). SCAR is the single international, interdisciplinary, non-
governmental organization which can draw upon the experience and
expertise of an international mix of scientists across the complete scientific spectrum. For over 30 years SCAR has provided scientific advice
to the Antarctic Treaty System. The membership of SCAR comprises
the National Committees of those national scientific academies or
research councils which are adhering bodies to ICSU and which are, or
plan to be, active in Antarctic research, together with the relevant Scientific Unions of ICSU. SCAR meets every two years to conduct its adminis-
trative business and agree policy and strategy. The majority of the
scientific work of SCAR, however, is carried out by the Working Groups
and Groups of Specialists. Groups of Specialists are created by SCAR in
response to specific scientific problems and their members are appointed
for the experience and expertise they can bring to the group. Laws (1986)
has reviewed these arrangements in relation to Antarctic conservation
and the Antarctic Treaty System.

The populations of Antarctic seals are believed greatly to outnumber
all other seal populations in the world (Laws, 1984) and following the
decline of the great whales are currently the largest group of animals
feeding on krill – the staple food of most vertebrates in the Southern
Ocean. The SCAR Group of Specialists on Seals, which developed from
a subcommittee of the SCAR Working Group on Biology, has a dual role
in research and management. It was formally constituted in 1974 to

enable SCAR to discharge its responsibilities under the Convention for the Conservation of Antarctic Seals (CCAS), part of the Antarctic Treaty System, for under this convention, SCAR has a significant role as an independent source of scientific advice (Articles 3–5). It is probably unique for a non-governmental body to be specified to fill such a role in an international convention of this kind.

This convention (the text of which is given in appendix 16.5) arose because the conservation measures (the Agreed Measures for the Conservation of Antarctic Flora and Fauna, 1964; text given in appendix 16.4) and other arrangements under the Antarctic Treaty could not give any protection to seals in the sea or on floating ice (Laws, 1986). This is owing to the fact that the Antarctic Treaty expressly reserves the rights of states under international law with regard to the high seas. CCAS was signed in 1972 and entered into force in 1978. It applies to all species of Antarctic seals and SCAR seal biologists played an important part in bringing it about.

The SCAR Group of Specialists on Seals has the following terms of reference:

1. To encourage the exchange of scientific data and information, to recommend research programmes, to co-ordinate seal research undertaken by SCAR nations, with particular reference to marking programmes, and to encourage standardization of techniques.
2. To scrutinize figures for the number of seals killed and to compile periodic summaries of these for publication.
3. To review the status of Antarctic seal stocks. If commercial sealing begins: to analyse catch returns, estimate and report to SCAR the dates by which the permissible catch limits are likely to be exceeded; to report to SCAR when the harvest of any species of seal is considered to be having a significantly harmful effect on the total stocks of such species, or on the ecological system in any particular locality.
4. To consider what statistical and biological information should be collected by sealing expeditions, and if necessary to make arrangements for its processing and analysis; to liaise with biologists accompanying sealing expeditions.
5. To report on methods of sealing and to make recommendations with a view to ensuring that the killing or capturing of seals is quick, painless and efficient.

6. To recommend amendments to the Annex to the Convention on the Conservation of Antarctic Seals.
7. To maintain liaison with international organizations concerned with marine mammals, such as IUCN and FAO.

Thus, through SCAR the group would, in the unlikely event of commercial sealing starting in the Antarctic, provide initial advice to control such exploitation, until the governments participating in CCAS established their own scientific advisory committee. The work of the group also has relevance to the aims of the Convention for the Conservation of Antarctic Marine Living Resources (CCAMLR) (appendix 16.6), because of the key role of seals in the Antarctic marine ecosystem. The group has responded to requests for advice from the scientific committee of CCAMLR, particularly in respect of ecosystem monitoring and the effects of marine debris.

In order to discharge these responsibilities the Group of Specialists on Seals has held eleven meetings since 1974 and has discussed problems by correspondence.

Thus, the group is concerned about research and conservation issues that focus on the seven Antarctic and sub-Antarctic seal species. Four of these species, the Ross, Weddell, leopard and crabeater seals, occupy the pack ice zone. The southern elephant seal and two species of fur seals have a more northerly distribution and breed mainly on the peripheral Antarctic and sub-Antarctic islands (Laws, 1984). The Antarctic seals are important in global terms for they are thought to be more abundant, in terms of both numbers and biomass, than all other seals in the world combined. The Group of Specialists is interested in research on all aspects of the biology of Antarctic seals: population ecology, behaviour, reproduction, growth and age, feeding and diet, and energetics. A major objective for Antarctic seal research is to use measures such as growth rates, reproductive rates, survival, foraging areas, feeding depths, energetics and general health, as indicators of ecosystem conditions. This is very relevant to the work of CCAMLR. Although a modern Antarctic sealing industry has not developed, the work of the group is particularly important in the face of commercial harvesting of fish and krill, and the increasing pressures of tourism and other human activities.

The present publication arises as a result of activities of the group relating to the first, fourth and fifth terms of reference, particularly the desirability of standardizing techniques and co-ordinating seal research. The summaries called for under the second term of reference have been published in the *SCAR Bulletin* (*Polar Record*, vol. 16 (101), pp. 343–5;

vol. 16 (105), pp. 901-2; vol. 18 (114), pp. 318-20; vol. 20 (125), pp. 195-8; and vol. 23 (146), pp. 622-7). These cover the period 1964-85. Subsequent summaries have been published in the reports of meetings of the SCAR Group of Specialists on Seals. The third term of reference – the review of the status of Antarctic seal stocks – is addressed at meetings of the Group and conclusions are published in the reports of these meetings.

The Group submitted a detailed report to the first Review Meeting held under the CCAS, in London, 12-16 September 1988. This had two purposes, first to describe how SCAR, through the Group of Specialists on Seals, has discharged its responsibilities under Article 5 of the Convention and paragraph 7 of its Annex. Secondly, the report offered information and views on scientific matters relating to the Convention and its Annex, and to the Agenda and papers for the Review Meeting.

Some published reviews of the biology of Antarctic seals are by Ridgway & Harrison (1981), King (1983), Laws (1984), and Reidmann (1990). Comprehensive information on diving physiology is given by Kooyman (1981), and Laws (1985, 1989) has reviewed the broad environmental and ecological background to studies of Southern Ocean ecosystems. The present publication does not address biological research on the Antarctic seals as such, but is specifically concerned with recommending tried, and where possible agreed, methodologies and techniques to be used in such research. There is an intentional bias towards field work, but relevant, sophisticated laboratory methods are also included. Thus, it is hoped that it will be helpful to non-specialists, who have the opportunity to make useful contributions. It also draws attention to new and developing techniques. We hope that it will encourage research, promote the adoption of standard techniques and the development of better ones. We expect that the information will also be useful to workers on other mammalian groups.

Following this introductory chapter, there is advice on the field identification of seven species: the southern elephant seal, *Mirounga leonina*; Weddell seal, *Leptonychotes weddellii*; Ross seal, *Ommatophoca rossii*; crabeater seal, *Lobodon carcinophagus*; leopard seal, *Hydrurga leptonyx*; Antarctic fur seal, *Arctocephalus gazella*; and the sub-Antarctic fur seal, *Arctocephalus tropicalis*. From the information and illustrations given, distance identification of adults should be possible; other information is provided to enable confirmation of the species in doubtful cases, if a closer approach is possible, if the animal is dead, or if only a skeleton or skull is found.

A knowledge of animal abundance is vital to all ecological studies and is particularly important for determining population trends and to formulate management and conservation policies; this is addressed in chapter 2. Ground, ship-based and aerial counting methods and population estimation are different for land-breeding and ice-breeding species. Mark–recapture techniques are also described in this chapter. Methods of extrapolating from density estimates to total population size are also treated, taking account of practical considerations relating to the polar environment.

For a number of purposes connected with research it may be necessary to restrain or immobilize a seal and this is dealt with in chapter 3. The categories of drugs available and the best methods of administering them are described, with special reference to both southern phocid seals and the fur seals. Other methods of capture, such as the use of a restraining sack may be adequate or appropriate for certain studies and are described.

Chapter 4 deals with marking techniques and programmes including: branding (hot-iron, freeze and explosive), punching and tattooing, tagging with plastic and metal tags, vital staining to establish rate of deposition of layers in teeth, the use of natural marks, paint, dyes and hair clipping to identify individuals. Some recommendations are made for standardizing numerical marking. A central seal tagging database is maintained through the Group of Specialists.

Rapid advances in the field of electronics in recent years, in particular the miniaturization of solid state circuitry through semiconductor technology, has made it feasible to deploy highly sophisticated instruments on free-ranging animals. Chapter 5 briefly outlines the types of instruments deployed in research. The main categories of instrument described are radio-telemetry (transmitters, receivers and recorders), sonic devices (recording vocalizations, sonic transducers and transponders), self-contained recorders (time-depth recorders, gastro-thermo recorders, geolocation by daily light levels, satellite linkage and transmission). A section on attachment methods is also included (harnesses, bracelets, glues and epoxy resins).

This naturally leads on to behaviour studies and in chapter 6 recommendations are made, mainly on the terrestrial aspects of seal behaviour (aquatic activities requiring telemetry are discussed in chapter 5). Behavioural observations are very relevant to interpreting population ecology as well as several other aspects of seal biology. Recommendations cover necessary background information that should be collected, recording vocalizations in air and underwater, detailed intraspecific behaviour, including territorial and reproductive activities, and inter-specific behaviour. Sampling methods and techniques for recording data are discussed.

For some research studies, however, seals have to be killed, and one of the terms of reference of the Group of Specialists was to make recom-mendations with a view to ensuring that the killing or capturing of seals by sealing expeditions or by research workers is quick, painless and efficient. Chapter 7 gives recommended methods of shooting, chemical euthanasia and clubbing. Under the Agreed Measures for the Conser-vation of Antarctic Flora and Fauna and the CCAS, seals can only be killed or captured under permits issued by national authorities, and the results are included in the exchange of information under Article XII of the Agreed Measures and Article 5 of CCAS.

If an animal is killed, whether for scientific or other purposes (for example for food for people or dogs in the Antarctic) it is important to try to make the fullest possible use of it that time and resources permit. However, there is little point in people collecting material which they do not intend to process themselves, without having made prior arrange-ments for its subsequent treatment. Some analyses are time-consuming or expensive, or both, and it may not be possible to find another research worker to process the material. Much time and effort can be saved by undertaking this preliminary advance planning. Chapter 8 advises on four basic collection needs most relevant to population ecology in research and management. The fundamental data to be recorded for each specimen include the recommended basic measurements and weights, the basic material to be collected and the recommended preservatives. The need to minimize the risk of infection (such as 'seal finger') is emphasized. Detailed instructions are given for collecting material for age determination (teeth and toe nails) which is fundamental to most studies. Priorities and methods of collection for skeletal material are recommended (skull, skeleton and baculum). Priority for collecting reproductive specimens is given to female material because it contributes more to the understanding of essential population parameters such as pregnancy rates and age at first reproduction. (More specialized require-ments are given in some later chapters, for example chapters 9 to 13.)

Chapter 9 is concerned with genetic-based studies which contribute to meeting the basic need of distinguishing between stocks, because both in terms of basic population biology and for management studies it is desirable or essential to establish the unity and integrity of the stocks concerned. Sample size and related considerations are discussed, and an outline of methods of collecting and preparing blood, plasma and serum samples is given. Gel eletrophoresis techniques are referenced. The analysis of polymorphic data is considered and an expression to estimate the genetic similarity of two populations is given.

With the increasing concern about pollutants in the environment and the demonstration of accumulation of such substances in the tissues of Antarctic animals, opportunity may be taken to collect samples for determination of pollutant levels, when seals are killed for other reasons. The contaminants include organochlorine residues and heavy metals, and recommendations for collecting and preserving samples are given in chapter 10.

As mentioned earlier, age determination is a basic tool in population biology research on mammals, particularly in studies of body growth rates, age-specific reproductive rates and other aspects of population dynamics. Chapter 11, on age determination, concentrates on the best proven method, depending on incremental lines in the teeth. The tooth structure and methods of preparing sections for examination are described, with literature references. There is a need to reduce confusion and ambiguity, particularly in relation to complex dentine structures; cementum has a much simpler structure. Aspects covered include: collection and storage, use of external and internal structures (direct sectioning, etching, decalcification and staining). Because of its importance the recommended methodology for preparation and reading is given for each species. The methods are not without error, and reliability and validation are specifically addressed. Other methods described more briefly are nail markings, laminated bones, body length, eye lens weight, suture closure, baculum development, ovarian structure (treated in greater detail in chapter 12), pelage and general appearance.

An understanding of the salient features of the reproductive cycle is fundamental to several aspects of seal biology, especially population ecology and behaviour. Reproductive status, pregnancy rate and reproductive success, and changes in the age at maturity are examples of the kind of quantitative data it is necessary to acquire. Chapter 12 provides an introduction to the seal reproductive cycle and organs of reproduction which should enable newcomers to the subject to find their way about. First the male reproductive system, including the gross and fine anatomy of the testis, epididymus, prostate, penis and baculum (the penis bone), is described. Then the fine and gross anatomy of the female reproductive tract – ovaries (and their follicles, corpora lutea, the corpus albicans and its regression) the uterus and vagina. Delayed implantation of the blastocyst, foetal growth and the estimation of foetal age, and the phenomenon of neonatal gonad hypertrophy are also covered. Quantitative aspects relate to the estimation of pregnancy rates, the secondary sex ratio, puberty and sexual maturity. A standard method of calculating the average age at sexual maturity is given so that meaningful comparisons between populations and time periods can be made. Recent developments in DNA technology have revealed very powerful methods of studying relatedness between individuals, with a resolution capable of accurately assigning a pup to one or both parents. This is known as DNA fingerprinting and is particularly important in determining paternity and hence reproductive success of individual males. It also has other applications such as the serial identification of individuals. These methods are summarized in this chapter. Finally some basic histological and analytical methods are given, including specific instructions for collecting material for electron microscopy.

Estimation of food consumption is a very important aspect of Antarctic seal studies because of the history of perturbations and interactions in the Southern Ocean ecosystem (Laws, 1985, 1989). It is usually approached by combining dietary data, dealt with in chapter 13, with metabolic energy requirements, considered in chapter 14, but much research still needs to be done before we have adequate quantitative information on the diet and its quality in energetic terms.

Chapter 13 recommends methods of obtaining, recording and preserving material suitable for these studies, from complete stomachs and intestines of killed seals, partial samples from stomachs of live seals, and the collection of faecal droppings or regurgitations. The ultimate aim is to record the weight and/or volume of individuals of each prey species, together with information on their size, age and reproductive status. Recommended methods of sorting and identification of food samples (including fragmentary material such as eyes, otoliths and beaks of crustaceans, fish and squid) are given with appropriate literature references. Biochemical techniques have also been used for very detailed studies. Finally, the problems and biases likely to be encountered in quantifying and interpreting the results of these studies, and ways of compensating for them, are outlined. An index of relative importance (IRI) and a modified volume index (MVI), which are useful compromises, are described and compared.

Chapter 14 is concerned with bioenergetics, which provides a common currency for comparative studies and is a powerful method of quantifying many aspects of animal life histories. The kinds of questions that may be addressed in such studies are outlined, namely: the costs of reproduction in the two sexes, the energetics of foraging and year to year changes in energy availability and use. The methods described include: estimation from change in mass (converted to energy terms); radioisotope dilution method, based on the difference in water content of lean and adipose tissue; measurement of correlates of energy expenditure, by means of activity budgets and heart rate telemetry; the relation between food intake and water influx, using radioisotopes; swimming energetics using time-depth recorders coupled to heart rate telemetry; the assessment of animal condition in terms of energy stores, using blubber layer thickness and radioisotope dilution to estimate water; and energy consumption, also using radioisotope methods.

The last chapter is concerned with the development of techniques for research. The objectives and the types of research programmes needed to meet them are outlined. They relate both to improved understanding of the basic biology of seals and to meeting the perceived needs of two international conventions within the Antarctic Treaty System, on the conservation of seals and other elements in Antarctic marine ecosystems (appendices 16.5, 16.6). The suggested research programmes are broadly framed so as to indicate where methodology and techniques particularly need to be improved. Emphases may well change as existing technical problems are solved, new research problems arise and new opportunities are grasped.

Finally, there are seven appendices covering the origins of the scientific names of the species of Antarctic seals, their vernacular names in different languages, and the scientific names of mammal species referred to, as well as the texts of the Agreed Measures for the Conservation of Antarctic Flora and Fauna, the Convention for the Conservation of Antarctic Seals, the Convention for the Conservation of Antarctic Marine Living Resources and the SCAR Guidelines on Animal Experimentation in Antarctica.

The changing membership of the Group of Specialists on Seals over the period of preparation of this publication included: J.L Bengtson, M.N. Bester, I.L. Boyd, H. Burton, J.P. Croxall, A.W. Erickson, R.M. Laws, A.W Mansfield, Y. Naito, T. Øritsland, L.A. Popov, D.B. Siniff, D. Torres Navarro and D. Vergani. R.M. Laws and D.B. Siniff were respectively Convenor and Deputy Convenor of the Group until 1988, when D.B. Siniff became Convenor; J.L. Bengtson was Secretary from 1988.

Acknowledgements

In addition to the authors of the chapters, other scientists who have contributed to the preparation of this publication, by participating in meetings or by correspondence are: W. Amos, G.A. Antonelis, B. Bergflodt, P.B. Best, I.L. Boyd, H.W. Braham, B. Mate, D. Miller, A.J. Nel, M.J. Oren, M.R. Payne, J. Ploetz, J.H. Prime, R.W. Davis, C.W. Fowler, P. Hammond, L. Hiby, R.L. De Long, D.P. De Master, H. Kajimura, K.R. Kerry, G. Kooyman, T.R. Loughlin, P.A. Prince, C.A. Repenning, G. Ross, W.R. Siegfried, C.F. Summers, J.W. Testa, W. Vaughan, A.E. York. The contributors and I wish to thank them all for their help. The assistance of P.D. Clarkson is much appreciated.

Participants in this project wish to express their gratitude for financial support from British Antarctic Survey, the Scientific Committee on Antarctic Research (SCAR), the International Union for the Conservation of Nature and Natural Resources (IUCN), the Marine Mammal Commission of the United States of America (MMC), the National Science Foundation, Division of Polar Programs, USA, the Ford Foundation, the Food and Agriculture Organisation of the United Nations (FAO), the United Nations Environment Program (UNEP) and the Trans-Antarctic Association.

References

King, J.E. 1983. *Seals of the World*. British Museum (Natural History), London.
Kooyman, G.L. 1981. *Weddell Seal: Consummate Diver*. Cambridge University Press, Cambridge.
Laws, R.M. 1984. Seals. In *Antarctic Ecology*, vol. 2, ed. R.M. Laws, pp. 621–715, Academic Press, London.
Laws, R.M. 1985. Ecology of the Southern Ocean. *Am. Sci*, **73**, 26–40.
Laws, R.M. 1986. *Animal conservation in the Antarctic*. Zoological Society of London, Symposia, no. 54, 3–23.
Laws, R.M. 1989. *Antarctica: the Last Frontier*. Boxtree, London.
Reidmann, M. 1990. *The Pinnipeds: Seals, Sea Lions and Walruses*. University of California Press, Berkeley.
Ridgway, S.H. & Harrison, R.J. 1981. *Handbook of Marine Mammals*, vol. 2, *Seals*. Academic Press, London.

1

Identification of species

R.M. LAWS

Introduction

The Order Pinnipedia is related to the Carnivora and was formerly classified within that Order. It includes three Families: the Otariidae or eared seals, including fur seals and sea lions; the Odobenidae, including the Atlantic and Pacific walruses; and the Phocidae or true seals, including the Phocinae (northern species) and the Monachtinae (Antarctic ice breeding seals and the elephant seals).

Of seven species of seal considered here, three (elephant seal and two species of fur seal) tend to be sub-Antarctic in distribution, but penetrate further south, particularly in the maritime Antarctic. The other four (Weddell, Ross, crabeater, leopard) are primarily fast ice or pack ice animals. The Weddell (at South Georgia only) and leopard seals also have sub-Antarctic populations. In elephant and fur seals, the males are much larger than the females; in the other species the females are slightly larger than the males (Laws, 1984). There is no sexual dimorphism of colour or pattern. The elephant and fur seals are grey brown, lighter ventrally but with no patterned markings (except small, lighter scars); the other species are all strikingly marked with spots, stripes or blotches, particularly when freshly moulted. Juvenile and adult coats are similar; pups of elephant and fur seals are born with uniform black fur or lanugo; in other species the newborn coat is mottled greyish brown (Weddell), pale 'milk coffee' coloured (crabeater) or strongly patterned like the adult (leopard, Ross). The degree of wetness of the fur or hair alters the colouration, generally making it darker; there may be a sharp colour contrast between wet and dry parts. With time elapsed since the moult the coat colour becomes browner (often creamier in the crabeater) and in patterned species this lessens the contrast of the pattern. From a distance,

1

body size and shape, the profile of head and body, are useful for distinguishing adults. The behaviour on disturbance is also characteristic – in the pack ice it may be necessary to wake sleeping seals by a blast from the ship's siren in order to identify them.

The seven species (Fig. 1.1) are dealt with below, with illustrations drawing attention to characteristic features, such as size, shape, colour and behaviour. Sufficient information is given so that distance identification of adults should be possible but this is difficult if the animals do not move. Other information is given, so that in doubtful cases if a

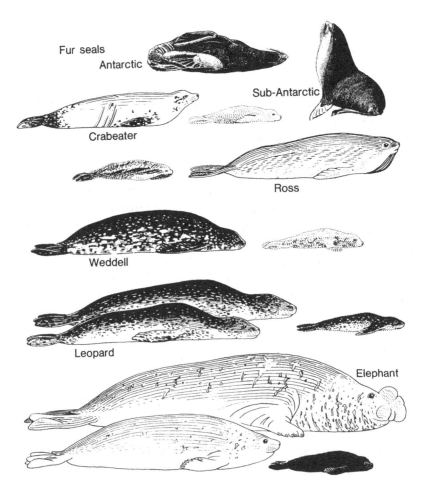

Fig. 1.1 Antarctic seals drawn to same scale for comparison of size, form and colour patterns.

close approach is possible, or if the seal is collected or has died naturally the identification may be confirmed, using additional criteria. Skulls are illustrated by species so that skeletons may be identified. (See Figs. 1.2–1.14.)

Finally, a condensed summary key is presented which may be helpful to observers with or without any biological training.

Elephant seal, *Mirounga leonina*, **(Linnaeus, 1758).**
Figs. 1.2, 1.3.

Fig. 1.2 Elephant seal characteristic attitudes.

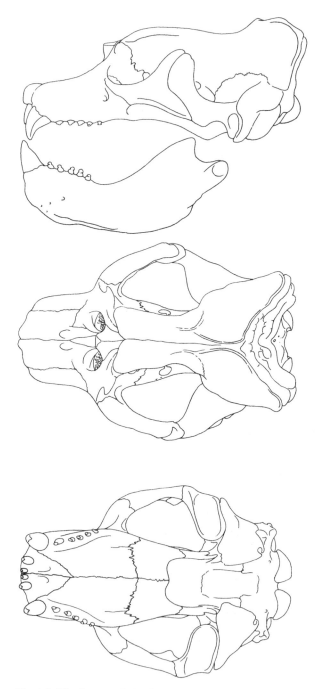

Fig. 1.3 Elephant seal skull.

Profile

Heavily built, neck inconspicuous, large thorax. The species is strongly sexually dimorphic – males after the first three years being conspicuously larger than females (Laws, 1953). The adult male has a conspicuous proboscis that enlarges during the breeding season. Females and younger males have a more pointed snout than the other Antarctic phocid seals.

Size

Newborn: length 127 cm; weight 40–46 kg.
Adult female: length 2.6–2.8 m; weight 400–900 kg.
Adult male: length 4.2–4.5 m; weight 3000–4000 kg.

Colour

Dark grey, lighter ventrally, fading to various shades of brown. No superimposed pattern of spots or other markings, but adults have scarring about neck and chest, adult females a lighter yoke around neck from bites during mating.

Pup

Black neonatal fur (lanugo), fading to very dark brown; in South Orkney Islands about 3% have prenatal moult of lanugo.

Yearling

Medium grey, lighter ventrally with yellowish staining.

Behaviour

Virtually never observed at sea even around South Georgia where the stock is about 360 000. In water near shore often floats with head and hind flippers clear of water. Submerges tail first; head pointed up and withdrawn vertically. Rarely seen in pack ice; if so, only solitary.

Breeding

On land in dense aggregations in spring. Dominant and subordinate males maintain position among breeding cows. Challenging roars, rear up in threat displays, fights.

Moulting

In summer in dense close packed aggregations, inactive for 30–40 days often in muddy wallows. From December to April; females and juveniles first, breeding bulls last.

Sand flipping

Reaction to heat and dryness; also a displacement activity.

Movement

Clumsy lumping motion, fore flippers spread out to lift body, then pelvic thrust to straighten out – like quasi-'looper' caterpillar; moves in short bursts of activity with frequent halts; often stops by falling on chest. Can back away also easily and quickly. To turn often arches tail and head upwards and pivots on belly, swivelling around with aid of fore flipper.

Vocalization

Large males have deep resonant pulsed roar; female barks, moans and howls. High pitched bark for short time after birth of pup. Both sexes produce a variety of belching noises.

Close-up

Teeth

Large canines especially in males; small peg-like post-canines (see Fig. 1.3).

Skull

Overall length up to: male, 561 mm; female, 333 mm.

Fore flippers

Relatively small in relation to body length; with large nails (to *c.* 5 cm × 1 cm).

Hair

Stiff, short; moult of epidermis and hair in large patches.

Fig. 1.4 Weddell seal characteristic attitudes.

Weddell seal, *Leptonychotes weddellii*, **(Lesson, 1826).**
Figs. 1.4 and 1.5.

Profile

Large, heavy 'barrel-shaped' body with relatively small head, moderate
snout and no distinct neck.

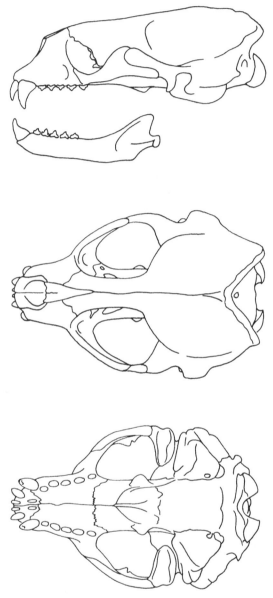

Fig. 1.5 Weddell seal skull.

Size

Newborn: length 120 cm; weight 25 kg.
Adult: length to 3 m; weight 500 kg.

Colour

Typically dark, slightly lighter ventrally, mottled with large darker and lighter patches, white patches predominant ventrally. Some individuals lighter generally, but never bleaches to whiteness of crabeater.

Pup

Soft grey, grey brown or golden, often with indistinct darker mottling.

Moult

December to March.

Behaviour

Inshore distribution on fast ice and pack ice. Forms large colonies (of many hundreds) on inshore fast ice during breeding season, September to November, often in hummocked ice and near tide cracks. Individuals relatively evenly spaced, not close. Small relict breeding population of about 100 based on Larsen Harbour, South Georgia. In pack ice found usually singly, normally on large smooth floes, occasionally in association with crabeater seals. Hauls out on beaches or fast ice during summer, usually singly or in small scattered groups, occasionally in close aggregations of up to 60.

In water

Floats vertically in leads with head pointing upward or level. Submerges by sinking tail first without showing back. May float at an angle, particularly in leads or at breathing holes, or horizontally in larger pools with much of the back exposed. Adult males defend three dimensional territories in water under breeding colonies.

Disturbed

Usually rolls on side with flipper raised in 'salute' (crabeaters also occasionally do this). Usually travels slowly on ice, jerking itself forward with small 'humping' movements with both fore flippers together, but pressed to sides; head usually slides along the ice on its chin. (But occasionally moves by lateral movements like crabeater.) When wishing to turn usually rolls until facing in right direction. May make 'ice sawing' movements of head and neck on ice – a displacement activity. Often makes dull glottal 'clapping' or gulping sound in throat while opening and closing mouth.

Close-up

Teeth

Incisors and canines often greatly worn; upper incisors procumbent for ice sawing; teeth blunt, molars with central prominent point, smaller one behind (Fig. 1.5).

Skull

Overall length up to 295 mm.

Fore flippers

Relatively small, centrally placed on body.

Underwater sound

Underwater calls readily audible to a person on surface of ice; highly vocal in repertoire and volume; pulsed trilling and explosive glottal sounds.

Ross seal, *Ommatophoca rossii*, Gray, 1844. Figs. 1.6 and 1.7.

Profile

Plump and rather shapeless; at a distance superficially like elephant seal and may be confused with it. Head short and wide; can be withdrawn into rolls of fat about neck; no external trace of neck. Short snout gives blunt pointed profile like elephant seal. Raises head near vertically, chest enlarged, back arched.

Size

Newborn: length 105–120 cm; weight 27 kg.
Adults: length to 2.6 m; weight to 225 kg.

Colour

Dark grey to chestnut dorsally, with little spotting and sharp line of demarcation from buff underside. Light and dark pattern about the eyes gives mask-like appearance. Often broad dark stripes from chin to chest, and on side of head; spotted or obliquely streaked on sides and flanks;

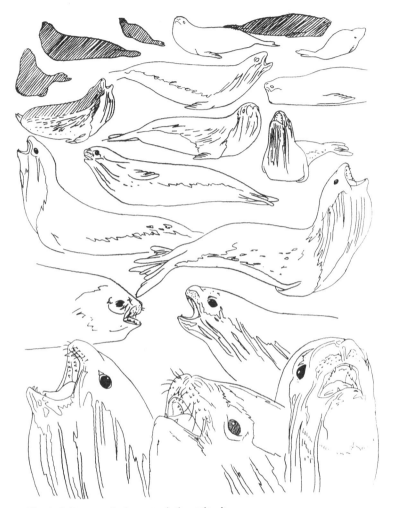

Fig. 1.6 Ross seal characteristic attitudes.

fore and hind flippers dark. Most adults have small pale scars about neck and shoulders.

Moult

In January and February.

Pup

Dark brown fur (Vallette, 1906), dark dorsally, light ventrally, with striping on the throat similar to adult (Thomas *et al.*, 1980), not white as described by Scheffer (1958). Pups rarely observed.

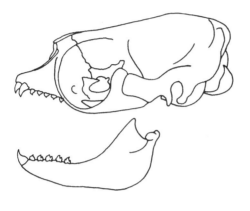

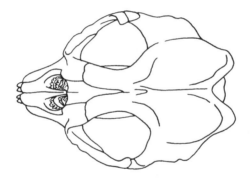

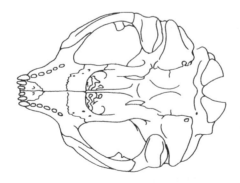

Fig. 1.7 Ross seal skull.

Behaviour

Usually in medium to close pack ice (see chapter 2), where they prefer smaller, hummocked floes in January and February; occasionally fast ice. Large concentrations described from 12°E to 6°W (Hall-Martin, 1974; Wilson, 1975). Usually solitary, but rarely in groups of up to five on single floe (Bonner & Laws, 1964). Placid disposition, easily approached even by helicopter. Pupping dates uncertain, probably November to December; mating late December to early January.

Disturbed

Raises head near vertically, back arched, inflates trachea and soft palate with air and with mouth open makes unique 'trilling, cooing, chugging' sounds. Usually travels slowly on ice, head raised and using fore flippers together, 'humping' like Weddell seal. (But can move fast, thrashing hind flippers from side to side with fore flippers pressed against body.)

Close-up

Teeth

Small, sharp canine teeth and very small post-canines, in great contrast to other species (see Fig. 1.7).

Skull

Overall length up to 265 mm.

Mouth

Relatively small, conspicuously pink inside, eyes large, protruding.

Fore flipper

Small size, well forward on body.

Hind flipper

Digits with long cartilagenous extensions, appear to drag behind.

Underwater sound

Can produce pulsed sounds, but characteristically unpulsed moan or 'buzzing' of varying frequency.

Hair

Short, velvety in texture, including flippers.

Crabeater seal *Lobodon carcinophagus*, (Hombron & Jacquinot, 1842). Figs 1.8, 1.9.

Profile

Relatively slim, lithe and streamlined, snout elongated, slightly tip-tilted or pig-like. Raises head and 'points' when disturbed, often open-mouthed.

Fig. 1.8 Crabeater seal characteristic attitudes.

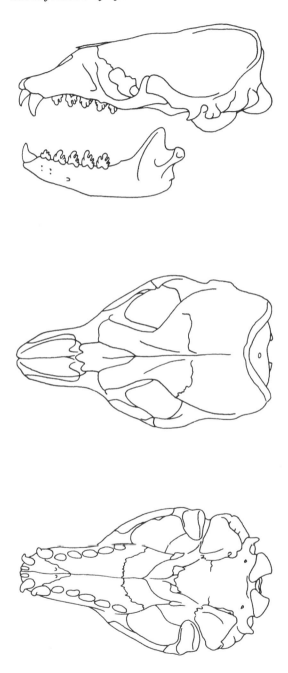

Fig. 1.9 Crabeater seal skull.

Size

Newborn: length 120 cm; weight 20 kg.
Adults: length 2.5 m; weight to 410 kg.

Colour

Background mainly silvery grey (newly moulted) to golden or creamy white (faded); sometimes brown; back darker than belly. Older animals progressively paler even when freshly moulted.

Markings

Reticulated chocolate brown marking and fleckings on shoulders, sides and flanks shading into predominantly dark hind and fore flippers and head.

Moult

January and February, some extending into March. Coat often conspicuously scarred with typically oblique parallel scars mainly on sides and flanks (in 60–85% of adults).

Pup

Light 'milk coffee' brown with darker hind flippers.

Behaviour

Usually found near periphery of pack ice in broken ice, particularly on 'cake' ice, but may be on large floes in open to close pack. Group size: usually single or small groups, average about one to two; larger groups of up to 28 individuals, particularly where floes are scarce; occasionally large aggregations on fast ice in bays or straits (Laws & Taylor, 1957).

On ice, fastest moving and most active of Antarctic true seals.

In September to October pupping season forms male/female pairs or triads with pup.

In water

Often swimming purposefully and directionally in groups of two to five or more, up to 500; between floes surfaces with head out and snorts. Shows back when submerging. When leaving water leaps out on floes. Occasionally uses Weddell seal breathing holes (Stirling & Kooyman, 1971).

Disturbed on surface

Raises head as if 'pointing' but remains belly down. Has 'neurotic' disposition, on closer disturbance resulting in jerky movements accompanied by open mouthed displays and 'hissing' and snorting noises. When turning rapidly, often pivots (like elephant seal).

If pressed, travels rapidly over ice with lateral swimming undulations and alternate strokes of fore flippers, but usually moves keeping fore flippers swept back against body (like Weddell seal) or occasionally with fore flippers spread out in unison like elephant seal.

Close-up

Teeth

Moderate sized canines with elaborate large four to five cusped sieve-like post-canines; cusps blunt (Fig. 1.9).

Skull

Overall length up to 306 mm.

Fore flipper

Moderate size, central position on body.

Faeces

Usually pink coloration from krill remains.

Underwater sound

Not yet precisely identified and described.

Leopard seal, *Hydrurga leptonyx* (Blainville, 1820). Figs. 1.10, 1.11

Profile

Large, very long and slender, with large 'shoulder', disproportionately large 'reptilian' head with long snout, marked neck constriction and flanks tailing off. Build more like crabeater than other species, but distinctive.

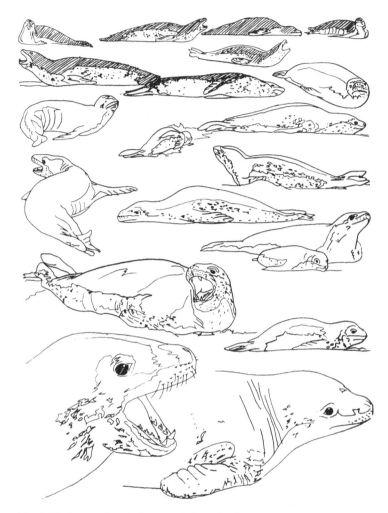

Fig. 1.10 Leopard seal characteristic attitudes.

Size

Newborn: length 150 cm; weight 35 kg.
Adults: length 4.5 m; weight to 600 kg.
Female markedly larger than male.

Colour

Usually very dark (black or dark grey) dorsally, silver ventrally. Liberally
spotted with light and dark grey, and black spots; relatively sharp line

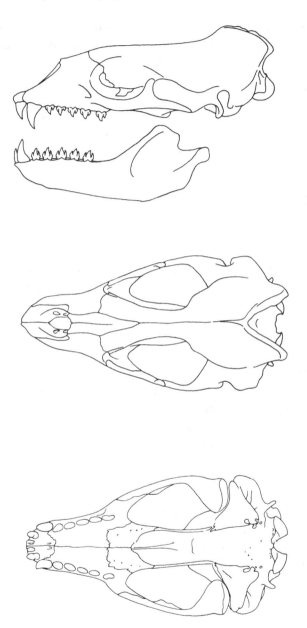

Fig. 1.11 Leopard seal skull.

of demarcation along sides; fore flippers light with dark spots. Another type has light grey ground colour dorsal and ventral (Brown, 1957). Head has broad silver grey band along upper lip.

Pup

Fur thicker and softer than adult but same length (*c.* 2 cm) and less dense than other species. Colour pattern similar to adult, unlike all other species except the Ross seal (Brown, 1957).

Moult

January to February.

Behaviour

Most widely distributed of all Antarctic seals – from continental shores to sub-Antarctic islands, and found in all densities of pack ice from open to closed. More often at edge of pack ice where distribution parallels that of crabeater seal. Usually solitary. Births take place from November to December.

In water

Often holds head out of water and characteristically shows back when submerging head first, like porpoise; if stationary in water sometimes submerges tail first. Follows small boats and will 'mouth' propeller or cone of inflatable float. Known to harass divers, but none have been attacked.

Disturbed

Often faces observer with head raised in deliberate direct gaze, flippers pressed to sides or used to raise fore part of body. Slow deliberate actions; sinuous lateral movements similar to crabeater but slower and with little use of fore flippers. Easily frightened by helicopters.

Close-up

Teeth

Incisors and canines large, pointed. Post-canines large, recurved pointed with main crown and two lateral cusps; cheek teeth interlock when jaw closed (as in crabeater) (see Fig. 1.11).

Skull

Overall length up to 431 mm.

Fore flipper

Large, otariid-like, well back on body.

Underwater sound

Pulsed, narrow band, about 300 c.p.s., also trilling and moaning. Unmistakable haunting quality once heard; sometimes heard when out of water.

Antarctic and Sub-Antarctic fur seals, *Arctocephalus gazella* (Peters, 1875), *A. tropicalis* (Gray, 1872). Figs. 1.12, 1.13, 1.14

Two species of fur seal are found south of the Antarctic Convergence; *A. gazella* is predominant, but there are occasional stragglers of *A. tropicalis* from islands further north. Field characters for *A. gazella* are given below with an indication of ways in which *A. tropicalis* differs.

Profile

Snout pointed. Small head, long neck, large fore flippers and 'shoulder' placed further back than in true seals; hind flippers elongated and can be turned forward. Can assume quadrupedal form. Smaller than true seals. Bulls have thickening of neck and shoulders due to presence of a mane.

Size

Newborn: length 60 cm; weight 5.5 kg.
Adult: male length to 200 cm; weight to 200 kg.
          female length to 145 cm; weight to 50 kg.
A. tropicalis is smaller than *A. gazella*.

Colour

Back and sides grey to brownish, depending on length of time ashore; throat and breast creamy; belly dark gingery colour. Adult male has heavy mane round neck, shoulder and on breast – grizzled appearance due to white hairs.

Fig. 1.12 Antarctic fur seal characteristic attitudes.

Yearlings

Grey dorsally, cream ventrally.

Pups

Very dark brown or black fur.

Pale form occasionally (0.1%) lacks pigmentation in guard hairs and appears creamy white or honey coloured (all ages and both sexes) but is *not* albino. *A. tropicalis* differs in that adult chest and throat yellow; mane less developed than in *A. gazella* with fewer white hairs; bulls with crest of longer hair on top of head (see Key at end of chapter).

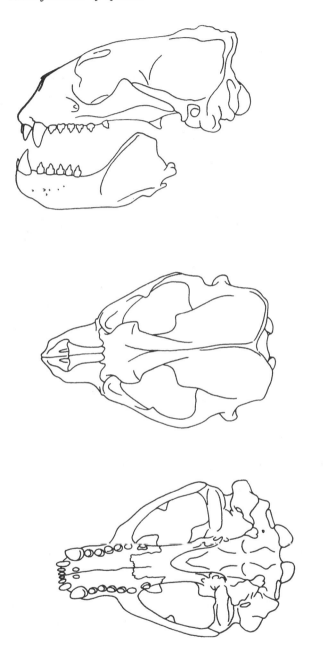

Fig. 1.13 Antarctic fur seal skull.

Fig. 1.14 Sub-Antarctic fur seal characteristic attitudes.

Behaviour

Breed on land, in November to December, in large aggregations; prefer rocky coasts. Form harems of 5–15 females, with harem bulls and subordinates. Non-breeding males occupy area peripheral to breeding beaches. Pups born November to December, moult in April to yearling coat. In summer, males and juveniles haul-out for moult and large numbers of males then swell South Orkney Islands and South Shetland Islands populations.

At sea

Swimming characteristic 'porpoising' in small groups, often leaping clear of water like penguins. At play in coastal waters adopt variety of positions at the surface – for example, head up or down, hind flippers fanned or fore flippers raised.

Disturbed

Very aggressive in breeding season, particularly males, whimpering roar; rapid walk or bounding run 'on all fours'.

Close-up

Teeth

Conspicuous canines and reduced cheek teeth; heavily stained dark brown. (*A. tropicalis* cheek teeth more prominent, less stained). (See Fig. 1.13.)

Skull

Overall length up to: male, 286 mm; female, 239 mm.

External ear

Obvious, over 1.5 cm long.

Snout

Pointed.

Fur

Long guard hairs conceal short thick underfur of commerce (sea lions have blunt snout, and short coarse guard hairs over very sparse underfur).

Testes

May be scrotal.

Key to field identification of adults

To use this key start at 1. If 1 applies, go to 2; if 1 does not apply go to 1(a) (= alternative); if 1(a) applies go to 6; if 6 applies go to 7; if 6 does not apply go to 6(a) and so on.

1. Colour grey weathering to brown; no spotting or patterned markings; pup fur black.　　　　　go to 2

   (a) Colour pattern of spots, blotches or stripes (may be

inconspicuous when faded); pup fur greyish-brown
or patterned markings.   6

2. Hind flippers capable of being turned forward and
   used in terrestrial locomotion in walking or running
   movements.   3

   (a) Hind flippers not capable of being turned forwards;
       not used in terrestrial locomotion; movements
       relatively slow.   5

3. Claws usually missing from fore flippers and small on
   hind flippers; underside of hind flippers hairless. Coat
   of guard hairs and underfur. Long neck. Small external
   ears over 1.5 cm long. Hind digits with cartilagenous
   extensions, united proximally. Testes often in scrotum.
   Teeth not lobed. Adult size 2 m, 200 kg.   4

4. Adult chest and throat yellow; adult females chest
   silvery-orange; younger animals with lighter chest. Short
   broad flippers; hind flippers relatively short. Adult
   male with crest of longer hairs on head, up to 5 cm in
   old animals. Ears short and fleshy. Teeth lightly
   stained. Clear sharp tones of vocalizations; growl like
   monosyllabic cough.

   Sub-Antarctic fur seal (Arctocephalus tropicalis)
   Fig. 1.14,

   (a) Adult chest grey to white, slight or no yellow tint.
       Adult male distinct grizzled or white head and back
       of neck. Long fore flippers; hind flippers long and
       awkward looking. No crest, but slight bunch of
       longer guard hairs between eyes. Ears obvious, long.
       Teeth heavily stained. Vocalizations nasal, some-
       times wheezy; growl deep and reverbatory.

   Antarctic fur seal (Arctocephalus gazella) Figs. 1.12,
   1.13.

5. Large claws on fore flippers. Coarse single hair coat. No
   obvious neck. Teeth not lobed. Size to 2.8 m, 900 kg
   (female); 4.5 m, 4000 kg (male).

   Elephant seal (Mirounga leonina) Figs. 1.2, 1.3

6. Profile lithe and streamlined with flanks tailing off.   7

   (a) Profile plump, barrel-shaped.   10

7. Very large head and thorax; distinct neck. Dark and light spots overall   8

   (a) Head moderate size; coat mainly silver-grey or cream with reticulated markings mainly on shoulders and near tail; snout upturned. Often scarred on flanks and belly with long parallel stripes.   9

8. Colour silver-grey with very dark dorsal band; fore flippers large, silvery with black markings, otariid-like, well back on body. Upper edge of mouth with broad grey band – reptilian appearance to head. Molars three-lobed, sharp, recurved, canines pointed. Adults to 4 m; 600 kg.

   Leopard seal (Hydrurga leptonyx) Figs. 1.10, 1.11.

9. Markings light circles surrounded by darker markings and spots; flippers dark, fore flippers centrally placed. When disturbed raises head as if 'pointing'. Adult males badly scarred about head and neck. Molar teeth 4–5 lobed blunt, canines blunt. Adults to 2.5 m; 410 kg.

   Crabeater seal (Lobodon carcinophagus) Figs. 1.8, 1.9.

10. Colour very dark dorsally, lighter ventrally, large spots or splashes of white, grey and black overall.   11

   (a) Colour very dark dorsally, lighter ventrally. Little spotting or oblique fleckings on sides. Dark streaks or bands on throat and chest.   12

11. Flippers dark, fore flippers centrally placed. Head small in relation to body. Rolls when disturbed. Adult males often scarred about belly. Upper incisors procumbent, blunt; incisors and canines often greatly worn; molar with central prominent cone, smaller one behind. Adults to 3 m; 500 kg.

   Weddell seal (Leptonychotes weddellii) Figs. 1.4, 1.5.

12. Flippers dark, well forward on body. Head short and broad, eyes protruding. When disturbed, raises head

near vertical, inflates neck and 'trilling' vocalization. Teeth very small, sharp pointed, recurved.

Ross seal (Ommatophoca rossii) Figs. 1.6, 1.7.

References

Bonner, W.N. & Laws, R.M. 1964. The true seals (Phocidae) In *Antarctic Research*, ed. R. Priestley, R.J. Adie and G. de Q. Robin, pp. 177-90. Butterworth & Co., London.

Brown, K.G. 1957. *The leopard seal at Heard Island, 1951-54*. Austral. Nat. Antarct. Res. Exped. Interim Rep., no. 16, 1-34.

Hall-Martin, A.J. 1974. Observations on population density and species composition of seals in the King Haakon VII Sea (Antarctica). *S. Afr. J. Antarct. Res*, **4**, 34-9.

Laws, R.M. 1953. *The elephant seal (Mirounga leonina* Linn.) *I. Growth and age*. Sci Rep. Falkland Is. Dep. Surv., no. 8, 1-62.

Laws, R.M. 1984. Seals In *Antarctic Ecology*, vol. 2, ed R.M. Laws, pp. 621-715. Academic Press, London.

Laws, R.M. & Taylor, R.J. 1957. A mass dying of crabeater seals, *Lobodon carcinophagus* (Gray). *Proc. Zool. Soc., Lond.* **129**, 315-24.

Scheffer, V.B 1958. *Seals, Sea Lions and Walruses. A review of the Pinnipedia*. Stanford University Press, California.

Stirling, I. & Kooyman, G.L. 1971. The crabeater seal (*Lobodon carcinophagus*) in McMurdo Sound, Antarctica, and the origin of mummified seals. *J. Mammal.*, **52**, 175-80.

Thomas, J., De Master, D., Stone, S. & Andriashek, D. 1980. Observations on a newborn Ross seal pup (*Ommatophoca rossi*) near the Antarctic Peninsula. *Can. J. Zool.*, **58**(11), 2156-8.

Vallette, L.H. 1906. Viaje a las Islas Orcades Australes. Republica Argentina, *Annales del Ministerio de Agricultura, Section de Zootechnica, Bacteriologi, Veterinaria y Zoologia*, **III**, 1-68, Buenos Aires.

Wilson, V.J. 1975. A second survey of seals in the King Haakon VII Sea, Antarctica. *S. Afr. J. Antarct. Res.*, **5**, 31-6.

2

Estimation of population sizes

A.W. ERICKSON, D.B. SINIFF AND J. HARWOOD

Introduction

This chapter provides a review of the methods which are currently available for estimating the size of Antarctic seal populations. Such estimates have a wide range of potential uses, ranging from the evaluation of the status of a species as part of a management or conservation plan, to the use of the species' abundance as an indicator of prey availability. However, there is no one estimation technique which is suitable for all purposes. The choice of a suitable technique will depend on the use which will be made of the results, the logistic constraints imposed on the survey team, and the basic biology of the species involved. The following sections have been designed to provide sufficient information about the available techniques to allow a choice between them to be made. They provide some advice on the general direction for subsequent analysis of results, but calculations of the precision and accuracy of the resulting estimates must be left to individual investigators. However, these latter calculations are essential to any estimate of abundance and anyone contemplating making such estimates should ensure that they will have access to adequate statistical advice during the design of a survey and the subsequent analysis of results.

Basic considerations

First it must be recognized that it is practically impossible to make a total census of all the age-classes in any seal population. Seals partition their time among areas such as beneath the surface of the sea, hauled out on land or ice, or at sea; the time spent in these habitats varies depending on sex, age and season. Survey techniques for seals have therefore attempted to estimate the number of animals that are hauled out, either on land or ice, at a particular time for a particular location. All species spend a greater proportion of their time hauled out at certain seasons and during certain times of day. Surveys should be timed to coincide with those times when the maximum proportion of the population is visible for counting.

Structured populations

When it is possible to estimate the size of some well-defined component of the population (such as the number of pups or breeding females), the ratio between this component and total population size can be used to estimate the size of the entire population in the area surveyed. Usually it is impossible to survey the entire range of a species and it is necessary to make a further correction for the number of seals in areas which have not been surveyed.

The choice of which component to survey is essentially determined by the biology of the species concerned. If a species pups at a limited number of well-defined sites (as is the case for Antarctic fur seals, southern elephant seals and Weddell seals) it is possible to estimate annual pup production, or the number of breeding females, from a series of counts or using capture–recapture techniques. Estimates for each colony can have high precision, but the estimates for the entire population can be seriously biased if all of the pupping sites are not identified (Myers & Bowen, 1988). Additional information on the species' demography is necessary to estimate total population size. The precision of the resulting estimate can be difficult to quantify.

Unstructured populations

If the species does not aggegate to pup, or if the number of aggregations is large, then some form of transect based survey will be required. If the entire area used by the species can be surveyed, the resulting counts will provide an estimate of the minimum size of the population which can then be corrected for the proportion of animals in the water at the time of the survey. However, this is rarely possible and it is usually necessary to survey some small fraction of the potential range of the species involved. In this case, good survey design to ensure representative coverage of all habitat types is essential. Estimates of total abundance from such surveys will not necessarily be of the population's minimum size, but they will be biased downwards, unless some correction is made for animals in the water at the time of the survey. Haul out behaviour is likely to be influenced by weather and ice conditions and it is important that information on these is recorded during the survey.

In most cases more than one seal species and more than one age-class of each species is likely to be encountered. It is therefore important that all observers are provided with objective criteria for classifying animals, that clear guidelines are provided on how to record animals whose species or age cannot be determined, and that checks for observer consistency are made regularly.

Total counts

Species such as Weddell seals, elephant seals and fur seals congegate in colonies during the breeding season. It is often possible to count all the seals hauled out, especially the pups, on foot, from fixed observation points, or from aerial photography. Counting techniques must be tailored to the physical topography of the breeding areas and the life history of the species.

However, even when species pup colonially at a limited number of sites, there is rarely a time when all of the pups born in a particular season can be counted on a single visit. Some pups will have died and their bodies disappeared, and some pups may have left before all the pups in a season have been born. If pup production is to be estimated reliably it is necessary to correct for these missing animals. This is best achieved using a series of counts made throughout the pupping season and spanning the time when the maximum number of pups is ashore. This series is then used to determine the parameters of an underlying statistical model of the way in which the number of pups, or breeding females, varies through the season. The method is described in Hiby, Thompson & Ward (1987) and Rothery & McCann (1987). If it can be assumed that the timing of the season and other parameters of the model remain constant from season to season and from site to site, it is then possible to estimate pup production from a single count if its timing in the pupping season can be established (Rothery & McCann, 1987).

Direct ground counts

For small colonies it is often possible to count the total number of pups or breeding females on the gound or from a vantage point. For example, McCann & Rothery (1988) used the top of the mast of a sailing boat to count elephant seals. In dense colonies it is often useful to mark animals which have been counted with a distinctive dye to ensure that they are not counted twice and to keep track of what parts of the colony have been censused. Where this is impractical, it is important that the observers do not take their eyes off the seals they are counting; this can be achieved by using hand tallies, small tape recorders or by assigning one member of a group of observers to record while the others count. The movement of observers through a colony can cause considerable disturbance. Not only may this lead to undercounting, because animals leave the colony, but it may disrupt some aspects of breeding behaviour and lead to estimates of pup production that are unrepresentative of an undisturbed colony.

In very large colonies, or where the ground is broken so that pups can be easily missed, direct counting becomes impossible and it is necessary to use aerial photography or capture-recapture analysis.

Direct aerial counts

Sometimes fixed wing aircraft or helicopters can be used as a counting platform. In such instances there are many factors that must be taken into account when designing surveys. The following are some considerations that need attention when total counts are taken from an aerial platform.

Type of aircraft

There are several criteria which should be considered in the choice of aircraft. Some of these in order of importance are availability, visibility, safety, cost and speed (both slow flight and cruise characteristics). Unless equipped with a nose bubble or otherwise projecting turret, few low-wing aircraft provide acceptable forward and lateral visibility, although removal of baggage hatches or doors may help. Some twin engine aircraft place the pilot and adjacent seat far enough forward of the wing. Both single and twin engine high-winged aircraft are available, but the cost of the latter is usually at least twice that of the former. The increased safety of twin engine aircraft is obvious and of particular importance when working in remote areas or over water. Some of these aircraft, however, are not capable of maintaining altitude on one engine, although in some this may be a matter of payload.

Altitude

This will vary depending on the local environment and topography of the breeding gounds. Altitudes reported in the literature vary from

100–500 m, but it is a matter of choice for the operator. Flight altitudes exceeding 300 m may result in difficulties identifying seals and species.

Disturbance

If relatively low altitude flying is required, aircraft noise must be considered since various seal species react differently to it. Except for fur seals, Antarctic pinnipeds appear moderately tolerant of such disturbances. However, some aircraft are probably not usable for pinniped work at low altitude because of the disturbance caused by their noise; helicopters often fall into this category. Pinnipeds appear to react mainly to loud noise, changes in volume and to certain frequencies. Most fast aircraft are noisy, but even slow aircraft can produce enough noise to disturb seals, and diving, climbing, steep turns and 'lugging down' should be avoided near the colonies. Certainly, the reaction of seals to aircraft sound can be a help in censusing certain seals. This is particularly so in the case of Antarctic ice seals. The reaction of seals to aircraft sounds aids in their identification. Most species appear more tolerant of noise during high seas or strong winds perhaps due to increased levels of background noise. Occasionally, an up-wind approach may make some difference in producing disturbance with the seals becoming 'attentive' well before any normal reaction distance is reached. However, on occasions when animals are 'attentive' to the presence of a disturbance, they raise their heads and this may make counting or age/sex discriminations easier. If animals start to move in apparent response to an approach, it is sometimes possible to ease away from the direct approach and animals' exodus will cease. However, a sudden turn will frequently increase the speed of animals moving toward the water. A recent history of disturbance may make some species more sensitive to further disturbance.

Aerial photography

In situations where an aircraft is available to conduct regular surveys and where the pups or adults can be readily distinguished from the backgound, aerial photography may offer an advantage over direct counting. Not only are the disturbance effects avoided but a number of different observers can be used to count animals in photographs, thus providing a valuable measure of inter-observer variability (e.g. Shaughnessy, 1987). In addition, the photographs provide a permanent quantifiable record of the distribution of animals in the colony.

The choice of a suitable camera and film depends on the terrain to be photographed and the type of aircraft which is available. Large format vertical photographs are easy to count for seals and minimize the problems of marking out areas which have been photographed twice. Hiby *et al.* (1987) provide some useful algorithms for defining non-overlapping polygons in aerial photographs. However, large format cameras are expensive and heavy, so that they may be impractical. If the overlap between sequential frames is increased it is possible to view the resulting photographs through a stereoscope. This is sometimes useful for identifying animals when the terrain is very broken. However, it substantially increases the amount of film which must be used and in many cases improving the resolution of the photographs is a better method for facilitating identification. If only a small aircraft is available, or if postural cues are important for identifying individuals or age classes, then oblique photography may be desirable. Hand-held 35 mm cameras fitted with a motor-drive are most useful for this. However, counting on photographs taken in this way can be difficult as it is sometimes difficult to identify areas of overlap between frames. In addition, pups are easily hidden by adults or rocks.

The choice of film will also depend on the circumstances but it should be of the highest resolution that is practical. Since most texts on aerial photography of wildlife recommend that shutter speeds of 1/250th to 1/500th should be used to avoid blurring of the image, this recommendation would lead to the use of fast speed black-and-white film under most Antarctic conditions. However, if the camera can be mounted on a motion-compensation platform, such as the one described by Hiby *et al.* (1987), it is possible to use lower shutter speeds and high resolution colour film. This can often make the identification and classification of animals substantially easier. In all cases, operation in the field is made much easier if the camera can be fitted with a back capable of holding large rolls of film.

Correct focus is also critically important in obtaining good quality photographs. Autofocus lenses can provide very good results with hand-held cameras but these are frighteningly expensive for large format cameras. If a large format camera is used it is important to calibrate the focus ring on the lens correctly before the survey, because the marked gradations are often imprecise. The lens can then be taped into the appropriate position for the altitude of the survey.

Most workers have used an altitude of 150–300 m on the grounds that a lower altitude causes disturbance and higher altitude surveys are vulnerable to the effects of cloud cover. However, with modern camera technology there is no need to abide by convention and the appropriate altitude should be determined from an analysis of the constraints of available flying time, required coverage and resolution, and likely weather conditions at the time of the survey.

The highest resolution is obtained by viewing negatives or colour transparencies directly. A microfiche reader is very useful for this purpose. A positive black-and-white image of good resolution on film can be obtained by 'printing' negatives onto sheet film (R.L. DeLong, pers. comm.). Areas of overlap can be marked directly onto large format film or onto a transparent overlay.

Ideally, surveys should be conducted when the sun is at its highest or obscured by cloud because shadows, particularly of ice hummocks, may resemble seals.

Estimates from sampling

Survey design

Where animals are widely dispersed over a large area – as is the case for most ice-breeding seals – some form of line transect survey is probably the most appropriate method for estimating abundance. The principles of line transect sampling are described in detail in Burnham, Anderson & Laake (1980). Recent modification to these techniques for use in surveys of large whales, which have some similarities to seal surveys, is described in detail in Hiby & Hammond (1989).

In general, all animals which are sighted are recorded and an estimate is made of their perpendicular distance from the track line being followed by the ship or aircraft which is conducting the survey. The strip transect is a special case where only those animals which are within a specified distance of the track line are recorded. The use of a strip transect simplifies data collection and analysis, but it makes a number of untested assumptions. The most important of these is that all animals within the strip are seen. When conducting a strip transect it is often necessary to estimate the actual distance of each group or animal from the track-line to determine whether it lies within the strip. In these circumstances it is usually preferable to use general line transect theory to analyze the data (Burnham & Anderson, 1984).

Surveys can cover only a small fraction of the region likely to be used by the target species. It is therefore important that the survey covers a representative part of the area for which the population estimate is desired. The survey design should take account of known variations in seal density with location, ice type and floe size. Transects should, in general, be placed at right angles to known gradients in abundance. Thus it is recommended that transect lines for Antarctic seals should, as far as possible, be oriented to run from the ice edge towards the consolidated pack. The ice pack is generally stratified into bands, tending from loose pack and small ice types at the ice edge to dense pack and large ice types as the continent is approached (Erickson *et al.*, 1971; Gilbert & Erickson, 1977). Different seal species prefer different ice types and survey design and analysis should take account of this.

Ideally the area to be surveyed should be divided into strata on the basis of prior knowledge about the distribution of seals and of ice types. Transects within each stratum are allocated on a random basis to ensure that each point within the stratum has an equal probability of being covered. However, ice conditions in the Antarctic often change so fast that this is impossible. The simplest solution is to post-stratify the observations on the basis of the ice conditions encountered. An alternative approach, which has yet to be applied to Antarctic seals, is to use the 'variable coverage probability' method of Cooke (described in Hiby & Hammond, 1989) which takes account of the fact that coverage probabilities within a stratum are inevitably non-uniform.

General considerations for data collection

Distinguishing species can be difficult, but species-specific behaviour patterns can provide important criteria for identification. Recognition of such patterns does, however, require considerable experience. Notes on identification are given in chapter 1.

The distinction between individual seals and groups should be agreed at the beginning of the survey. It is recommended that all seals within 20 m of each other on one floe should be considered as a group. All those lying on one small floe should be considered as the same group, while on large floes judgement is required. However, this seldom produces problems as most animals in a group are usually within a few metres of each other.

Most reports indicate that out of the breeding season (i.e. January to March) there is a general increase in the number of seals hauled out on the floes between 10.00 h and 16.00 h, although this does not necessarily apply to Weddell seals. Ideally, surveys should only be conducted during this period, but such a restriction could seriously reduce area coverage which, in the case of ship surveys, is already restricted by the ship's speed capabilities through the pack ice. In these circumstances it is usually assumed that all seals are hauled out and visible during the survey period. However, studies can be undertaken to determine a suitable correction factor (see section 'correction factors').

Information on weather conditions, which may influence the sightability of seals, can be collected and used to stratify or correct analyses. Water depth, air and sea temperature, and details of inter- and intra-specific relationships can be recorded at the same time as seals are observed. Such information is often useful for interpreting species' distributions and behaviour. However, it should be recognized that if observers are asked to record large quantities of information this will increase fatigue and will reduce the length of time that observers can operate at maximum efficiency.

Data collection from a ship

Counts should be made from a high platform. Generally, the ship's bridge is used, but higher positions should be considered if they can be made comfortable for work. If a strip transect survey is being conducted Erickson & Hanson (1990) recommend that all animals within 200 m on either side of the ship's track are recorded, as they come abeam of the ship. If standard line transect methodology is being followed, the perpendicular distance of an animal from the track-line can be estimated if its distance and bearing from the track are recorded. However, in pack ice the track-line may be too erratic for such estimates to be meaningful. It is probably more appropriate to record the animal's distance from the vessel as it comes abeam. Distance from the track line can be estimated using commercial range-finders, inclinometers, or specially-constructed 'sightings boards' (Siniff, Cline & Erickson, 1970).

Ice type and concentration (see section on ice classification), and the ship's position should be recorded at least once every 15 mins. Conventional navigation combined with dead reckoning is usually not accurate enough, but must be relied upon in the absence of other navigational aids. However, most modern ships are equipped with navigational aids which make use of satellite systems (such as GPS), or networks of beacons (such as Decca, Loran or Omega). Some of these aids provide a continuous record of the ship's position. In these cases, positions can be noted more frequently.

Data collection from an aircraft

Aerial surveys enable a far greater area to be covered than shipboard surveys and they can be conveniently carried out during the optimal haul out period (see section on general considerations for data collection). However, poor visibility or high winds can severely curtail flying time. The choice of a suitable aircraft has already been discussed in section 'Direct aerial counts'

Maintenance of the selected altitude and the agreed track-line is critical. Pressure altimeters are usually satisfactory for controlling height over flat terrain, but they are less reliable than radar altimeters over ice. Ice type and concentration should be recorded approximately every 5 km (e.g. at 5 min intervals when flying at 60 knots).

Accurate measurement of the distance of individual seals from the track-line is often not possible during aerial surveys, and some form of strip transect methodology is normally employed. However, an estimate of the variation in sightings probability with distance from the track-line can be obtained by dividing the strip into a number of sectors. The boundary of each sector can be defined by a series of markings on the windscreen of the aircraft (Scott & Gilbert, 1982; Hiby *et al.*, 1987), or by streamers attached to the aircraft's wing struts. In both cases it is important that observers do not alter the elevation of their head relative to the markers, as this can change the strip width.

If it proves difficult to count large groups of seals accurately on a single pass, the precise size of the group can be determined later from a photograph, or series of photographs, taken at the time (Norton-Griffiths, 1974).

An accurate estimate of ground speed is essential for the analysis of aerial survey results. It is therefore important that the aircraft's exact position is determined at regular intervals. This can be achieved by linking transects to a series of well-defined way-marks, or by using the navigational systems described in the previous section, where these are operational. The calculation of ground speed should take account of 'crabbing' of the aircraft due to side winds, and variation in attack angle when flying upwind or downwind.

Analysis of results

Estimates of seal density within each stratum are easily calculated if the strip transect method has been used. The number of seals of each species sighted is divided by the total length of the track-line within the stratum multiplied by two times the strip width. If line transect methods have been used it is necessary to calculate an 'effective search width', equivalent to the strip width in strip transect calculations. A generalized distribution is fitted to the observed distribution of sightings with distance from the track-line (see Burnham *et al.*, 1980 and Hiby & Hammond, 1989 for details). Different theoretical distributions can be used for different strata and weather conditions, if there are obvious differences in the sightability of seals.

The estimates of seal density for each stratum may be used to estimate population size for the area covered by the survey, and may even be extrapolated to cover a wider geographic area. This calculation requires an estimate of the total area occupied by each of the stratum types at the time of the survey. If the strata are based on ice types it may be possible to use satellite imagery to determine these areas. However, it is important to ensure that the ice types identified in the satellite images correspond with the classifications used during the surveys.

The variance associated with each density estimate should be calculated where possible. Variance can be estimated either by applying jack-knife or bootstrap procedures to the raw survey data (Efron, 1981) or by conducting replicate surveys within strata. The latter approach is desirable but expensive.

Calculation of variance for overall abundance estimates when all strata are pooled is a complicated procedure and investigators are urged to consult a statistician at an early stage of analysis.

Correction factors

Visibility bias

Line transect theory assumes that an animal on the track-line is certain to be seen. For a strip transect survey the assumption is that all animals within the chosen strip are certain to be seen. These assumptions should be tested whenever this is practical. Available methods are reviewed in Pollock & Kendall (1987). For Antarctic seal surveys the most appropriate method is to carry out at least part of the survey with two observers who record their sightings independently and who do not communicate with each other. The subsequent analysis of these independent observer data to estimate the probability that an animal on the track-line or within the strip will be seen, is discussed in Graham & Bell (1989), Hiby &

Hammond (1989) and Marsh & Sinclair (1989). The abundance estimates can then be divided by this probability to correct for visibility bias.

Diurnal activity

Diurnal fluctuations in numbers of seals hauled out occur in at least some Antarctic seals (e.g. crabeater seals). As already mentioned in section on general considerations for data collection, the main haul out period in the pack ice occurs between 10.00 h and 16.00 h with peak haul out occurring about mid-day local time (Erickson, Bledsoe & Hanson, 1989). Fur seals may also show diurnal haul out patterns, even during the breeding season. Elephant seals, which remain on land during the breeding and moulting seasons, do not show this tendency. In order to correct for this variation with time of day, it is desirable to carry out activity studies periodically throughout the survey. This entails making sequential counts (e.g. every 30 min) throughout a 24 h period with the ship stationary in the pack ice (i.e. drifting with the ice with the engines stopped). These counts can be used to calculate a set of correction factors to compensate for the effects of surveying outside the main haul out period. Erickson *et al.* (1989) have used such correction factors to revise survey data for crabeater seals.

Capture–recapture methods

In circumstances where ground counts or aerial surveys are inappropriate, the number of pups or adults can be estimated by marking a number of animals uniquely and more-or-less permanently, and then determining the proportion of marked animals in subsequent samples.

These techniques are particularly appropriate for seals which utilize land or fast ice as the haul-out platform during the breeding season. These seals are often easy to approach and examine, which makes them especially suitable organisms on which to use capture–recapture techniques. This technique has been widely used for seals in the region of the pupping sites (Payne, 1977; Siniff *et al.*, 1977). A review of recent applications, including examples of seal studies, is provided by Seber (1986).

Petersen estimator

In this simplest form of capture–recapture a random sample of the population or age-class is tagged and the proportion of tagged animals in subsequent samples is recorded (these samples need not be random, provided there is complete mixing of tagged and untagged animals, and tagging does not influence the chances of being sampled). Population size is then estimated by multiplying the number of animals originally marked by this proportion. Seber (1982) describes the statistical basis for the technique and a variance estimator. The resulting estimate of abundance is susceptible to bias due to small sample size if the number of recaptures is less than 20. Different estimators are available if sample sizes fall below this figure.

The method relies on a number of restrictive assumptions which are described in Seber (1982). Bowen & Sergeant (1983) describe a study of harp seals where the validity of all these assumptions was tested. The method is also vulnerable to differences in catchability between classes or individuals. Modified estimators for use with closed populations which take account of this problem are described in Burnham & Overton (1979).

Jolly-Seber technique

A more general form of capture–recapture analysis was independently developed by Jolly and Seber in 1965 for open populations. A series of censuses (at least three but preferably five to seven) is required. New animals may be marked during each census period to increase the size of the marked population. The method and its assumptions are described in Seber (1982). Again it is necessary to check on the validity of the assumptions before the method is used on a large scale because violation of some of the assumptions can seriously bias the estimates. A long-term capture–recapture study can provide estimates of demographic parameters such as survival, immigration and emigration as well as estimates of abundance (e.g. Siniff *et al.*, 1977; Croxall & Hiby, 1983).

Practical considerations

Capture–recapture methods are vulnerable to the effects of differential catchabilities; it is therefore important to minimize the disturbance and discomfort caused by the tagging process, to treat all of the obvious sub-divisions of the population (by age, sex and reproductive condition) separately in analysis, and to quantify the rate of tag loss by double-tagging experiments.

General considerations

Ice classification

While pack ice occurs in many forms, there are two important considerations affecting the distribution of seals:

(a) The suitability of ice as a hauling out surface.
(b) The amount of open water access in the pack ice.

Ice types suitable as hauling out surfaces are ice cakes and ice floes, and the amount of free water access is generally inversely proportional to the concentration of the pack ice. The distribution of the various types of ice is variable, but as a rule the pack progressively grades from very consolidated cake ice along the edge of the pack to large floe ice as the distance into the pack ice increases.

The pack ice inhabiting seals are not evenly distributed within the pack ice; crabeater and leopard seals are usually more abundant in the outer pack where access to water is more uniformly available. Ross seals appear to favour mid-pack areas and Weddell seals occur in the areas of deep-pack or in fast ice close to land area (Erickson *et al.*, 1971; Condy, 1977). These considerations highlight the importance of relating census results to the pack ice characteristics (Gilbert & Erickson, 1977).

The following classification based upon the *WMO Sea-Ice Nomenclature* (World Meteorological Organisation, 1970), and the *Illustrated Glossary of Snow and Ice* (Armstrong, Roberts & Swithinbank, 1973) is recommended:

> Brash ice (B) – The wreckage of other forms of ice, too small to support seals.
>
> Cake ice (C) – Capable of supporting a seal, but smaller than 10 m across.
>
> Small floes (SF) – Ice floes measuring 10–100 m across.
>
> Medium floes (MF) – Ice floes measuring 100–500 m across.
>
> Large floes (LF) – Ice floes greater than 500 m across.

To standardize floe-size classification it is suggested that the average of the two greatest widths be estimated and used as the basis for assessing floe size.

Open water (W): Ice-free water
Pack ice
    concentration: recorded in tenths,
                        e.g. open water (0/10th)
                                open pack ice (1/10th to 3/10th),
                                medium pack ice (4/10th to 6/10th),
                                close pack ice (7/10th to 9/10th)
                                solid pack ice (10/10th)
Ice surface
    topography: smooth (1), rough (2), hummocked (3)

A convenient method of recording ice type and cover is illustrated by the following examples: C5/B3/W2 (5/10th cake ice, 3/10th brash ice, 2/10th open water), SF6/B3/W1 (6/10th small floes, 3/10th brash ice, 1/10th open water).

During ship and aerial surveys, ice type, concentration and ice surface topography should be recorded every 15 min (ship) or 3 min (aerial). During ship surveys it is also desirable to record these data at each seal observation such that:

(a) The ice concentration and type within 100 m of the occupied floe is described.

(b) The surface topography and size of the ice floe occupied by a seal is described in the following example: SF6/B3/W1 – MF/1 which indicates that the occupied ice is a medium-sized floe (MF) with a smooth surface (1), and that the local ice coverage (within 100 m) is 6/10th small ice floes, 3/10th brash ice and 1/10th open water.

Navigation accuracy is improved if the ship's track through pack ice is maintained in as straight a line as possible, a practice more convenient to ice breakers than to research or supply ships. Visible icebergs or prominent land points serve as good navigational aids, especially for the maintenance of a straight course, and if they are not visible radar direction may assist.

In the case of aerial surveys the plotting and measurements of transect lines may be more difficult depending on the navigational aids available. In the case of ship-supported aerial surveys, the transect can be plotted by radar from the ship to distances of 60 nautical miles (111 km) from the ship. In the absence of ship support, sample strip plots must for the most part be accomplished by using in-flight speed and bearing calculations. Again, visible markers (i.e. icebergs or land features) are useful for orientation and course maintenance.

Standard units

It is recommended that the following standard units be used:

(a) Distance and area – given in nautical miles and nautical square miles:

$$1 \text{ nautical mile} = 1.852 \text{ km}$$
$$1 \text{ square nautical mile} = 3.4299 \text{ km}^2$$

(b) Ocean depth – given in fathoms:

$$1 \text{ fathom (6 feet)} = 1.8288 \text{ m}$$

(c) Time – given as local time and indicating the variation from GMT.

References

Armstrong, T., Roberts, B.B. & Swithinbank, C.W.M. 1973. *Illustrated Glossary of Snow and Ice* (2nd edn). Scott Polar Research Institute, Cambridge, England, Special Publication No. 4. Scolar Press Ltd., Menston, Yorkshire.

Bowen, W.D. & Sergeant, D.E. 1983. Mark–recapture estimates of harp seal pup (*Phoca groenlandica*) production in the Northwest Atlantic. *Can. J. Fish. & Aquat. Sci.* **40**, 728–42.

Burnham, K.P. & Overton, S. 1979. Robust estimation of population size when capture probabilities vary among animals. *Ecology* **60**, 927–36.

Burnham, K.P., Anderson, D.R. & Laake, J.L. 1980. Estimation of density from line transect sampling of biological populations. *Wildl. monogr.* **72**, 1–202.

Burnham, K.P. & Anderson, D.R. 1984. The need for distance data in transect counts. *J. Wildl. Mgmt.* **48**, 1248–54.

Condy, P.R. 1977. Results of the fourth seal survey of the King Haakon VII Sea. *S. Afr. J. Antarctic Res.* **7**, 10–13.

Croxall, J.P. & Hiby, A.R. 1983. Fecundity, survival and site fidelity in Weddell seals, *Leptonychotes weddelli. J. appl. Ecol.* **20**,19–32.

Efron, B. 1981. Nonparametric estimates of standard error: the jackknife, the bootstrap and other methods. *Biometrika* **68**, 589–99.

Erickson, A.W. & Hanson, M.B. 1990. Continental estimates and population trends of Antarctic ice seals. In *Antarctic Ecosystems*, ed. K.R. Kerry and G. Hempel, pp. 253–64. Springer-Verlag, Berlin.

Erickson, A.W., Siniff, D.B., Cline, D.R. & Hofman, R.J. 1971. Distributional ecology of Antarctic seals. In *Symposium on Antarctic Ice and Water Masses*, ed. G. Deacon, pp. 55–75. Heffer and Sons, Ltd, Cambridge.

Erickson, A.W., Bledsoe, L.J. & Hanson, M.B. 1989. Bootstrap correction for diurnal activity cycle in census data for Antarctic seals. *Mar. Mamm. Sci.* 5(1), 29–56.

Gilbert, J.R. & Erickson, A.W. 1977. Distribution and abundance of seals in the pack of the Pacific sector of the Southern Ocean. In *Adaptations Within Antarctic Ecosystems*, ed. G.A. Llano, pp. 703–40. Proceedings Third SCAR Symposium on Antarctic Biology, Smithsonian Institute.

Graham, A. & Bell, R. 1989. Investigating observer bias in aerial survey by simultaneous double counts. *J. Wildl. Mgmt.* 53, 1009–16.

Hiby, A.R. & Hammond, P.S. 1989. Survey techniques for estimating abundance of cetaceans. *Rep. int. Whal. Commn.* (Special Issue 11), 47–80.

Hiby, A.R., Thompson, D. & Ward, A.J. 1987. Improved census by aerial photography – an inexpensive system based on non-specialist equipment. *Wildl. Soc. Bull.* 15, 438–43.

Marsh, H. & Sinclair, D.F. 1989. Correcting for visibility bias in strip transect aerial surveys of aquatic fauna. *J. Wildl. Mgmt.* 53, 1017–24

McCann, T.S. & Rothery, P. 1988. Population size and status of the southern elephant seal (*Mirounga leoninea*) at South Georgia, 1951–85. *Polar Biol.* 8, 305–9.

Myers, R.A. & Bowen, W.D. 1988. Estimating bias in aerial surveys of harp seal pup production. *J. Wildl. Mgmt.* 53, 361–72.

Norton-Griffiths, M. 1974. Reducing counting bias in aerial censuses by photography. *E. Afr. Wildl. J.* 12, 245–8.

Payne, M.R. 1977. Growth of a fur seal population. *Proc. Royal Soc. Lond. Series B.* 279, 67–79.

Pollock, K.H. & Kendall, W.L. 1987. Visibility bias in aerial surveys: a review of estimation procedures. *J.Wildl. Mgmt.* 51, 502–10.

Rothery, P. & McCann, T.S. 1987. Estimating pup production of elephant seals at South Georgia. *Symp. zool. Soc. Lond.* 58, 211–23.

Scott, G.P. & Gilbert, J.R. 1982. Problems and progress in the U.S. BLM-sponsored CETAP surveys. *Rep. int. Whal. Comm.* 32, 587–99.

Seber, G.A.F. 1982. *The Estimation of Animal Abundance*. Hafner Publ. Co., Inc., New York.

Seber, G.A.F. 1986. A review of estimating animal abundance. *Biometrics* 42, 267–92.

Shaughnessy, P.D. 1987. Population size of the Cape Fur Seal *Arctocephalus pusillus. 1. From aerial photogaphy. Investl. Rep. Div. Sea Fish. Rep. S. Afr.* 130, 1–56.

Siniff, D.B., Cline, D.R. & Erickson, A.W. 1970. Population densities of seals in the Weddell Sea, Antarctica, in 1968. In *Antarctic Ecology*, vol. 1, ed. M.W. Holdgate, pp. 377–94. Academic Press, London.

Siniff, D.B., DeMaster, D.P., Hofman, R.J. & Eberhardt, L.L. 1977. An analysis of the dynamics of a Weddell seal population. *Ecol. Mon.* 47, 319–35.

World Meteorological Organisation 1970. *WMO Sea-Ice Nomenclature: Terminology, Codes and Illustrated Glossary*. Secretariat of the World Meteorological Association, Geneva, Switzerland. WMO No. 259, TP. 145.

3

Immobilization and capture

A.W. ERICKSON AND M.N. BESTER

Introduction

Studies of wild animals often require the capture and immobilization of representative specimens for such purposes as making accurate body measurements, determination of sex and age, collecting of blood or milk samples, or attaching recorders and transmitters. Although great progress has been made in the development of techniques for the capture and immobilization of terrestrial animals using drugs, their use on seals is still in the experimental stage and further work is needed before definitive recommendations can be made for most species.

The use of drugs for the field anaesthetizing of wild animals is considerably more difficult than working with animals in the laboratory or in normal animal practice where information usually exists on the history of the subjects and where greater support facilities and expertise are normally available. Characteristically, the field researcher is faced with the problem of having to immobilize free-ranging animals in difficult situations and often under adverse conditions. The researcher is further required to estimate the body size and physiological state of the animal to be immobilized in order to approximate dosage rates, and not infrequently must proceed without knowledge of the physiological effect of the various drugs on a particular species.

Investigators should also be aware that the distribution or availability of some immobilizing drugs may be restricted so that they are not generally available. Sufficient lead time should be allowed to obtain such things as permits. (See Appendix 16.7.)

Desired properties of immobilizing drugs

An ideal immobilizing agent should have the following characteristics:

(a) A wide safety margin.
(b) Be suitable for most species.
(c) Allow the retention of essential body functions.
(d) Possess high activity and be non-irritating.
(e) Be highly soluble and miscible in a solute, preferably water, and effective in small amounts.
(f) Be relatively short-acting or readily reversible by an antagonist.
(g) Be readily administered.

Unfortunately, few of the immobilizing agents available today possess even half of these characteristics. For field use, an immobilizing agent with a wide safety margin is necessary because of difficulties associated with correct weight and physiological assessment, especially in the case of moulting, lactating or fasting seals. Retention of essential body functions, such as respiration and temperature control, is of particular importance. In general, pinnipeds can tolerate fairly long periods of apnoea (cessation of respiration), but compounds which have a markedly depressant effect on respiration can lead to anoxia (oxygen deprivation), secondary depression of vital functions, shock and death. Body temperature control, especially in the case of seals on dry land, deprived of their normal methods of heat dissipation, is an important consideration. Drugs (such as phencyclidine) which increase muscle tonus tend to induce a rise in body temperature, while other drugs (such as the promazines) disrupt the heat regulating mechanisms and cause a rise or fall in body temperature depending on ambient temperature. In zero or sub-zero conditions, animals tend to become hypothermic although severe disruption of the heat-regulating centre can lead to death from hyperthermia. Vision can be impaired by drugs such as atropine and hyoscine which dilate the pupil of the eye so that sunlight becomes painful and may damage the retina. The eye is also dilated because of adrenaline secretion resulting from fear, and from the administration of amphetamine and similar sympathetic stimulants. Generally, the eyes of anaesthetized seals should be shaded without obstructing air flow through the nostrils and mouth. Hyperthermia can be partly alleviated by shading or cooling the animal with water and snow.

Harthoorn (1976) discusses in much greater detail the physiological processes which are affected by immobilization, and suggests ways and means of avoiding or alleviating many undesirable side effects.

A relatively quick-acting drug is important to prevent the escape of animals and to obviate the multiple injection of individual animals during times of multiple captures as often happens with seals. A drug with small bulk is a prerequisite since drugs must usually be administered by a projectile or pole-mounted syringe. A further desirable characteristic of the immobilizing drug is a relatively short immobilization period in order to limit the period of the adverse effects noted above and to ensure that the animals have recovered and regained self-sufficiency.

Suitable immobilizing drugs and their effects

Immobilizing drugs can be divided into two categories according to their effects: Paralyzing compounds and centrally acting compounds. Both types of drugs have been used to immobilize seals.

Paralyzing compounds (neuromuscular blockers)

Neuromuscular blockers acting at the muscle endplate cause paralysis of voluntary muscles. Acetylcholine is the normal transmitter chemical at the neuromuscular junction and on release is immediately broken down by cholinesterase.

(a) Depolarizing neuromuscular blockers (i.e., succinylcholine chloride) act by simulating the action of acetylcholine but are broken down less freely. The effect is to cause unco-ordinated contraction of isolated muscle groups, and then paralysis which occurs progressively, affecting distal limb muscles before proximal ones. Muscles of the face and tail are affected first, intercostal muscles and those of the diaphragm last.

(b) Competitive (curariform) neuromuscular blockers (e.g. nicotine) act by combining with the endplate so that the action of acetylcholine is prevented. The removal of this block is hastened if the concentration of acetylcholine is increased by inhibiting cholinesterase with physostigmine or neostigmine methylsulphate.

Effective immobilization with neuromuscular blockers depends on paralyzing the locomotory muscles without seriously affecting breathing. This is difficult to perform in the field, since correct dosage depends on accurate weight assessment. Some competitive neuromuscular blockers (e.g. gallamine triethiodide) are reversible but others can be extremely dangerous to handle (e.g. nicotine) and in general use are not as safe or successful as the depolarizing neuromuscular blockers.

It should be noted that there has been some criticism of the use of neuromuscular blockers as being inhumane (e.g. Haigh, 1978). This is a matter of opinion and researchers may still consider a neuromuscular blocker the drug of choice in individual circumstances, particularly in situations requiring short periods of immobilization.

Centrally-acting compounds

Centrally-acting compounds act entirely or mostly on the central nervous system, the effect ranging from tranquilization to anaesthesia. Their main advantages in comparison with neuromuscular blockers are:

(a) Wide safety margin.
(b) Sparing action on basic physiological mechanisms of the body.
(c) Easy handling of immobilized animal.
(d) Possibility of reversal of the drug effect.
(e) Minimal infliction of fear, distress and pain.
(f) Possibility of partial reversal to render the animal mobile but tractable.

Their disadvantages are:

(a) Excitement before immobilization which may induce considerable movement rendering the animal difficult to capture, and which may cause death from exhaustion and hyperthermia.
(b) Legal restrictions which limit the availability and use of some drugs, and problems in handling scheduled compounds.

Administration of drugs

Equipment

Several forms of applicator have been used to capture seals using liquid immobilizing drugs. The simplest has been direct injection with needle and syringe, a syringe mounted on a hand-held pole (Ling, Nicholls & Thomas, 1967; Seal & Erickson, 1969) and tubing between needle and syringe to avoid spillage and so increase the volume of injectate (Ryding, 1982; Gales & Burton, 1987; Bester, 1988). Projectile syringes have included devices fired from air- or gas-powered guns, crossbows, longbows and blowpipes (Boyd *et al.*, 1990). A small hole drilled in the side of the needle prevents blubber coring. The applicator of choice depends in part on the circumstances of use and the preference of the researcher. Inasmuch as Antarctic seals are readily approached under most circumstances, the pole applicator can usually be used on less active species such as the Weddell, Ross and elephant seals, and on the active but docile crabeater seal (Cline *et al.*, 1969). It is less successful with the active and potentially dangerous leopard and fur seals. Crabeater and leopard seals are usually encountered in broken pack ice where the problem of preventing the escape of the seals exists. Conversely, there is an element of danger when working in a fur seal rookery and great social disruption may result from a close approach to the animals. In these circumstances, success is most likely to be obtained using projectile syringes. An additional advantage is that projectile syringes can be marked and a number of seals simultaneously drugged and monitored, provided one uses barbed needles in the syringe. Further, the successful administration of the drug can be ascertained from examination of the syringe, which is less easily done with pole syringes.

Site of administration

Regardless of the applicator used to deliver the drug, there are several points to be borne in mind. A first consideration is the site of administration. This is particularly critical with seals, because they are heavily invested with fat. Consequently, care must be taken to inject the drug into the muscle rather than into the blubber. Hence, the needle used needs to be long (particularly early in the season when the seals are fatter) and should be of large diameter, 16 or 18 gauge, to limit breakage and prevent back pressure on injection. As a rule a 5 cm (2 inch) needle will suffice for all but the largest leopard, Weddell or elephant seals. Shorter needles, 2.5 cm or 3.75 cm (1 or 1.5 inch) can be satisfactorily used on younger or smaller seals.

The recommended areas for injection are the hind upper back just lateral of the mid-line about 10 cm anterior of the insertion of the hind flippers or the anterior or 'upper' shoulder region. The hind upper back is preferred, although the tendency of an active animal to face an antagonist renders access to this area difficult if one is working alone. Frontal shoulder injections work well on facing animals but injections to this area with the pole syringe must be administered with a quick thrust because of the tendency of the seals to bite the syringe head. It is equally important to ensure that the injection is made more or less perpendicular to the surface of the seal's body. A tangential injection is unlikely to be successful – firstly, because of bending or breaking of the needle or syringe and secondly, because of the failure to penetrate the fat layer.

Prevention of drug freezing in needles or syringes

One of the problems associated with attempts to immobilize Antarctic seals using liquid drugs is the tendency of the drug solution to freeze in the needle or syringe. This is particularly a problem with the pole syringe, which must be loaded on the sea ice, and with a projectile syringe if several minutes elapse between loading the syringe into the gun and firing. It is helpful to pre-load syringes before going into the field and to place them in a container heated with a hand-warmer or in a carrying case placed next to the body. In this procedure the loaded syringe will remain effective at freezing temperatures for several minutes after being loaded in the syringe gun. For most captures of Antarctic seals the syringe pistol is to be preferred to the rifle. Carried in a shoulder holster under the outer garment the pistol remains warm and can be quickly loaded and used. For a similar reason, gas-operated projectors are less desirable because of the marked loss of operating pressure at low temperatures.

Estimating body weight

Critical to the successful and responsible drug immobilization of free-ranging animals is accurate estimation of body weight to determine dosage. This initially requires visual assessment which is normally aided by knowledge of the body weight range of recognizable elements of the population, that is newborn pups, yearlings and adults by sex. Only after anaesthetizing can determination of the actual dosage rate administered be determined. Ideally, the weight of an immobilized seal should be determined by weighing the animal. This procedure is for obvious reasons largely limited to seals weighing less than 150 kg unless a special weighing apparatus is devised.

The senior author and his colleagues developed a tripod constructed of a one and a half inch (c. 3.8 cm) aluminium tubing that could be assembled in the field. The tripod legs measured 8 feet (c. 2.5 m) in length and consisted of two coupled sections for ease of transport. The three upper leg sections had holes drilled through them and together with an attachment hook were joined together by a loose-fitting pin. To assemble the tripod the lower leg sections simply needed to be screwed into place.

To complete the weighing equipment a stretcher type sling consisting of two lengths of tubing and an interconnecting web liner was devised. This procedure was preferable to net slings which tend to bag and require greater height clearance than permitted by our eight foot-high tripod. A compact scale is a critical part of the equipment; the scales that we used were PIAB ramometers (Renfroe and Sons, Jacksonville, Florida, USA) with capacities to 1100 pounds (*c*. 500 kg).

For weighing, the seals were rolled onto the sling and the tripod was positioned above. To weigh smaller seals one leg of the tripod was pulled outwards to lower the scale suspended from the tripod head, which was then attached to the shackle of the weighing stretcher. The seal was then raised for weighing by moving the extended tripod leg back toward the seal. A small three-sheave rope pulley was used to lift larger seals. With these devices leopard seals which exceeded the 1100 pound (500 kg) weight capacity of the scale were successfully lifted from the ground.

If animals cannot be weighed directly, fairly reliable dosage rates can be calculated from body measurements of seals (McLaren, 1958; Bryden, 1969, 1972). Ling & Bryden (1981) used the relationships: W = 131.49 + 0.00002079 L and W = 195.51 + 0.00000400 L to estimate the body weight of male and female elephant seals respectively from length, where W = body weight (kg) and L = the cube of standard body length in cm. However, while length is a good indicator of size it does not effectively express the fatness of an animal. Furthermore, the relationship between bodyweight (kg) and standard length (cm) for male elephant seals (W = 9.98 L − 2.317.63) determined by Gales & Burton (1988) differed by 15% from those calculated by Ling & Bryden (1981) for males of similar lengths. Usher & Church (1969) found that girth was better than length as a single indicator of weight of ringed seals (*Pusa hispida*), but the most precise indicator involved the combined parameters of length and girth. They developed the formula:

$$X_1 = 0.023 \ X_2^{1.024} \ X_3^{1.857}$$

(X_1 = body weight in pounds; X_2 = length in inches; X_3 = maximum girth in inches) and suggested that the formula X_1 is a function of (X_2X_3) and would probably be the best predictor of weight for all phocid seals. A rule of thumb approximation proposed for field use is:

$$X_1 = \frac{3X_2X_3^2}{2000}$$

Hofman (1975) performed an analysis of the relationship between the length, girth and weight measurements of Antarctic phocid seals and developed the formula:

$$W = 2.9 \frac{LG^2}{1728}$$

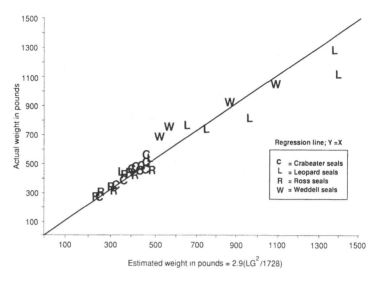

Fig. 3.1 Comparison of scale weights versus estimated weights (Hofman 1975).

where 2.9 is the slope of the regression line, length (L) and axillary girth (G) are measured in inches, and divided by 1728 to obtain the estimated weight in pounds. Figure 3.1 presents the linear regression of the weight to volume index for four species of Antarctic phocids. There were no significant differences between the slopes of the regression lines for the four species. Plotted against the known weights of seals the average difference between the known and estimated weights (from this equation) of Antarctic phocid seals was 8.1% (Hofman, 1975).

Investigators are encouraged to test the validity of these formulae for estimating the weights of seals whenever they have the opportunity to take accurate weight measurements of seals.

Substances used to immobilize seals

(see Tables 3.1 and 3.2)

Succinylcholine chloride

Generally, injection is followed by a regular sequence of muscle paralysis, limpness of hind flippers and then the fore flippers, sagging of the lower jaw, closing of the eyes, intercostal muscular paralysis, and apnoea for

Table 3.1 *Immobilizing agents and dose rates used on phocid seals*

Species and references	Method and route of administration	Immobilizing agent	Sex and age	Dose rate (mg drug/ kg body mass)	Comments
Southern elephant seal, *Mirounga leonina* Ling *et al.*, 1967	Intramuscular in longissimus dorsi anterior and lateral to tail by syringe pole	Succinylcholine chloride	83 (M) 31 (F) all classes to 1900 kg	0.50–1.50 1.51–2.00 2.01–2.50 2.51–3.00 3.01–3.50 3.51–5.50	Excellent paper, an essential review for researchers using this drug on seals 18 of 29 immobilized 22 of 27 immobilized 35 of 36 immobilized 9 of 9 immobilized, 1 mortality 4 of 4 immobilized 7 of 9 immobilized, 4 mortalities Recommended dose 2.5 mg/kg, induction time unstated, immobilization period 6–27 min. Author evaluation: a successful workable drug for elephant seals.
Southern elephant seal, *M. leonina* Ling & Nicholls, 1963	Intramuscular in longissimus dorsi anterior and lateral to tail by syringe pole	Succinylcholine chloride	M and F adults to 2500 kg	0.5–1.0 2.0	Unreliable. Effective, paralysis usually in 1 min, immobilization up to 45 min. Apnoea common.
Southern elephant seal, *M. leonina* Shaughnessy, 1974	Intramuscular anterior to hind flipper	Succinylcholine chloride	All classes	2.5	Not as reliable as phencyclidine but with shorter immobilization period
Weddell seal, *Leptonychotes weddelli*	Intramuscular in gluteal region by projectile syringe	Succinylcholine chloride	Unstated 181–227 kg (E) 91–295 kg (E) 28–340 kg (E)	2.2–6.6 (E), x̄ 1.4 1.8–3.4 (E), x̄ 2.8 12.5–3.9(E), x̄ 3.3	2 of 2 sub-paralytic 9 of 9 paralytic 3 of 3 mortalities

Species, Reference	Route of administration	Drug	Animals	Dose (mg/kg)	Comments
Flyger et al., 1965					Recommended dose 2.8 mg/kg, latent period, x̄ 9 min, immobilization x̄ 34 min.
Crabeater seal, *Lobodon carcinophagus* Hofman 1975	Intramuscular in gluteal region by syringe pole	Succinylcholine chloride	6 adult M & F	1.1–2.0	Unpredictable results, slight body temperature depression. 1.1 mg/kg fatal in one instance. Latent periods 8–18 min. Immobilization 7–41 min. Author evaluation: unsatisfactory for species.
Weddell seal Flyger et al., 1965	Intramuscular in gluteal region by projectile syringe	Nicotine alkaloid	Unstated 227–331 kg (E), 127–295 kg (E), 91–286 kg (E)	3.0–6.2(E), x̄ 4.6; 1.7–3.9(E), x̄ 3.0; 5.0–9.8(E), x̄ 7.4	2 of 2 sub-paralytic; 3 of 3 paralytic; 4 of 4 mortalities. Author evaluation: Not recommended for species; causes vomiting and other undesirable side effects
Southern elephant seal Shaughnessy, 1974	Intramuscular anterior to hind flippers	Phencyclidine hydrochloride	All classes	0.5	Induction times, 20 + min, Immobilization, 1 + h. Author evaluation: preferred to succinylcholine chloride
Southern elephant seal Ross & Saayman, 1970	Intramuscular	Phencyclidine hydrochloride	5 + yr (M)	0.2	Latent period 5 min, 2 additional doses of 100 mg at 55 and 85 min to maintain effect.
Weddell seal Flyger et al., 1965	Intramuscular in gluteal region by syringe pole	Phencyclidine hydrochloride	All classes 2 (286–367 kg(E)), 11 (45–304 kg (E)), 2 (164–181 kg (E))	0.1–0.3, x̄ 0.2; 0.2–1.1, x̄ 0.5; 1.2–4.4, x̄ 2.8	Sub-paralytic; Paralytic; Mortalities.

Table 3.1 *contd*

Species and references	Method and route of administration	Immobilizing agent	Sex and age	Dose rate (mg drug/ kg body mass)	Comments
					Latent period x̄ 18 min. Immobilization to 9 h. Effective dosage rate 0.5 mg/kg. Mild convulsions in some animals
Crabeater seal Ross *et al.*, 1976	Intramuscular	Phencyclidine hydrochloride	Immature (F)	0.5	Latent period 60 min, muscular spasms, twitching and profuse salivation. Immobilization period 6 h. Died from hyperthermia.
Crabeater seal Hofman, 1975	Intramuscular in gluteal region by syringe pole	Phencyclidine hydrochloride	150 kg (F)	0.7	Immobilization trials with captive seals Induction period 25 min, incomplete immobilization.
			245 kg (F)	0.4	Induction period 13 min, recovery 5 h.
			153 kg (F)	0.8	Induction period 10 min, recovery 5.75 h. Author evaluation: used alone as an immobilizing agent for Antarctic seals was unpredictable and had undesirable side effects.
Leopard seal *Hydrurga leptonyx* Hofman, 1975	Intramuscular in gluteal region by syringe pole	Phencyclidine hydrochloride	580 kg (F)	0.5	Induction period 13 min, immobile period not reported

Species / Reference	Route of administration	Drug	Body weight / class	Dosage	Remarks
Ross seal *Ommatophoca rossi* Hofman, 1975	Intramuscular in gluteal region by syringe pole	Phencyclidine hydrochloride	203 kg (F) 193 kg (F)	0.5 0.8	Induction period 16 min, immobile period 3 h, not completely immobile Induction period 8 min, recovery 2.5 h
Ross seal Condy, 1979	Intramuscular by blowgun and projectile syringe	Phencyclidine hydrochloride	All classes 7 individuals 6 individuals	0.32–0.80 Total dose 70–120 mg	Effective immobilization for 20–60 min, recovery period up to 6 h, not weighed, no dose rate.
Southern elephant seal Cline et al., 1969	Intramuscular in gluteal region by projectile syringe and syringe pole	Phencyclidine hydrochloride plus equal amount of promazine hydrochloride	2 imm. (M) 159 kg (E) 3 imm. (F) 102–13 kg (E)	0.16–0.47, $\bar{x}$ 0.32 phencyclidine 0.22 phencyclidine	Latent periods 25 and 28 min. Latent periods 15–33 min, $\bar{x}$ 18, immobilization periods unreported. No mortalities, F more tolerant than M. Recommended dosage: immatures 0.2 mg/kg, adults 0.4 mg/kg. Author evaluation: drug suitable for immobilizing this species.
Southern elephant seal Hofman, 1975	Intramuscular in gluteal reg on by syringe pole	Phencyclidine hydrochloride plus equal amount of promazine hydrochloride	5 M, 167–1624 kg 4 F, 159–461 kg	0.20–0.60, $\bar{x}$ 0.34 0.10–0.60, $\bar{x}$ 0.23	Latent period 9–30, $\bar{x}$ 22 min. Latent period 8–20, $\bar{x}$ 14.5 min. No mortalities, recommended dosage 0.30 mg/kg Author evaluation: highly effective drug combination for species, no side effects.
Weddell seal Cline et al., 1969	Intramuscular in gluteal region by projectile syringe and syringe pole	Phencyclidine hydrochloride plus equal amount of promazine hydrochloride	8 adult M, 181–363 kg (E) 2 imm. (M), 57–91 kg(E)	0.66–1.38, $\bar{x}$ 0.96 Phencyclidine 0.55–0.88, $\bar{x}$ 0.72 Phencyclidine	Latent period 4–38, $\bar{x}$ 23 min Latent period 15–40, $\bar{x}$ 28 min.

Table 3.1 *contd*

Species and references	Method and route of administration	Immobilizing agent	Sex and age	Dose rate (mg drug/ kg body mass)	Comments
Weddell seal Cline *et al.*, 1969	Intramuscular in gluteal region by projectile syringe and pole syringe	Phencyclidine hydrochloride plus equal amount of promazine hydrochloride	2 imm. F, 91 kg each	0.49–0.55, x̄ 0.52	Latent period 5–26, x̄ 16 min., 2 mortalities Author evaluation: drug suitable for species at rates of 0.25 and 0.50 mg/kg for immature and adult, respectively.
Weddell seal Siniff *et al.*, 1971	Intramuscular in gluteal region by projectile syringe and pole syringe	Phencyclidine hydrochloride plus equal amount of promazine hydrochloride	1 yr 1 yr	0.21 0.66	Both killed.
Crabeater seal Cline *et al.*, 1969	Intramuscular in gluteal region by projectile syringe and pole syringe	Phencyclidine hydrochloride plus equal amount of promazine hydrochloride	12 adult M, 113–227 kg (E) 10 adult M, 125–227 kg 8 imm. M, 57–113 kg 7 imm. F, 57–159 kg	0.69–2.94, x̄ 1.44 0.55–1.57, x̄ 1.08 0.66–2.20, x̄ 1.26 0.44–4.39, x̄ 1.71	Latent periods 10–106, x̄ 48 min, immobilization periods not reported. Latent periods 4–100, x̄ 34 min. Latent periods 10–82, x̄ 36 min. Latent periods 4–40, x̄ 17 min. No mortalities. Recommended dosage: 0.5 mg/kg for imm., 1.0 mg/kg for adults. Author evaluation: highly effective for species.
Crabeater seal	Intramuscular in gluteal region by	Phencyclidine hydrochloride	23 M, 104–245 kg	0.31–1.50, x̄ 0.73	Latent periods 5–18, x̄ 12.6, Recovery periods not reported

Reference	Method	Drug	Subjects	Dose	Comments
Hofman, 1975	projectile syringe and pole syringe	plus equal amount of promazine hydrochloride	18 F, 109–248 kg	0.30–1.30, x̄ 0.75	Latent periods 3–18, x̄ 11.8, Recovery periods not reported. Of 183 immobilized, including the above, 7 mortalities due to procedures. Indicated effective dose: 0.70 mg/kg; lethal dose: 1.5 mg/kg. Author evaluation: An appropriate immobilizing drug combination for species. Untoward side effects with phencyclidine alone.
Leopard seal Hofman, 1975	Intramuscular in gluteal region by syringe pole	Phencyclidine hydrochloride plus equal amount of promazine hydrochloride	5 M, 246–420 kg 7 F, 296–534 kg	0.47–0.81, x̄ 0.63 0.44–2.06, x̄ 0.74	Latent periods 7–21, x̄ 13 min. Latent periods 4–15, x̄ 11 min. Indicated effective dosage, 0.70 mg/kg suitable for species. Author evaluation: tolerated in same manner as crabeater seal.
Ross seal Hofman, 1975	Intramuscular in gluteal region by syringe pole	Phencyclidine hydrochloride plus equal amount of promazine hydrochloride	4 adult M, 123–203 kg 3 adult F, 148–204 kg	0.50–0.70, x̄ 0.65 0.50–0.70, x̄ 0.56	Latent periods 4–15, x̄ 9.3 min. Latent periods 8–15, x̄ 11.0 min. Indicated effective dose: 0.60 mg/kg. No mortalities. Author evaluation: Suitable dose for species, tolerated in same manner as crabeater seal.
Leopard seal Hofman et al., 1975	Intramuscular in gluteal region by syringe pole	Phencyclidine hydrochloride plus diazepam (Valium)	10 F, 261–445 kg 9 M, 189–323 kg	0.17–0.80, x̄ 0.30 +5–10 mg Valium 0.19–0.93, x̄ 0.40 +5–10 mg Valium	Latent period approx. 20 min, Immobilization period 3.8 h. Latent period approx 20 min, Immobilization period 3.4 h. December anaesthetizing of sleeping seals required extremely low doses; 4 mortalities due to initiation of diving reflex.

Table 3.1 *contd*

Species and references	Method and route of administration	Immobilizing agent	Sex and age	Dose rate (mg drug/ kg body mass)	Comments
Weddell seal Hofman, 1975	Intramuscular in gluteal region by syringe pole	Phencyclidine hydrochloride plus diazepam (Valium)	19 M, 123–359 kg 18 F, 107–518 kg	0.29–0.86, $\bar{x}$ 0.45 +5–10 mg Valium 0.23–0.99, $\bar{x}$ 0.48 +5–10 mg Valium	Latent periods 6–34, $\bar{x}$ 16.8 min. Latent periods 3–37, $\bar{x}$ 22.1 min. 6 of 7 mortalities in 70 anaesthetizations were adult males thus risks primarily to this age group. Indicated dosage: 0.40 mg/kg.
Crabeater seal Condy, 1977	Intramuscular anterior and lateral to tail	Phencyclidine hydrochloride (250 mg) and diazepam (2.5 ml)	Adult M moulting	1.3	Latent period 20 min, no convulsions but excessive salivation. Died suddenly 3.5 h later while recovering.
		Phen. 300 mg and diazepam 3.0 ml	Adult, F pregnant	1.5	Latent period 5 min., head and flipper shaking during recovery, died 3.0 h after injection.
Ross seal Condy, 1977	Intramuscular anterior and lateral to tail by projectile syringe	Phencyclidine hydrochloride (300 mg) plus diazepam (2.0 ml)	Adult M moulting	1.5	Latent period 80 sec, full recovery in 3 h. No convulsions or salivation.
Southern elephant seal Vergani, 1985	Intramuscular by projectile syringe	Xylazine hydrochloride (Rompun)	68 pups (M & F)	1.0–4.9	Variable results, hyperthermia and vomiting, 3 mortalities. recommended dosage: 2–3.9 mg/kg. Author evaluation: effective immobilizing agent with reversibility of undesirable side effects.

Species / Reference	Administration	Drug	Animals	Dosage	Comments
Crabeater seal Vergani, Spairani & Aguirre, 1986	Intramuscular by projectile syringe	Xylazine hydrochloride (Rompun)	4 adult M (185–217 kg) 1 adult F (148 kg) 5 not recorded	1.6–3.9, $\bar{x}$ 2.8	Good immobilization with no undesirable secondary effects.
Crabeater seal Erickson & Denny (unpub. data) Erickson et al., 1974	Intramuscular gluteal region (projectile syringe)	Xylazine hydrochloride (Rompun)	3 F 124–226 kg 4 F 157–226 kg 2 M 195–220 kg	1.3–1.6, $\bar{x}$ 1.5 2.2–2.5, $\bar{x}$ 2.4 2.3–2.5, $\bar{x}$ 2.4	Incomplete sedation. Latent period, $\bar{x}$ 9 min. Immobilization period, $\bar{x}$ 1.7 h. Latent period, $\bar{x}$ 8 min. Immobilization period, $\bar{x}$ 1.2 h. 5 exhibited slight to moderate salivation. Author evaluation: Excellent, no mortalities, recommended dosage 2.5 mg/kg.
Weddell seal Erickson & Denny (unpub. data) Erickson et al., 1974	Intramuscular gluteal region (projectile syringe)	Xylazine hydrochloride (Rompun)	2 adult F 240–330 kg 1 adult M 33 kg 4 M 110–80 kg	1.5 and 1.7 0.9 1.7–2.7, $\bar{x}$ 2.2	Immobilization period, 0.3 and 2.7 h. No effect. Latent period, $\bar{x}$ 8 min. Immobilization period, $\bar{x}$ 1.3 h. 2 chewed at snow. Author evaluation: Excellent, no mortalities, recommended dosage 2.0 mg/kg
Ross seal Erickson & Denny (unpub. data) Erickson et al., 1974	Intramuscular gluteal region (projectile syringe)	Xylazine hydrochloride (Rompun)	5 adult M 184–98 kg	2.3–3.3, $\bar{x}$ 2.7	Latent period: 20 min to 2 hr, $\bar{x}$. 55 min. 2 had slight mucus at nares. Author evaluation: no mortalities but highly varied response; appears more resistant to drug than crabeater and Weddell, recommended dosage 3.0 mg/kg.

Table 3.1 contd

Species and references	Method and route of administration	Immobilizing agent	Sex and age	Dose rate (mg drug/ kg body mass)	Comments
Grey seal *Halichoerus grypus* Parry, Anderson & Fedak, 1981	Intramuscular in lumbar region by blowgun and projectile syringe	Fentanyl citrate plus nalorphine hydrochloride (antidote)	13 adults M and F (E)	0.20–0.40	8 of 13 immobilized in $\bar{x}$ 7.8 min. 5 of 13 required 2nd or 3rd injection. Author evaluation: erratic results and untoward side effects. Not recommended for species.
Hooded seal *Cystophora cristata* Haigh & Stewart, 1979	Intramuscular in lumbar area (longissimus lumborum muscles) by projectile syringe	Fentanyl citrate plus nalorphine hydrochloride (antidote)	12 adults M and F	0.3–0.6 (E)	Rapid sedation. Time elapsed to handling 3.5–27.5 min; the longer periods after additive doses. Apnoea ensued in 50% of cases, fine muscle tremors usually. No mortalities.
Southern elephant seal Ryding, 1982	Intramuscular in dorsal hip area by hand syringe	Ketamine hydrochloride	3 adult M (1102–650 kg) 1 adult F, (279 kg) 1 pup F, (212 kg)	4.6–7.56 13.6 4.85	Dosage delivered over 2 injections in 2 males, minimal side effects with satisfactory immobilization from 1–2 h.
Northern elephant seal *Mirounga angustirostris* Briggs *et al.*, 1975	Intramuscular in dorsal hip region by hand syringe	Ketamine hydrochloride	9 lactating F 500–675 kg (E) 8 1–1.5 yrs 116–141 kg	1.7–2.9, $\bar{x}$ 2.5 1.4–6.9, $\bar{x}$ 3.8	Latent period 6–15, $\bar{x}$ 10.0 min. Recovery 23–79, $\bar{x}$ 51 min. Latent period 2–6, $\bar{x}$ 4.5 min, Recovery 10–128, $\bar{x}$ 55 min. Author evaluation: effective for species, recommended dosage 2.5–3.5 mg/kg

Species / Reference	Route	Drug	Subjects	Dosage	Remarks
Northern elephant seal Antonelis et al., 1987	Intramuscular in pelvic area by hand syringe	Ketamine hydrochloride	59 sub-adult M plus adult F	2.01–5.0	Good immobilization, slight muscle tremors in association with apnoea in some cases, increased lacrimation and salivation commonly observed. Author evaluation: optimal dose rate 2.51–3.5 mg/kg for smaller M and adult F, 2.5–3.0 mg/kg larger males. Relatively safe drug, good results, one mortality.
Harp seal *Phoca groenlandica* Engelhardt, 1977	Intravenous injections	Ketamine hydrochloride	10 pups, 10–21 days old, 15–27 kg	(3) 0.5–3.3, $\bar{x}$ 1.7 (6) 4.3–6.5, $\bar{x}$ 5.3 (4) 7.4–11.1, $\bar{x}$ 9.2	2–12, $\bar{x}$ 52 sec. 10.3 min. 15–30, $\bar{x}$ 25 sec. 18.2 min. 3–30, $\bar{x}$ 16 sec. 40.3 min. Author evaluation: Excellent drug and procedure.
Leopard seal P. Murrell (pers. comm.)	Intramuscular	Ketamine hydrochloride	Sex?, 127 kg (E)	3.5–3.7(E)	Incomplete anaesthetization but tractable 20–40 min. post injection. Entered water 80 min after injection.
Ringed seal *Phoca hispida* Geraci, 1973	Intramuscular in dorsal hip area by hand syringe	Ketamine hydrochloride	18 M and F, 15 months old, 20–52 kg	4.5–11.0 (24 trials)	Response dose related, latent period usually 2–16 min. Recovery 10–150, $\bar{x}$ 66.5 min. Tremoring common, 1 mortality. Author evaluation: effective for species; cardiovascular and Recommended dosage 4–6 mg/kg. Contraindicated when in poor condition.
Southern elephant seal Bester, 1988	Intramuscular by projection injection	Ketamine hydrochloride plus xylazine hydrochloride	5 adult F, pre-post-partum 244–254 kg (E)	Ket. 3.59–6.38, $\bar{x}$ 4.4 Xyl. $\bar{x}$ 0.9	Latent period, 10–27 min. Recovery period, 25–98 + min. $\bar{x}$ 58 min.

Table 3.1 *contd*

Species and references	Method and route of administration	Immobilizing agent	Sex and age	Dose rate (mg drug/kg body mass)	Comments
Grey seal Parry *et al.*, 1981	Intramuscular in lumbar region by blowgun, projectile syringe	Ketamine hydrochloride plus xylazine hydrochloride	14 adult M and F, (E)	Ket. 3.93 plus Xyl. 0.76	6 of 14 immobilized in $\bar{x}$ 12 min, 8 required 2nd or 3rd injection. Author evaluation: drug effective but results erratic, indicated dosage 4.0 mg/kg + 0.75 xylazine.
Grey seal Baker & Gatesman, 1985	Intramuscular in posterior lumbar region by projectile syringe	Ketamine hydrochloride plus xylazine hydrochloride	1980: 12 adult M 42 adult F 1981: 4 adult M 46 adult F	2.0–8.8, $\bar{x}$ 5.0 +1.0 xylazine 3.6–7.4, $\bar{x}$ 5.1 +1.0 xylazine	Latent periods for 27 single darted seals 13.5 min. Latent periods for 27 single darted seals 13.9 min. Author evaluation: reasonably reliable immobilization agent but some adverse side effects, mortalities in 3.8% of cases
Crabeater seal Shaughnessy, 1991	Intramuscular by hand syringe	Ketamine hydrochloride plus diazepam	7 adult M 10 adult F	Ket. 2.8–7.7, $\bar{x}$ 6.0 Diaz. 0.11–0.25, $\bar{x}$ 0.18	No adverse reactions except muscle tremors (one case), time to handling 20–40 min, 1 mortality probably unrelated to drugs.
Grey seal Anderson (pers. comm.)	Intramuscular by blowgun	Ketamine hydrochloride plus diazepam	All classes	5.9 Ket. 0.2 diazepam	This dosage used as a standard procedure to achieve a good plane of anaesthesia.
Common seal *Phoca vitulina* Anderson (pers. comm.)	Intramuscular by blowgun	Ketamine hydrochloride plus diazepam	All classes	5.9 Ket. 0.2 diazepam	This dosage used as a standard procedure to achieve a good plane of anaesthesia.

Species / Reference	Method	Drug	Animals	Dose	Comments
Southern elephant seal Gales & Burton, 1987	Intramuscular in lumbar region by projection syringe (initial dose), intravenously for additional doses	Ketamine plus diazepam or xylazine plus atropine	13 pre- & post-partum F 33 adult & sub-adult M 1 weaned pup M 897–1932 kg x̄ 465 M and F (E) 192–1102 kg x̄ 1477	Ket. 5.4–12.5, x̄ 8.7 Diaz. 0.03–1.02, x̄ 0.09 Xyl. 0.24–0.85, x̄ 0.4	Latent periods 5–70 min. Recovery periods not stated. One mortality. Author evaluation: excellent drug combination for species.
Weddell seal Gales & Burton, 1987	Intramuscular in dorsal hip area by projection syringe	Ketamine hydrochloride plus diazepam plus atropine	15 M, 137–402 kg	Ket. 4.76–11.31, x̄ 7.99 + Diaz. 0.03–0.07, x̄ 0.05	Latent periods 1–60 min, variable results, 10 showed no side effects, apnoea occurred and 3 mortalities. Author evaluation: not safe for this species.
Southern elephant seal Condy, 1980	Intramuscular 10 cm dorsal and lateral to tail by hand injection	Suxamethonium chloride (Scoline)	30 M, pups, 18–40 days 45 F pups 18–40 days	x̄ 1.91 x̄ 1.84	Latent period x̄ 3.07 min. Recovery period x̄ 12.24 min. Latent period x̄ 3.05 min. Recovery period x̄ 11.65 min. No mortalities, apnoea with dose over 3.0 mg/kg. Recommended dose 1.9 mg/kg.
Grey seal Parry et al., 1981	Intramuscular in lumbar region with blowgun projectile syringe	Etorphine hydrochloride	23 adults M & F (E)	0.005–0.018	10 of 23 immobilized in x̄ 14.6 min. 12 required 2nd or 3rd injection. Lower dosages induced deep narcosis and had to be reversed with an antagonist. Higher dose gave satisfactory anaesthetizing. Author evaluation: unsatisfactory for species.
Hooded seal Haigh & Stewart, 1979	Intramuscular	Etorphine hydrochloride plus nalorphine hydrochloride (antidote)	3	0.012–0.015 + 0.08–0.18	Convulsions and apnoea, variable results.

Table 3.1 *contd*

Species and references	Method and route of administration	Immobilizing agent	Sex and age	Dose rate (mg drug/kg body mass)	Comments
Grey seal Baker & Gatesman, 1985	Intramuscular in posterior lumbar region by projectile syringe, antagonist intravenously	Carfentanil with Naloxene hydrochloride as antagonist	1980: 42 adult M 2 adult F 1981: 34 adult M	6.7–15.3, $\bar{x}$ 9.9 μg/kg + 0.5 mg/kg Naloxone 4.2–24.9, $\bar{x}$ 10.2 μg/kg + 1.7 mg/kg Naloxone	Latent periods for 30 single darted seals $\bar{x}$ 6.5 min. Latent periods for 22 single darted seals $\bar{x}$ 10.1 min. Mortality rate 5.1% of cases. Author evaluation: reasonably good immobilizing agent for species, somewhat unpredictable.
Southern elephant seal Baker et al., 1990	Intramuscular in posterior lumbar region by projectile syringe, additional doses intravenously	Titelamine/ zolazepam mixture (1:1)	3 adult M 90 adult plus subadult F (195 trials)	Not given $\bar{x}$ 0.95 (1 dose) $\bar{x}$ 1.15 (>1 dose)	Immobilized satisfactorily, low initial doses followed by boosters, immobilizations per seal $\bar{x}$2.2, ≤ 9 repeats. Apnoea in 6 animals, no mortalities. Latent period $\bar{x}$ 10.4 min, but $\bar{x}$ 28.1 when additional doses were needed. Author evaluation: best immobilization agent available for this species.
Grey seal Baker et al., 1990	Intramuscular in posterior lumbar region by projectile syringe, additional doses intravenously	Titelamine/ zolazepam mixture (1:1)	25 adult 19 adult (84 trials)	$\bar{x}$ 1.01	Immobilized satisfactory, $\bar{x}$ 1.9 immobilizations per animal. Latent period $\bar{x}$ 12.2 min, minor tremoring in 12 cases, with apnoea in 3. No mortalities. Author evaluation: best immobilization for this species.

M = male; F = female; E = estimate

Table 3.2 *Immobilizing agents and dose rates used on otariid seals*

Species and references	Method and route of administration	Immobilizing agent	Sex and age	Dose rate (mg drug/ kg body mass)	Comments
Northern fur seal *Callorhinus ursinus* Peterson, 1965	Intramuscular by Projectile syringe	Succinylcholine	93 territorial bulls	0.26–0.40 (wts estimated)	13 sub-paralytic, 58 immobilized, 18 mortalities (mostly in trials). Effective dosage 0.30–0.39, latent periods 12 min, immobilization periods 3–10 min.
Northern fur seal Peterson, 1965	Intramuscular by projectile syringe	Succinylcholine + hyaluronidase	22 territorial bulls	0.24–0.35 plus 150 TR units hyaluronidase	Immobilization in 66%, one mortality. Latent periods reduced to 5 min. Author evaluation: Best drug for brief field immobilization but low therapeutic ratio (effective dose/lethal dose) a major drawback.
Northern fur seal Peterson, 1965	Intramuscular by projectile syringe	Nicotine alkaloid	6 adults M 5 adults M 3 adults M	< 4.0 4.0–5.0 5.6, 6.2, & 7.0	No observable reaction. Ataxia and tremors but not sedated. All mortalities. Author evaluation: Drug not successful for species.
Northern fur seal Peterson, 1965	Intramuscular by projectile syringe	Phencyclidine hydrochloride	9 adult F	0.09–0.30	5 animals immobilized, latent periods 5 min., immobilization period 1–7 hrs, 2 mortalities at 0.26 and 0.30 mg/kg. Effective dose range 0.2–0.3 mg/kg. Author evaluation: dosage rate remarkably low.
California sea lion *Zalophus californianus* Seal & Erickson, 1969	Intramuscular by pole syringe	Phencyclidine hydrochloride plus promazine hydrochloride	12 adult M & F	0.5	Author evaluation: suitable drug formulation for species

Table 3.2 (cont.)

Species and references	Method and route of administration	Immobilizing agent	Sex and age	Dose rate (mg drug/ kg body mass)	Comments
Northern fur seal Peterson, 1965	Intramuscular: pups by hand; adults by projectile syringe	Propriopromazine hydrochloride	30 pups through 5 years 11 harem bulls	1.0 2.0 3.0–4.0 2.0	Produced calmness Slight tranquilization Sedation and ataxia, latent period 15 min; effects as long as 24 h. Post breeding season use to calm for capture and tagging. Author evaluation: Bulls too aggressive to be handled with tranquilizer during breeding season.
South American sea lion, *Otaria flavescens* Cardenas & Cattan, 1984	Intramuscular	Xylazine hydrochloride	6 adult M	0.7 0.8 1.2–2.0	1 animal – no effect. 1 animal – partial immobility. 4 animals – no effect.
South American sea lion Cardenas, 1984	Intramuscular	Xylazine hydrochloride	18 mixed sexes and ages	0.4–4.87	Effective dosage for pups, 1.4 mg/kg; for adults, 2.0.
South American sea lion Ramirez, 1986	Intramuscular by projectile syringe	Xylazine hydrochloride	26 adult and subadult M 100–300 kg (estimated)	0.65–16.0	Highly variable results. Tremors, facial ticks and respiratory depression in some cases. Latent period 15–55 min, recovery 15–50 min.

Species / Reference	Route	Drug	Sample	Dosage (mg/kg)	Comments
Juan Fernandez fur seal, *Arctocephalus philippii* Cardenas, 1984	Intramuscular	Xylazine hydrochloride	6 mixed ages	0.40–4.87	Effective dosage for pups, 1.4 mg/kg; for adults, 2.0.
South African fur seal *Arctocephalus p. pusillus* David *et al.*, 1988	Intramuscular in neck and shoulder region by projectile syringe	Ketamine hydrochloride	20 adult M	4.3–7.8	5 immobilized, 5 partly immobilized, 5 mortalities. Procedure discontinued on species.
South American sea lion Ramirez, 1986	Intramuscular by projectile syringe	Ketamine hydrochloride	26 F and M 100–300 kg (E)	1.0–2.0 2.1–4.0 4.1–10.0	3 of 10 immobilized 1 of 9 immobilized 1 of 7 immobilized Highly erratic results. Author evaluation: Agitated animals resistant to drug.
New Zealand sea lion *Phocarctos hookeri* Cawthorn, 1975, in Gales, 1989	Intramuscular	Ketamine hydrochloride	11	2.5–4.5	Variable results.
Antarctic fur seal *Arctocephalus gazella* Boyd *et al.*, 1990	Intramuscular in neck and shoulder region by projectile syringe	Ketamine hydrochloride	30 adult M	$\bar{x}$ 6.94	Response roughly dependent on dose. Minor tremoring common, convulsions in one case. Dose required for this species greater than for most other species. No adverse long-term effects. No mortalities.

Table 3.2 (*cont.*)

Species and references	Method and route of administration	Immobilizing agent	Sex and age	Dose rate (mg drug/ kg body mass)	Comments
Antarctic fur seal Boyd *et al.*, 1990	Intramuscular in neck and shoulder region by projectile syringe	Ketamine hydrochloride plus xylazine hydrochloride	45 adult M	Ket. x̄ 7.32 Xyl. 0.62	Variable results. No particular advantage in using xylazine combination. 3 mortalities (one overdose, the others within normal range).
Antarctic fur seal Bester, 1988	Intramuscular injection by hand	Ketamine hydrochloride plus Xylazine hydrochloride	2 M 21.3 & 22.4 kg 7 M 16.5–22.0 kg 5 F 11.0–13.5 kg	Ket. 7.9 & 10.7 plus Xyl. 2.0 Ket. 4.4–5.6, x̄ 4.7 plus Xyl. x̄ 0.9 Ket. 3.8–6.5, x̄ 5.0 plus Xyl. x̄ 0.9	Lethal dose Latent period 4.5 min, recovery 22 min. Latent period 4.2 min, recovery 29 min. Author evaluation: stressed animals require doses above 5.6 mg/kg
South American fur seal *Arctocephalus australis* Gentry & Kooyman, 1986	Intramuscular	Ketamine hydrochloride plus Xylazine hydrochloride	Adults	Ket. 2–4 plus Xyl. 0.3–0.4	Light dosage to reduce co-ordination for handling
South African fur seal David *et al.*, 1988	Intramuscular	Ketamine hydrochloride plus Xylazine hydrochloride	2 adult	4.2 & 5.2, Xyl. not reported	Both mortalities with lung conditions, hyperthermic. Procedure discontinued on fur seals.

Species / Reference	Drug	Route	Animals	Dosage	Comments
Galapagos sea lion *Zalophus californianus wollebaeki* Gentry & Kooyman, 1986	Ketamine hydrochloride plus Xylazine hydrochloride	Intramuscular	Adult F	Ket. 2–4 plus Xyl. 0.3–0.4	Light dosage to reduce co-ordination for handling
Galapagos sea lion Trillmich & Weisner, 1979	Ketamine hydrochloride plus Xylazine hydrochloride	Intramuscular	9	Ket. 1.5–5.0 Xyl. 0.085–1.0	Problems with hyperthermia
Galapagos sea lion Trillmich, 1983	Ketamine plus Xylazine hydrochloride	Intramuscular projectile injection by blowgun	3 subadult M 60–70 kg (E) 7 adult F	Ket. 2.5–4.6, $\bar{x}$ 3.3 plus Xyl. 0.5 Ket. 2.1–7.1, $\bar{x}$ 3.6 + Xyl. 0.3–1.43, $\bar{x}$ 5.4	Latent period unstated. Recovery period 53–86, $\bar{x}$ 66 min. Latent period unstated. Recovery period 40–50, $\bar{x}$ 49 min. 1 mortality. Author evaluation: a safe immobilizing drug @ 3.0–5 mg/kg Ketamine plus 0.3–0.5 mg/kg Xylazine. Long anaesthetic period a problem.
Galapagos fur seal *Arctocephalus galapagoensis* Trillmich, 1983	Ketamine plus Xylazine hydrochloride	Intramuscular projectile injection by blowgun	1 adult F 6 adult M 1–4 day pup	Ket. 3.1–9.3 $\bar{x}$ 4.8, Xyl. 0.5–1.2, $\bar{x}$ 0.5 Ket. 18.7	Latent period not stated. Recovery 30–120, $\bar{x}$ 68 min. Mortality. Author evaluation: suitable for species @3.0–5 mg/kg Ketamine plus 0.3–0.5 mg/kg Xylazine. Long anaesthetic period a problem.
Subantarctic fur seal *Arctocephalus tropicalis* Bester (unpublished data)	Ketamine plus Xylazine hydrochloride	Intramuscular by Projectile syringe in neck and rump area	5 adult F plus 10 subadult M (21–32 days)	Ket. 4.8–7.1, $\bar{x}$ 5.66 Xyl. 0.3–0.9, $\bar{x}$ 0.62	Varied response from little effect to total immobilization, steady body temperature, no obvious negative effects. One mortality possibly due to manhandling.

Table 3.2 (*cont.*)

Species and references	Method and route of administration	Immobilizing agent	Sex and age	Dose rate (mg drug/kg body mass)	Comments
South American sea lion Ramirez, 1986	Intramuscular projectile injection by blowgun	Ketamine plus Xylazine hydrochloride	3 adult M	Ket. 0.4–1.2, Xyl. 0.8–0.65 Ket. 2.1, Xyl. 12.6	2 animals, little effect. 1 animal, mortality after 127 min.
Australian sea lion *Neophoca cinerea* Gales, 1989	Intramuscular	Ketamine hydro-chloride + diazepam	12	4.0–6.5 0.02	Good immobilization
Antarctic fur seal Boyd et al., 1990	Intramuscular by projectile syringe	Ketamine hydro-chloride + diazepam	23 adult M	$\bar{x}$ 6.35 (Ket.) $\bar{x}$ 6.3 µg (diaz.)	Response highly variable. Convulsions in two cases, minor tremoring common. One mortality due to excessive 2nd dose. No adverse long-term effects observed.
South African fur seal David et al., 1988	Intramuscular by projectile and hand-held syringe	Xylazine hydro-chloride plus Azaperone (ratio 1:1)	3 adult M 12 subadult M plus adult F	0.57–2.0 (weight estimated)	12 immobilized, 2 partly immobilized, 1 mortality. Procedure discontinued for species.
South African fur seal David et al., 1988	Intramuscular by projectile syringe	Carfentanyl plus Xylazine (plus azaperone)	12 adult M (in 7 trials)	0.006–0.018 Xylazine not stated	6 immobilized, 2 partly immobilized, 3 mortalities. Recovery times 13 min to 2.5 h. Procedure discontinued for species.

Species / Reference	Route	Drug	Animals	Dose	Comments
Northern fur seal Peterson, 1965	Intramuscular with projectile syringe–adults; rectally–subadults	Thiopental sodium	5 adult M 4 2–4 yr subadults	40–200 26–80	Only 200 mg/kg dose caused slight ataxia in adults. This near-lethal dose in most species of younger seals showed no sedation effects after 2 h. Author evaluation: Even massive doses too slow for field work.
Northern fur seal Peterson, 1965	Intramuscular by hand and projectile syringe	Insulin	6 2–5 yrs	80–150 USP units. 3 animals given glucagen as an antagonist	Long delays before coma (to 3.5 h). Large dose in 3 animals reversed by glucagen, 3 of which later died of secondary coma. Author evaluation: technique worth further testing.
Antarctic fur seal Boyd et al., 1990	Intramuscular in lumbar and shoulder region by projectile syringe	Tiletamine hydrochloride plus zolazepam (ratio 1:1)	33 adult M 139 adult F	$\bar{x}$ 1.48 $\bar{x}$ 1.58	Variable results. The only adverse effect was respiratory depression. 3 M and 2 F mortalities. Author evaluation: Optimum dose is 1.2–1.7 mg/kg, measurement of dose critical and more scope for error when used on small seals of the species.

M = male, F = female, E = estimate.

up to 10 min; recovery is in the reverse order. The length of apnoea which elephant seals can survive is sometimes alarming and characterized by cyanosis of the tongue, but breathing starts again, initially being slow and irregular. In smaller species showing cyanosis or respiratory stress, respiration may be aided by applying rhythmic thoracic pressure at a rate of 8–10 applications per min. The relatively short active period of this drug usually limits the required respiratory assistance to several minutes, unless severe over-dosage has occurred. In this case even prolonged assistance with a respiratory device may not prevent death. Dosage rate is critical and depends on accurate weight assessment.

Dose rates vary with sex and age, and possibly even season, but in general this drug has been used successfully on elephant and Weddell seals but is unpredictable and not recommended for crabeater seals (Table 3.1). Recommendations for drugging otariid seals with succinylcholine chloride are presented in Tables 3.2 and 3.4(a)(b).

Second administrations should not be attempted until the effects of the first have been overcome (Flyger *et al.*, 1965; Hofman, 1975). Some authors (Hofman, 1975; Shaughnessy, 1974) consider phencyclidine hydrochloride to be a better immobilizing agent, although Flyger *et al.* (1965), appeared to prefer succinylcholine chloride in the case of Weddell seals.

Phencyclidine hydrochloride

Injection is usually followed by progressive ataxia and loss of co-ordination. Muscular convulsions, especially of the neck and flippers, salivation and hyperthermia commonly occur (Seal & Erickson, 1969). Animals may also have difficulty breathing when under the influence of this drug. This can be alleviated by forcing air from the lungs by rhythmically pressing on the upper back region at a rate of 8–10 times/min. Calm, lethargic and docile seals tend to be more susceptible than agitated or excited ones (Cline, Siniff & Erickson, 1969; Hofman 1975). Lactating and non-lactating females appear to be equally sensitive to the drug but adult males may require larger than normal doses during the breeding season. Dose rates vary with age, but Siniff, Tester & Kuechle (1971) and Siniff *et al.* (1979), have successfully used very low doses on Antarctic phocid seals in conjunction with the sack-restraining device (Stirling, 1966).

Dose rates, when used without the sack method, vary with age, but the drug has been widely and successfully used on both phocid and otariid seals. Dosage rates for Antarctic phocid and otariid seals of phencyclidine hydrochloride are presented in Tables 3.3(a)(b) and 3.4(a)(b).

Inclusion of a tranquilizer with phencyclidine is effective in reducing muscular convulsions, excessive salivation and hyperthermia associated with phencyclidine (Seal & Erickson, 1969). The tranquilizers promazine hydrochloride and diazepam hydrochloride used with half to equal amounts of phencyclidine hydrochloride, have proved to be effective. Cline *et al.* (1969), administered an additional 1–2 ml of promazine hydrochloride whenever prolonged or excessive excitation and minor convulsions occurred following initial doses. Phencyclidine hydrochloride used in conjunction with promazine hydrochloride or diazepam hydrochloride used to be the drug of choice for immobilizing Antarctic seals. However, since the introduction of ketamine hydrochloride, phencyclidine has been used very little (Gales, 1989). Immobilization experiences and drug dosages for these drug combinations are presented in Tables 3.1 and 3.2.

Nicotine alkaloid

Flyger *et al.* (1965), used nicotine alkaloid as an immobilizing drug on Weddell seals and found it to be unsatisfactory because of the violent side effects, although immobilization was achieved. They noted severe muscular spasms, groaning and vomiting following administration, but suggest that these effects could be reduced or eliminated by addition of an anticonvulsant such as pentobarbital sodium. See Tables 3.1 and 3.2 for a summary of immobilization experience with nicotine.

Ketamine hydrochloride

Ketamine produces an increase in heart rate and blood pressure together with mild depression in most animals (Green, 1979) and has been widely used with success to immobilize seals. The desirable features of using ketamine hydrochloride in both ringed and elephant seals were the rapid induction of anaesthesia, a wide margin of safety, little interference with the cardiovascular and thermoregulatory mechanism, rapid and uncomplicated recovery and intramuscular administration (Geraci, 1973; Briggs, Henrickson & Le Boeuf, 1975). Quivering and violent muscular shaking, occasionally occur (Geraci, 1973) but slight muscle tremoring is a more common side effect (Briggs *et al.*, 1975; Antonelis *et al.*, 1987; Boyd *et al.*, 1990) which may be associated with apnoea and increased lacrimation and salivation (Antonelis *et al.*, 1987). Very few mortalities

Table 3.3(a) *Recommendations for immobilizing drugs administered by intramuscular injection – phocid seals*[1]

Species	Drug, dosage (mg/kg bodyweight), and efficacy rating[2]							
	Succinylcholine chloride	Nicotine alkaloid	Phencyclodine hydrochloride	Phencyclodine hydrochloride + promazine hydrochloride	Phencyclodine hydrochloride + Valium	Xylazine hydrochloride	Fentanyl citrate	Ketamine hydrochloride
Mirounga leonina	EF-1 2.0 mg/kg	–	EF-1 0.5 mg/kg	EF-1 0.3 mg/kg each	–	–	–	EF-1 2.5–3.5 mg/kg
Leptonychotes weddellii	EF-2 2.8 mg/kg	EF-4	EF-2 0.5 mg/kg	EF-2 0.3 mg/kg each	EF-1 0.4 mg/kg + Valium 0.1 mg/kg	EF-1 2.0 mg/kg	–	–
Lobodon carcinophagus	EF-4	–	EF-2 0.5 mg/kg	EF-1 0.7 mg/kg each	–	EF-1 2.5 mg/kg	–	–
Hydrurga leptonyx	–	–	EF-2 0.5 mg/kg	EF-1 0.7 mg/kg each	EF-2 0.3 mg/kg + Valium 0.1 mg/kg	–	–	–
Ommatophoca rossii	–	–	EF-2 0.5 mg/kg	EF-1 0.6 mg/kg each	EF-1 1.5 mg/kg	EF-2 3.0 mg/kg	EF-4	–
Halichoerus grypus	–	–	–	–	–	–	EF-3 0.2–0.4 mg/kg	–
Phoca groenlandica	–	–	–	–	–	–	–	EF-1 50 mg/kg
Phoca vitulina	–	–	–	–	–	–	–	–

[1] Includes only drugs with sufficient use for evaluation.
[2] Efficacy rating (EF) of drug: 1 = effective with low mortality; 2 = effective but with undesirable side effects or 5–10% mortality; 3 = relatively ineffective or erratic; 4 = not recommended due to high mortalities.

Table 3.3(b) *Recommendations for immobilizing drugs administered by intramuscular injection – phocid seals* [1]

	Drug, dosage (mg/kg bodyweight), and efficacy rating [2]						
Species	Ketamine hydrochloride + xylazine	Ketamine plus diazepam	Ketamine + diazepam + xylazine or atropine	Suxamethonium chloride (Scoline)	Etorphine hydrochloride	Carfentanyl + antagonist Naloxone	Titelamine + Zolazepam
Mirounga leonina	EF-1 4.5 + 0.9 xylazine	—	EF-1 8.0 + 0.1 Diazepam + 0.4 Xylazine	EF-1 1.9	—	—	EF-1 1.0 + 1.0 Zolazepam
Leptonychotes weddellii	—	—	—	—	—	—	EF-1 0.5-1.0 + 0.25-0.5 Zolazepam
Lobodon carcinophagus	—	EF-1 6.0 + 0.2 Diazepam	—	—	—	—	—
Hydrurga leptonyx	—	—	—	—	—	—	—
Ommatophoca rossi	EF-3 4.0 + 0.75 xylazine	—	—	—	EF-4	—	—
Halichoerus grypus	EF-2 5.0 + 0.7 xylazine	EF-1 5.9 + 0.2 Diazepam	—	—	EF-3 0.005-0.018	EF-2 10-15 + 0.5 Naloxone	EF-1 1.9 each
Phoca groenlandica	—	—	—	—	—	—	—
Phoca vitulina	—	EF-1 5.9 + 0.2 Diazepam	—	—	—	—	EF-1 0.5-1.3 + 0.25-0.7 Zolazepam

[1] Includes only drugs with sufficient use for evaluation.
[2] Efficacy rating (EF) of drug: 1 = effective with low mortality; 2 = effective but with undesirable side effects or 5-10% mortality; 3 = relatively ineffective or erratic; 4 = not recommended due to high mortalities.

Table 3.4(a) *Recommendations for immobilizing drugs administered by intramuscular injection – otariid seals*[1]

Species	Succinylcholine chloride	Nicotine alkaloid	Phencyclidine hydrochloride	Phencyclidine + Promazine	Promazine hydrochloride	Xylazine hydrochloride	Xylazine hydrochloride + Azaperone
				Drug, dosage (mg/kg body weight) and efficacy rating[2]			
Arctocephalus gazella	–	–	–	–	–	–	–
Arctocephalus phillippi	–	–	–	–	–	EF-2 pups 1.4, adults 2.0	–
Arctocephalus australis	–	–	–	–	–	–	–
Arctocephalus pusillus	–	–	–	–	–	–	EF-2 0.6–0.2
Arctocephalus galapagoensis	–	–	–	–	–	–	–
Callorhinus ursinus	EF-2 0.3–0.4	EF-4	EF-2 0.2–0.3	–	EF-3 0.3	–	–
Otaria flavescens	–	–	–	–	–	EF-2 pups 1.4, adults 2.0	–
Zalophus californianus	–	–	–	EF-1 0.5 + 0.5 Promazine	–	–	–

[1] Includes only drugs with sufficient use for evaluation.
[2] Efficacy rating (EF) of drug: 1 = effective with low mortality; 2 = effective but with undesirable side-effects or with 5–10% mortality; 3 = relatively ineffective or erratic; 4 = not recommended due to high mortalities.

Table 3.4(b) *Recommendations for immobilizing drugs administered by intramuscular injection – otariid seals*[1]

Species	Drug, dosage (mg/kg body weight) and efficacy rating[2]					
	Ketamine	Ketamine + xylazine	Carfentanyl + xylazine	Thiopental sodium barbiturate	Insulin	Titelamine + Zolazepam
Arctocephalus gazella	—	EF-2 4.0–5.0 + 2.0 xylazine	—	—	—	EF-2 1.2–1.7 + 1.6 Zolazepam
Arctocephalus phillippi	—	—	—	—	—	—
Arctocephalus australis	—	EF-2 3.0–4.0 + 0.4 xylazine	—	—	—	—
Arctocephalus pusillus	EF-4	—	EF-4	—	—	—
Arctocephalus galapagoensis	—	EF-2 3.0–5.0 + 0.5 xylazine	—	—	—	—
Callorhinus ursinus	—	—	—	EF-3 200	EF-4	—
Otaria flavescens	EF-3 4.0–10	—	—	—	—	—
Zalophus californianus	—	EF-3 3.0–5.0 + 0.5 xylazine	—	—	—	EF-4

[1] Includes only drugs with sufficient use for evaluation.
[2] Efficacy rating (EF) of drug: 1 = effective with low mortality; 2 = effective but with undesirable side-effects or with 5–10% mortality; 3 = relatively ineffective or erratic; 4 = not recommended due to high mortalities.

occurred when using ketamine in phocids (Table 3.1), except for Weddell seals which showed high mortalities (Hammond & Elsner, 1977). Otariids have been drugged with variable results and while satisfactory in Antarctic fur seals with no mortalities (Boyd *et al.*, 1990) the use of ketamine has been rejected as unsafe for South African (Cape) fur seals (David *et al.*, 1988).

Sedatives such as diazepam and xylazine hydrochloride are widely used in combination with ketamine to minimize undesirable side effects. In conjunction with diazepam it again proved to be unsafe for immobilizing Weddell seals (Gales & Burton, 1988) but was relatively safe and practical with few side effects in crabeater seals and elephant seals (Gales & Burton, 1987; Shaughnessy, 1991), satisfactory in sea lions (Gales, 1989) and male Antarctic fur seals without severe respiratory depression being observed (Boyd *et al.*, 1990). Boyd *et al.* (1990) experienced no particular advantage by using diazepam or xylazine hydrochloride with ketamine in Antarctic fur seals and suggested that its presence may increase the mortality rate. Nevertheless, the ketamine–xylazine mixture proved a reasonably reliable immobilizing agent for a number of fur seals (Table 3.2) and phocids (Table 3.1), and excellent for southern elephant seals (Gales & Burton, 1987; Bester, 1988).

Tiletamine hydrochloride–zolazepam hydrochloride combination

Marketed as Zoletil; or Telazol, the 1:1 mixture of tiletamine and zolazepam has successfully been used to immobilize over 800 walruses (*Odobenus rosmarus*) with no unwanted side effects (Stirling & Sjare, 1988; I. Stirling & B. Sjare, pers. comm. in: Gales, 1989). Described as having the smoothest induction and fewest side effects of all the drugs they used in harbour and Weddell seals (Hammond & Elsner, 1977), the tiletamine–zolazepam combination was also highly successful in the immobilization of both grey seals and southern elephant seals (Baker *et al.*, 1990). Only grey seals showed a minor degree of tremor, but in a few instances both species became apnoeic and artificial respiration was given. However, use of this drug combination at levels that were successful in walruses, resulted in death in two of five southern elephant seals (P. Mitchell, pers. comm. in: Gales, 1989).

It has been used with less success in otariids where the best results were obtained at doses between 1.8 and 2.5 mg/kg in sea lions (Loughlin & Spraker, 1989) and 1.2 to 1.7 mg/kg in Antarctic fur seals (Boyd *et al.*,

1990). Respiratory depression was the only adverse side effect (Boyd *et al.*, 1990), and seen together with the high mortalities in sea lions (six of 29 and two of five in *Eumetopias jubatus* and *Zalophus californianus*, respectively) recorded by Loughlin & Spraker (1989) and Grey, Bush & Heck (1974), it is not ideal for otariids. Boyd *et al.* (1990) suggested its use as a replacement for ketamine in Antarctic fur seals with its main advantages being a smaller volume of injectate which overcomes some practical problems when immobilizing large seals, and the virtual absence of the side-effects associated with ketamine anaesthesia. This drug combination deserves further investigation for use in Antarctic seals.

Other drugs

Various other drugs such as xylazine hydrochloride, fentanyl citrate, etorphine hydrochloride and carfentanil (Tables 3.1 and 3.2) have been used singly or in combination with other drugs as immobilizing agents for seals. Xylazine hydrochloride at doses of 1.5 to 3.9 mg/kg provided for satisfactory immobilization in Antarctic phocids (crabeater, Ross and Weddell seals), with varied response in southern elephant seals, where vomiting and hyperthermia were recorded side effects (Vergani, 1985). Greater experience with these drugs is necessary before firm recommendations can be made relative to their use on seals.

Resuscitation and some precautions

Apnoea accompanied by alarming cyanosis of the mouth and tongue is not infrequent in immobilized seals. Prolonged apnoea or sudden collapse may call for treatment by artificial respiration as noted earlier.

When administering artificial respiration one should straighten the seal's neck and see that the tongue or mucous fluids are not blocking the throat. The naral plates should also be opened, as the animal may be in dive reflex, and the larynx opened with a probe. To apply artifical respiration one should press on the thorax at a rate of 8–12 times per min to ventilate the lungs and stimulate natural breathing. These procedures are generally effective with drugs of short duration such as succinylcholine chloride. When possible, longer acting drugs should be counteracted with an antagonist in addition to application of artificial respiration. Tracheal intubation and ventilation by mouth as described by Baker *et al.* (1990) may be necessary until spontaneous breathing has been restored.

A further precaution to be taken during anaesthetizing is to ensure that immobilized seals are not left with their head and thorax inclining downward. Because seals do not possess venous valves, this coupled with relaxation of the skeletal musculature accompanied by bradycardia may allow blood pooling (Seal & Erickson, 1969). Blood pooling and fluid build up in the nares, mouth and throat can also be reduced by periodically moving immobilized seals.

Researchers are also cautioned that immobilization and restraint can cause haemodilution and other blood effects which may affect recovery responses and affect other baseline data (Seal *et al.*, 1972).

Other capture techniques

In Antarctica, crabeater, Weddell, Ross and elephant seals can be easily approached. However, because of their size and strength, even two persons may not be able to restrain some seals without the aid of either immobilizing drugs or some form of restraining apparatus.

Many capture methods have been devised that do not use drugs (Ronald *et al.*, 1970; Smith, Beck & Steno, 1974), but Stirling (1966) devised the simplest technique for capturing and restraining Antarctic seals. His apparatus consisted of a sack with four ropes attached to each quarter round the sack mouth. Two sizes of sack were used, one 45 cm in diameter and 61 cm deep, the other 66 cm in diameter and 91 cm deep. Two persons each holding two ropes stood on either side of the seal and pulled the sack over its head. Once secured, the seal could be handled by one person, provided enough tension was put on the ropes to keep the sack in position. The larger sack fitted over the fore flippers and held them immobile, so that the seals were more restricted than when the smaller sack was used. Most seals restrained in this way became calm within a few minutes. Doses of 0.23 mg/kg phencyclidine hydrochloride may be used in concert with the sack restraining device (Stirling, 1966). Siniff *et al.* (1979) used 0.2–0.4 mg/kg on adult crabeater seals on ice floes to reduce mobility so as to allow capture by the sack technique. Following capture, anoxia caused by the rebreathing of air in the confining sack seemed to increase drug effectiveness. Seals handled in this manner recovered quickly after the sack was removed.

Among physical capture techniques employed for otariids, such as nets, hoop nets, noose poles and portable fence with restraint cages, restraining boards or restraint bars (Gentry & Johnson, 1978; Gentry & Holt, 1982; David, Meyer & Best, 1990) the long-handled hoop net described by

David *et al.* (1990) is particularly useful. Modified from the already developed aluminium hoop net (Gentry & Holt, 1982) it is fitted distally with a tapering, cone-shaped sleeve of opaque PVC material or terylene sailcloth. Open and reinforced with a circular piece of rope sewn into the plastic, a thick-walled, plastic cylinder with a layer of fine wire mesh at the distal end is anchored in the open end with a hose clamp. Once trapped and lightly wedged inside, the PVC cone acts as an effective blind-fold of the fur seal. By placing a mask over the plastic cylinder in close proximity to the nose of the seal, chloroform can be administered with excellent results (David *et al.*, 1990). The use of inhalation anaesthetics has been reviewed by Gales (1989) and it appears that chloroform provides the best results under field conditions, notably a finely controlled tranquilizing effect coupled with rapid and complete recovery without mortalities (David *et al.*, 1990).

Recording results

Although a large amount of drug testing and routine immobilization of southern seals has occurred, considerable variation exists in the results reported. It appears that as yet only phencyclidine hydrochloride and ketamine hydrochloride produce reasonably predictable results, although the dosage rates for all drugs tried vary, depending on species, sex, age, and physiological status. There is a need for further experimentation and exchange of data between researchers. A narcotic such as fentanyl citrate or M99 (etorphine hydrochloride) with a tranquilizer such as Azaperone (fluperidol) may yet prove to be the most satisfactory immobilizing mixture. A drug combination of this nature would have the advantage that the narcotic effect could be partly or totally reversed using either nalorphine hydrobromide for fentanyl citrate and nalorphine hydro-chloride or disreorphine for etorphine hydrochloride.

Whether drug-testing or routine immobilization is being performed, it is very important to record data such as dose rates, time to first effects, time to complete immobilization, symptoms of induction and their progression and recovery time. Additionally, since these factors can be influenced by age, sex, season, social status, pregnancy, lactation and condition, such information should also be noted. Also, every effort should be made to obtain the body weight of immobilized seals. A data sheet such as that illustrated in Fig. 3.2 (modified from Harthoorn, 1976) should be used.

SPECIES: .... PLACE OF CAPTURE: ....
DATE: ....
DRUGS USED (and concentrations): ....
....
....
ENVIRONMENTAL CONDITIONS (temperature, wind, etc): ....
....

No. or name	Sex	Age	Estimated weight (kg)	Time of injection	Immobilizing drug (mg)	Tranquilizer (mg)	Additional drug (mg)	Dart size and route	Induction time (first effects) (min)	Full induction time (min)	Restraint method	Antidote (mg)	Reaction time (min)	Subsequent procedure (marking, crating, sampling)	Heart rate	Respiration rate	Rectal temperature °C	Remarks (including viability, mortality)

Fig. 3.2 Drug immobilization data Form

Conclusions

Immobilization as a technique for handling seals still requires much more research, and success is highly dependent on the use of correct procedures and documentation of results. In theory, a reversible drug such as etorphine should be most acceptable, but in practice it seems that ketamine hydrochloride in combination with a suitable tranquilizer, and the tiletamine HCL-zolazepam HCL mixture are the current drugs of choice. Dose rates are still largely uncertain and the summaries provided in Tables 3.1 and 3.2 should be regarded as records of use rather than firm recommendations.

Tables 3.3 and 3.4 present summaries of the drugs used to immobilize seals, efficacy ratings of the suitability of various drugs for the chemical immobilization of seals, and indicated dosages of these drugs for the several species of Antarctic seals and other representative seal species.

References

Antonelis, G.A., Lowry, M.S., De Master, D.P. & Fiscus, C.H. 1987. Assessing northern elephant seal feeding habits by stomach lavage. *Mar. Mamm. Sci.* **3**, 308–22.

Baker, J.R., Fedak, M.A., Anderson, S.S., Arnbom, T. & Baker R. 1990. Use of tiletamine-zolazepam mixture to immobilise wild grey seals and southern elephant seals. *Vet. Rec.* **126**, 75–7.

Baker, J.R. & Gatesman, T.J. 1985. Use of carfentanil and a ketamine-xylazine mixture to immobilize wild grey seals (*Halichoerus grypus*). *Vet. Rec.* **116**, 208–10.

Bester, M.N. 1988. Chemical restraint of Antarctic fur seals and southern elephant seals. *S. Afr. J. Wildl. Res.* **18**, 57–60.

Boyd, I.L., Lunn, N.J., Duck, C.D. & Barton, T. 1990. Response of Antarctic fur seals to immobilisation with ketamine, a ketamine-diazepam or ketamine-xylazine mixture, and zoletil. *Mar. Mamm. Sci.* **6**, 135–45.

Briggs, G.D., Henrickson, R.V. & Le Boeuf, B.J. 1975. Ketamine immobilization of northern elephant seals. *J. Am. Vet. Med. Assoc.* **167**, 546–8.

Bryden, M.M. 1969. Growth of the southern elephant seal, *Mirounga leonina* (Linn). *Growth* **33**, 69–82.

Bryden, M.M. 1972. Body size and composition of elephant seals (*Mirounga leonina*): Absolute measurements and estimates from bone dimensions. *J. Zool. Lond.* **167**, 265–76.

Cardenas, J.C. 1984. *Evaluation de un metodo de inmovilizacion quimica a distancia en pinnipedos otaridos en condiciones naturales.* Tesis para optar all titulo de medico veterinario y al grado de licenciado en ciencias pecuarias y medico veterinarias. Univ. de Chile, Coll. Veterinarias y Forestales, Santiago.

Cardenas, J.C. & Cattan, P. 1984. Administracion de chlorhidrato de xilacina en otaridos en condiciones naturales mediante el emplo de cerbatana. *Primera Reunion de Trabozo de Expertos en Mamiferos acuaticos de America del Sur.* Resumes 25–9 junio 1984. Buenos Aires, Argentina. Pages 14. (English trans. B. Roderic & A.M. Mast.)

Cline, R.D., Siniff, D.B. & Erickson, A.W. 1969. Immobilizing and collecting blood from Antarctic seals. *J. Wildl. Mgmt.* **33**, 138–44.

Condy, P.R. 1977. Fourth seal survey in the King Haakon VII Sea,

Antarctica – January/February 1977. Unpubl. report, Mamm. Res. Inst., University of Pretoria, Pretoria, South Africa.

Condy, P.R. 1979. Population biology, population dynamics and feeding ecology of the Ross Seal, *Ommatophoca rossi* (Gray, 1844). Unpub. Rept, Mamm. Res. Inst., University of Pretoria, Pretoria, South Africa.

Condy, P.R. 1980. Postnatal development and growth in southern elephant seals (*Mirounga leonina*) at Marion Island. *S. Afr. J. Wildl. Res.* **10**, 118–22.

David, J.H.M., Hofmeyr, J.M., Best, P.B., Meyer, M.A. & Shaughnessy, P.D. 1988. Chemical immobilization of free-ranging South African (Cape) fur seals. *S. Afr. J. Wildl. Res.* **18**, 154–6.

David, J.H.M., Meyer, M.A. & Best, P.B. 1990. The capture, handling and marking of free-ranging adult South African (Cape) fur seals. *S. Afr. J. Wildl. Res.* **20**, 5–8.

Engelhardt, F.R. 1977. Immobilization of harp seals (*Phoca groenlandica*), by intravenous injection of ketamine. *Comp. Biochem. Physiol.* **56C**, 75–6.

Erickson, A.W., Denny, R., Brueggeman, J.J., Sinha, A.A., Bryden, M.M. & Otis, J. 1974. Seal and bird populations of Adelie, Claire, and Banzare coasts. *Antarct. J. United States* **9**, 292–6.

Flyger, V., Smith, M.S.R., Damm, R. & Peterson, R.S. 1965. Effects of three immobilizing drugs on Weddell seals. *J. Mammal.* **46**, 345–7.

Gales, N.J. 1989. Chemical restraint and anesthesia of pinnipeds: a review. *Mar. Mamm. Sci.* **5**, 228–56.

Gales, N.J. & Burton, H.R. 1987. Prolonged and multiple immobilizations of the southern elephant seal using ketamine hydrochloride-xylazine hydrochloride or ketamine hydrochloride-diazepam combinations. *J. Wildl. Disease* **23**, 614–61.

Gales, N.J & Burton, H.R. 1988. Use of emetics and anaesthesia for dietary assessment of Weddell seals. *Aust. Wildl. Res.* **15**, 423–33.

Gentry, R.L. & Holt, J.R. 1982. *Equipment and Techniques for Handling Northern Fur Seals*. NOAA Techn. Rep. NMFS 758, pp. 1–15.

Gentry, R.L. & Johnson, J.H. 1978. Physical restraint for immobilizing fur seals. *J. Wildl. Mgmt.* **42**, 944–6.

Gentry, R.L. & Kooyman, G.L. 1986. Methods of dive analysis. In *Fur Seals: Maternal Strategies on Land and at Sea*, ed. R.L. Gentry & G.L. Kooyman, pp. 28–40. Princeton: Princeton Univ. Press.

Geraci, J.R. 1973. An appraisal of ketamine as an immobilizing agent in wild and captive pinnipeds. *J. Amer. Vet. Med. Assoc.* **163**, 574–7.

Green, C.J. 1979. *Animal Anesthesia*. Laboratory Animals Ltd., London.

Grey, C.W. Bush, M. & Heck, C.C. 1974. Clinical experience using CI 744 in chemical restraint and anesthesia of exotic species. *J. Zoo Anim. Med.* **5**, 12–21.

Haigh, J.C. 1978. Use and abuse of drugs for chemical restraint of wildlife. *Vet. Clinics of North America* **8**, 343–52.

Haigh, J.C. & Stewart, R.E.A. 1979. Narcotics in hooded seals (*Cystophora cristata*): a preliminary report. *Can. J. Zool.* **57**, 946–9.

Hammond, D. & Elsner, R. 1977. Anesthesia in phocid seals. *J. Zoo. Anim. Med.* **8**, 7–13.

Harthoorn, A.M. 1976. Application of pharmacological and physiological principles in restraint of wild animals. *Wildl. Monogr.* **14**, 1–78.

Hofman, R.J. 1975. *Distribution Patterns and Population Structure of Antarctic Seals*. Ph.D. Thesis, Univ. Minnesota.

Hofman, R.J., Reichle, R. Siniff, D. & Muller-Schwarze, D. 1975. Behavior and movements of leopard seals at Palmer Station, Antarctica. In *Proc. Third Symp. on Antarct. Bio., Washington, DC, Aug. 1974.* Washington, DC: Smithsonian Institution.

Ling, J.K. & Bryden, M.M. 1981. Southern elephant seal (*Mirounga leonina* Linnaeus, 1758). In *Handbook of Marine Mammals*, ed. R.J. Harrison & S.H. Ridgway, pp. 297-327. New York, Academic Press.

Ling, J.K. & Nicholls, D.E. 1963. Immobilization of southern elephant seals using succinylcholine chloride. *Nature* **200**, 1021.

Ling, J.K., Nicholls, D.E. & Thomas, C.D.B. 1967. Immobilization of southern elephant seals with succinylcholine chloride. *J. Wildl. Mgmt.* **31**, 468-79.

Loughlin, T.R. & Spraker, T. 1989. Use of Telazol to immobilize female northern sea lions (*Eumetopias jubatus*) in Alaska. *J. Wildl. Dis.* **25**, 353-58.

McLaren, J.A. 1958. The biology of the ringed seal (*Phoca hispida* Schreber) in the eastern Canadian Arctic. *Bull. Fish. Res. Bd. Can.* **118**, 1-97.

Parry, K., Anderson, S. & Fedak, M.A. 1981. Chemical immobilization of gray seals. *J. Wildl. Mgmt.* **45**, 986-90.

Peterson, R.S. 1965. Drugs for handling fur seals. *J. Wildl. Mgmt.* **29**, 688-93.

Ramirez, G.D. 1986. Rescue of entangled South American sea lions (*Otaria flavescens*). Report for the Center for Environmental Education, Argentina.

Ronald, K., Johnson, E., Foster M.E. & Vander Pol, D. 1970. The harp seal, *Pagophilus groenlandicus* (Erxleben 1777). I. Methods of handling, molt, and diseases in captivity. *Can. J. Zool.* **48**, 1035-40.

Ross, G.J.B., Ryan, F., Saayman, G.S. & Skinner, J. 1976. Observations on two captive crabeater seals *Lobodon carcinophagus* at the Port Elizabeth Oceanarium. *Int. Zoo Yearb.*, **16**, 160-4.

Ross, G.J.B. & Saayman, G.S. 1970. A young bull elephant seal immobilized. *Afr. Wildl.* **24**, 331-6.

Ryding, F.N. 1982. Ketamine immobilization of southern elephant seals by a remote injection method. *Br. Antarct. Surv. Bull.* **57**, 21-6.

Shaughnessy, P.D. 1974. An electrophoretic study of blood and milk protein of the southern elephant seal *Mirounga leonina*. *J. Mammal.* **55**, 796-808.

Shaughnessy, P.D. (1991). Immobilisation of crabeater seals *Lobodon carcinophagus* with ketamine and diazepam. *Wildl. Res.* **18**, 165-68.

Seal, U.S. & Erickson, A.W. 1969. Immobilization of Carnivora and other mammals with phencyclidine and promazine. *Federation Proceedings* **28**, 1410-19.

Seal, U.S., Ozoga, J.J. Erickson, A.W. & Verme, L.J. 1972. Effects of immobilization on blood analysis of white-tailed deer. *J. Wildl. Mgmt.* **36**, 1034-40.

Siniff, D.B., Stirling, I. Bengtson, J.L. & Reichle, R.A. 1979. Social and reproductive behaviour of crabeater seals (*Lobodon carcinophagus*) during the austral spring. *Canad. J. Zool.* **57**, 2243-55.

Siniff, D.B., Tester, J.R. & Kuechle, D.B. 1971. Some observations on the activity patterns of Weddell seals as recorded by telemetry. In *Antarctic Pinnipedia*, ed. W.H. Bert, pp. 173-80. *Ant. Res. Ser. No. 18.* American Geophysical Union.

Smith, T.G., Beck, B. & Steno, G.A. 1974. Capture, handling and branding of ringed seals. *J. Wildl. Mgmt.* **37**, 579–83.

Stirling, I. 1966. A technique for handling live seals. *J. Mammal.* **47**, 543–4.

Stirling, I. & Sjare, B. 1988. Preliminary observations on the immobilization of male Atlantic walruses (*Odobenus rosmarus rosmarus*) with Telazol. *Mar. Mamm. Sci.* **4**, 163–8.

Trillmich, F. 1983. Ketamine xylazine combination for the immobilization of Galapagos sea lions and fur seals. *Vet. Rec.* **112**, 279–80.

Trillmich, F. & Weisner, H. 1979. Immobilization of free-ranging Galapagos sea lions Zalophus californianus wollebaeki. *Vet. Rec.* **105**, 465–6.

Usher, P.G. & Church, M. 1969. On the relationship of weight, length, and girth of the ringed seal (Pusa hispida) of the Canadian Arctic. *Arctic*, 120–9.

Vergani, D.F. 1985. *Comparative Study of Populations in Antarctica and Patagonia of the Southern Elephant Seal*, Mirounga leonina *(Linne, 1758) and its Methodology.* Publication No 15. Direccion Nacional del Antartico, Instituto Antartico Argentino.

Vergani, D.F., Spairani, H.J. & Aguirre, C.A. 1986. *Immobilization of Crabeater Seals*, Lobodon carcinophagus, *with the use of Xilazine Hydrochloride at 25 De Mayo Island (Antarctica) and Identification of Polymorphism in Transferrins.* Contribution No 317. Direccion Nacional del Antartico, Instituto Antartico Argentino.

4

Marking techniques

A.W. ERICKSON, M.N. BESTER AND R.M. LAWS

Introduction

In studies of population ecology or behaviour it is essential to be able to mark individuals in such a way that they can be positively identified when they are re-observed. For certain purposes, however, it is sufficient to identify year classes or a specific group of seals. Even then, if it can be done, individual marking should be attempted because it has the potential of giving so much more information.

The techniques adopted will depend upon the objectives of the research. Long-term marking by brands, tags or intra-vitam staining is necessary for validating age criteria by means of known-age animals; for stock identification; for studies of population dynamics and longevity; for dispersal and migration studies; and for assessing breeding success (e.g. the frequency of pupping). Also in behavioural studies, temporary conspicuous marks can often be used individually or in combination with a routine tagging programme.

The more permanent marking methods usually require a measure of restraint of the animal. This usually presents no problem where pups are concerned, and even adults of some species can be tagged without first restraining them. Usually, however, adults must be immobilized, particularly if reliable weights and measurements are to be taken, if a tooth is to be extracted for age determination, or if the marking procedure is complex. Methods of immobilization and capture are described in chapter 3. The bagging technique devised by Stirling (1966) is adequate for several of the ice-breeding seals with or without partial drugging to slow them down. Rand (1950) and J.K. Ling (pers. comm.) used nets or bags to catch Cape fur seals, Australian sea lions and southern elephant seals to 500 kg in weight, the bags being more successful. The secret of the bagging or net technique is to pinion the animals' flippers to their sides, thus inhibiting their mobility. For methods to restrain otariid seals, the reader is referred to Gentry & Holt (1982) and David, Meyer & Best (1990). Drug immobilization may be necessary when marking large elephant seals, leopard seals and fur seals.

Marking techniques

In studies of the population ecology and behaviour of animals, scientists have devised many ways of identifying or marking individual animals (Taber & Cowan, 1969). These range from naturally occurring coat patterns and body scars, to artificial scarring, such as brands and tattoos, and applied tags, collars and other devices.

Natural markings

The use of natural body markings and scars is largely limited to narrowly defined, short-term behaviour studies of specific groups of animals, as recognition of the subject animals seldom extends beyond a single group of investigators. Notable exceptions are studies of humpback whales (Katona *et al.*, 1979), southern right whales (Payne *et al.*, 1981) and killer whales (Bigg, MacAskie & Ellis, 1976; Balcomb *et al.*, 1980) in which international catalogues of photo-identifiable whales are maintained.

The relatively few studies of seals based on natural body patterns or scars have been limited to colonial species. These studies have included grey seals (Fogden, 1961), northern elephant seals (LeBoeuf, 1972), Steller sea lions and California sea lions (Gentry, 1979).

To date, none of the Antarctic ice-breeding seals have been studied extensively in this way. The difficulty lies in the large numbers and the pack ice distribution of all species except the inshore breeding populations of Weddell seals. However, this species could be studied in this way using ventral pelage markings (I. Stirling pers. comm.).

Temporary marking techniques

On occasion, researchers may wish to apply temporary visual markers to seals, either as an adjunct to a permanent marking programme or in short-term studies. These marking techniques are often more easily performed than permanent marking methods, as in many instances the seals do not need to be restrained during marking. Furthermore, temporary markers often permit the remote identification of seals without disturbance.

The disadvantage of temporary markers is that the continuity of investigations beyond a single season is usually not possible unless coupled with a permanent marking programme.

Among the various temporary marking techniques are paint markings, dyes and bleaches, pelage clippings, and streamers attached to permanent tags and collars.

Paints, bleaches and dye marking

Gentry (1979) presents a perceptive analysis of the foibles attendant on the myriad approaches researchers have used to mark seals with paints, dyes and bleaches. Difficulties in using these marking techniques include the short duration of the marks and problems associated with applying discernible marks, particularly if the animals are wet. Normally, the maximum duration of the applied marks is until the following moult.

Paint marking Painting has proved to be an effective marking procedure in a few seal studies, involving a limited number of animals. The kinds of paints used have included marine paint, rubber-based highway paint, quick-drying cellulose paint, aerosol spray paint and general house paint. It naturally follows that unless a quick drying paint is employed, the applied markers will be subject to considerable smearing. Although evidence of a paint mark may persist until moult, most marks fade rapidly and are usually unreadable after about one month.

Examples of paint marking programmes on seals include Laws (1956), who used quick-drying cellulose paint to mark southern elephant seals, which have relatively sparse hair. The paint marks were applied using interchangeable brushes on a bamboo pole. Cows were painted with a number, while adult bulls were marked by spots or other marks, often on the bare skin of the proboscis. The majority of bulls did not need remarking over an eight-week period, but cows required repainting at two-week intervals.

Similarly, LeBoeuf (1971) painted northern elephant seals by filling plastic bags with marine paint in bright colours, which burst when thrown against a seal. Identification of individual animals was by location and colour of the mark – for example, YLN = yellow left neck, CLS = crimson left side. The marks were conspicuous and lasted several months but the number of combinations was limited. A paint roller on a long stick offers an alternative method of marking elephant seals.

LeBoeuf also used a carbon dioxide operated pistol, the Nel-Spot colour marking gun (Nelson Paint Company, Box 907, Iron Mountain, Michigan 49801, USA) to mark northern elephant seals. The gun fires capsules of non-toxic paint which break on impact to leave a bright mark 5–10 cm in diameter. The gun loads 14 capsules at a time and is accurate up to about 15–20 m. However, it should be noted that the capsules often did not break when fired at a seal except at point-blank range and even then the small amount of paint was inadequate for reliable marking (Gentry, 1979).

More recently, Gribben (1979), Gallucci & Gribben (1979) and Gribben *et al.* (1984), marked sub-adult male northern fur seals using a fluorescent plastic resin, naphtha-based paint manufactured by Lenmar Lacquers, Inc. (Baltimore, Maryland, USA). The paint markers remained visible on all marked seals for two months and were visible on some seals for 12 months. Furthermore, even when the paint disappeared, the pattern of the marks frequently remained identifiable. Apparently, the guard hairs were matted by the paint and broke off leaving an outline of the symbol. Not unexpectedly, the marks persisted even after the pelts were processed.

M.W. Cawthorn (pers. comm.) used high-gloss marine enamel paints in aerosol cans to mark Hooker's sea lion. Applied against the nap of the hair the paint allowed identification of individual seals for three months including sea trips of 200 n.m. (370 km).

In the Arctic, tagged harp seals and hooded seals have been paint-marked to warn sealers not to take them (Sivertsen, 1941).

Dye marking Dyes offer an alternative procedure for the short-term marking of seals but the results reported to date are variable, probably because of the methodologies employed. Among the difficulties associated with the use of dyes to mark seals is that for the dyes to be effective the animals must be dry and remain out of the water for variable periods of time following treatment. It is also important that the dyes be permanent. Furthermore, because dye solutions tend to be thin, they spread easily, thus making it difficult to obtain small discrete marks. It is also difficult to dye dark-furred seals.

Boyd & Campbell (1971), successfully used green, pink and purple dyes to colour-mark grey seal pups and Pitcher (1979) used red 'Woollite' branding liquid, a pelage dye, to mark harbour seals. The Woollite markers lasted up to four months. Gentry (1979) reported that black Nyanzol D dyes persist for at least three months on California sea lions and Steller sea lions and he thought that this dye would work well on any light-coloured pinniped.

Conversely, Laws (1956) had little success with a series of similar dyes on southern elephant seals. Stirling (in Gentry, 1979) reports that the addition of a little absolute alcohol to Nyanzol D dissolves fur oils and thus leaves a clearer dye mark. The alcohol also prevents the solution from freezing in cold weather.

Picric acid (yellow) has also been used to mark seals. Hoek (1979) marked grey seal pups for counting using picric acid in a saturated solution of full strength alcohol. He applied the dye with a back-pack tree sprayer and was able to mark both wet and dry pups successfully. The marks brightened with exposure to sunlight. Furthermore, Beck (in Pitcher, 1979) reports that the picric acid marks on grey seals not only last through the pup moult but show up on the adult pelage as well. An interesting marking variation reported by S.S. Anderson (pers. comm.) consists of mixing a small amount of epoxy resin with a fluorescent dye. She reports that the mixture works well as a dye marker, lasts well and is highly visible.

Bleach marking Another pelage marking procedure for seals is bleaching. The principal bleach tested to date has been Lady Clairol Ultra Blue, which produces a white or cream-coloured mark especially visible on dark pinnipeds (Gentry, 1979). To obtain consistent results, a developer, proteinator and hydrogen peroxide must be mixed together to form a paste. Gentry reports that the thick consistency of the resulting paste allows one to write clear, thin lines on immobilized or tractable animals. Bleach marks on northern fur seals were sometimes visible for two seasons after application.

LeBoeuf & Peterson (1969) and LeBoeuf (in Gentry, 1979) successfully marked northern elephant seals by mixing 35% hydrogen peroxide with Lady Clairol Ultra Blue and a commercial emulsifier. The bleach mixture was applied with a plastic squeeze bottle as the animals slept. The pelage is bleached in just a few minutes and the marks remain visible until the moult six months later. The procedure requires that the animals be kept out of the water for about 20 min after treatment. Wilkinson (1991) used a mixture of 35% hydrogen peroxide and a household cleaning agent (Handy Andy, Lever Brothers) on southern elephant seals, which produced a cream mark on adults, visible until the moult, and a bright orange mark on the lanugo of the pups (the mixture was applied from a plastic squeeze bottle). Kaufman, Siniff & Reichle (1975), also had good success

Fig. 4.1 Bleach- and dye-marked Weddell seal at McMurdo Sound. (Courtesy of D.B. Siniff.)

using Clairol Ultra Blue to mark Weddell seals at McMurdo Sound for underwater monitoring by television (Fig. 4.1).

The use of appropriate bleaches and dyes as temporary marking techniques for seals appears to be very efficacious. The dyes are best suited to light-coloured species while the bleaches work best on dark-coloured species. Dyes and bleaches may also be used in combination as reported by M.W. Cawthorn (pers. comm.). He used hydrogen peroxide with a small amount of domestic washing detergent added as a wetting agent to bleach-mark male Hooker's sea lions and Lady Clairol jet black dye was used to dye the females.

Research workers should be aware that the chemicals in dyes and bleaches are toxic and may irritate and even burn the skin and eyes of both animals and people. Consequently, these formulations must be used with care.

Hair clipping

This marking procedure involves clipping the hair coat or brand singeing the pelage to produce a distinctive patch or number. The procedure is most useful where the underfur is a different colour from the guard hairs. The method is short-term, lasting only to the next moult, but is easy to perform, is painless to the animal and is distinctive. Scheffer (1950) described marking northern fur seals and Steller's sea lions, and Payne (1977) has marked Antarctic fur seal pups in this way for mark-recapture estimates of numbers.

An alternative procedure for animals with like-coloured guard hairs and underfur is to clip the guard hairs and dye, paint-mark or stain the underfur.

Streamer markers

Still another procedure for temporarily marking animals is to apply colour-coded or numbered streamers to applied tags. S. Jeffries (pers. comm.) has experimented with a number of flag materials attached to tags on harbour seals and reports that nylon cloth reinforced with vinyl is very effective. Flagging attached as shown in Fig. 4.2 commonly lasts

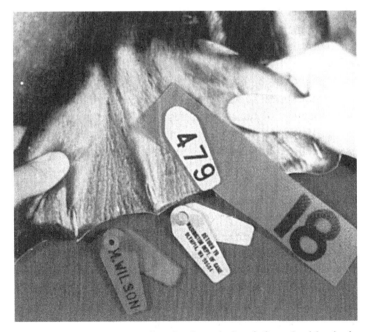

Fig. 4.2 Streamer tag made of nylon cloth reinforced with vinyl and numbered with a permanent marker pen. Here the tag is attached to a hard plastic Dalton Jumbo Rototag. Illustrated below right is the reverse side of the Jumbo tag and at the left is a soft plastic Dalton Reise tag.

for a year or more, as do marker numbers written with a permanent felt marker.

Scar marking techniques

Scar marking procedures have been used extensively to mark both wild and domestic animals. Unfortunately, many of the procedures have been criticized as being inhumane, particularly branding. However, the efficacy of many of the procedures argues for their continued use, particularly in view of the fact that most seals regularly sustain injuries significantly more painful than those induced by marking. Further, as pointed out by Gentry & Holt (1982), slightly injuring an animal once in its life seems preferable to capturing an animal repeatedly to refresh temporary marks.

Tattoos

Tattoos have proved to be an effective procedure to mark animals (Neal, 1959; Craighead *et al.*, 1960) and are routinely used to mark domestic animals and dogs for registration. Tattoos applied to thinner-furred areas such as the inner ears, thighs or lips of animals are especially effective.

Surprisingly, this highly effective and easily performed marking procedure does not appear to have been employed by seal biologists. Although the identification of tattoos requires close examination of animals, the technique offers a reliable auxiliary marking technique particularly useful in assessing the reliability of other marking methods. The site most suited for tattooing seals would be the inner lip area.

Vital stains

Another form of marking uses vital stains to mark hard tissues (bones and teeth). The use of these techniques is virtually limited to calibrating age criteria, however, since with the exception of extractable teeth, the bony tissues suitable for reading require the killing of the marked animals. In essence, the vital stain markers establish a checkpoint, or checkpoints, in bony tissue against which growth, often in the form of annulations, can be assessed. Thus, the technique is useful for marking animals for growth or age-related studies but not for large scale marking of individuals.

Among the vital stains that have been used for marking are certain of the antibiotic drugs, notably the tetracyclines, alizarine red and lead acetate (Yagi, Nishiwaki & Nakajima 1963), which is toxic at high doses. Only examples of the tetracyclines will be discussed here.

Tetracycline chelates with calcium ions deposited in the body and fluoresces when bones and teeth are exposed to ultraviolet (UV) light. Inasmuch as tetracycline deposition is limited to areas of new bone growth, a discernible checkmark occurs in growing bony tissue following administration of tetracycline. A number of workers have used the technique in age determination studies of mammals with varying success. This may be due to species differences, to the amount of the drug administered, or to the mode and period of administration.

Best (1976) conducted an experiment on captive dusky dolphins in which he administered tetracycline daily for eight days in their food at a dosage rate of 58–80 mg/kg body weight. Though the dosage was effective at this rate, he was of the opinion that a lower dosage rate might have been adequate. Fluorescent layers were detected in the dentine but none occurred in the cementum. Cementum tissue is less mineralized than dentine and the fluorescent layer may have been too fine or faint to be discernible. Conversely, Spinage (1973) was able to identify fluorescing lines in both the cementum and dentine of the water buck using demethylchlortetracycline administered intraperitoneally at a dosage of 12 mg/kg body weight.

The use of this technique on seals has apparently been fairly limited. Yagi *et al.* (1963), dosed a captive northern fur seal with tetracycline hydrochloride both orally and by injection. The animal survived captivity 162 days and when the teeth were examined under UV illumination, conspicuous tetracycline lines were observed. Condy & Bester (1975) performed intramuscular injections on 300 sub-Antarctic fur seals with oxytetracycline hydrochloride at a dosage of 10 mg/kg body weight. However, no follow-up work has been done. Also, the British Antarctic Survey and collaborators have since 1977 marked fur seals and elephant seals using this technique. Bengtson (1988) made periodic intramuscular injections (20 mg/kg body weight) of oxytetracycline to mark specific layers of dentine being laid down in the teeth of 21 fur seal pups. The position of the fluorescent bands in the dentine was successfully examined under UV illumination to determine the timing of layer deposition in the pup.

The marking of seals using vital stains has limited application in field studies of seals, as fairly sophisticated equipment and procedures are required to make histological tissue preparations and to perform the readings under UV illumination. Nonetheless, the procedure is useful in validating age-related criteria and as a redundancy marking technique.

Punch marks and amputations

Punch marks and amputations have long been in the repertoire but these techniques have been criticized because they maim animals and may affect behaviour or survival. However, most of the procedures are easily performed with a minimum of equipment and punch marks or notching do not harm the animals.

Scheffer (1950) marked northern fur seals by punching holes in the webs of flippers, and by varying the positions and number of holes, several hundred individual codes could be applied. However, the procedure proved unreliable as a long-term marking technique as the holes often become occluded and inconspicuous. Bonner (1968) experienced the same general result with Antarctic fur seals in which the hind flippers were punched with 6 mm diameter holes to estimate tag losses. The holes usually healed leaving only a 1–2 mm aperture surrounded by a dense ring of scar tissue and were difficult to identify visually, though they were easy to discern by palpating. Hole punching of the flippers appears not to have been tried in phocid seals. The experience with fur seals suggests that the procedure would prove even less effective applied to the hair-covered flippers of these seals (the webs of fur seal flippers are devoid of hair).

Other forms of scar marking on seals have included ear clips and the surgical notching of the flippers of northern fur seals (Roppel, 1979). Ear clippings of 1400 fur seals were successful as a cohort marking technique but the practice was discontinued because of concern that the loss of portions of the ear pinnae might destroy the deep-diving ability of the animals, since the external ears of otariids may function as valve-like organs during dives.

Surgical notchings of the flippers of over 200 000 fur seals were also performed but the procedure was found to be unreliable for marking individual seals because of tissue regrowth.

Surprisingly, seal workers have apparently failed to attempt marking seals by the toe clipping technique, with the exception of Johnson (1971), who used the technique to mark northern fur seal pups. This technique has been used extensively on a variety of multi-toed species ranging from small mammals (Baumgartner, 1940; Blair; 1941) to bears (Erickson,

1957) and would appear to be workable on seals. Using two toe clips removed at the distal joint of the digit over 1000 seals could be individually marked by this method (Melchior & Iwen, 1965).

Branding

Hot iron branding

History Hot iron branding was one of the first procedures used to mark seals and remains one of the most effective methods of permanently marking seals. The procedure appears to have been first employed in marking northern fur seals on the Pribilof Islands in 1912 (Osgood, Preble & Parker, 1915; Scheffer, 1950), and by the 1970s over 50 000 fur seals had been branded for scientific studies (Roppel, 1979). Some brands remained discernible for up to 20 years. Subsequently, Rand (1950) successfully used hot brands to mark Cape fur seals.

Over 2000 southern elephant seals were branded at South Georgia during the late 1920s (Matthews, 1929), but without result. Another 8000 or more southern elephant seals were branded at Heard and Macquarie islands from the 1940s to the 1960s (Chittleborough & Ealey, 1951; Carrick & Ingham, 1960; Ingham, 1967).

The first branding of ice-breeding seals appears to have been performed by Lindsey (1937) in the Bay of Whales, Antarctica. He branded 243 cow and pup Weddell seals to follow their movements. Later, Chittleborough & Ealey (1951) marked several transient leopard seals at Heard Island.

Hot branding studies, while continuing, appear to have waned during the late 1950s to early 1970s as seal workers experimented with other marking techniques, particularly tags, which were easier to apply and thought to be more humane. Recently, however, there has been a resurgence in the use of hot branding as a reliable permanent marking technique for seals (Hoek, 1979; Gentry & Holt, 1982; Anderson, 1985; J.H.M. David pers. comm.; J.K. Ling pers. comm.). However, because of the bulky apparatus necessary, the method is largely limited to colonial or at least aggregated seals.

Apparatus for hot branding The basic apparatus required for hot branding is an appropriate branding tool and a suitable heat source. Most of the current researchers recommend a branding head formed from a round mild steel rod (S.S. Anderson pers. comm.), although Hoek

(1979) preferred a stainless steel and brass rod as these metals do not corrode and thus lose circumference. A rounded rod is preferred to a flat rod because all the surfaces will be burnt evenly instead of only at the edges.

Research workers have used various procedures to heat branding irons, although some form of forge has generally proved most successful. Unfortunately, many workers have experienced difficulty in developing reliable forges or heating apparatus and one would do well to study Hoek (1979) for recommendations as regards apparatus, fuels and general techniques. Figure 4.3 illustrates a simple coke-fired forge used by Mirtha Lewis to mark elephant seals on the Valdez Peninsula in Argentina. The forge can be carried by two persons even while hot and in that environment it worked well to mark several hundred recently weaned seals during the years 1981–83.

S.S. Anderson (pers. comm.) and her colleagues have had good success using a furnace designed for glass-blowing (Fig. 4.4). It consists of a thin metal cylinder with ceramic fibre lining; it is fixed to a metal stand so that it lies on its side. The furnace is fired by a propane burner inserted through a hole in the side wall towards the back. Irons are inserted for heating through its open front. With this furnace, branding irons can be heated to a cherry-red in about five minutes. The unit is quite light and is attached to a pack frame for easy transport.

Fig. 4.3 Coke-fired forge used to heat branding irons to mark elephant seals on the Valdez Peninsula, Argentina. (Courtesy of M. Lewis.)

Fig. 4.4 Hot-branded grey seal and ceramic glassblower furnace used to heat branding irons in the United Kingdom. (Courtesy of S.S. Anderson.)

Hoek (1979) reports that the TECO branding iron heater manufactured by NASCO (Fort Atkinson, Wisconsin, USA, catalogue C3832N) is an ideal furnace for use where the seals can be brought to the apparatus. The unit is 2 ft square (60 cm square) and weighs about 73 lbs (33 kg) without the liquid propane fuel tank. It is capable of heating mild steel branding irons to a cherry-red in three minutes. Since the irons can be interchanged, it affords maximum flexibility in the choice of marks.

Over the years, seal branders have devised many forges and furnaces to heat branding irons and this will undoubtedly continue to be the case. The important requirement is that the heat source is sufficient to heat the branding irons to a cherry-red colour in a relatively short time in the particular environment in which the work is to be performed.

Branding recommendations Most attempts to brand seals appear to have initially met with problems in achieving clear, readable brands. Experience is the key and seasoned branders are able to discern when just the right mixture of heat and pressure has been achieved. Some (S.S. Anderson, pers. comm.) note a distinctive 'squeak-like' sound at the time a suitable branding has occurred. Responses to similar clues are undoubtedly a key to the success of other branders.

Although the techniques of individual branders vary considerably, almost all agree that the branding irons should be kept clean with a wire brush and heated to a cherry-red colour. Hoek (1979) expressed the view that an inadequately heated branding iron requires a longer application of the brand to achieve a mark, which causes the skin to cook rather than sear and results in open wounds rather than the formation of a simple scab. It is also important not to break the skin during branding.

The techniques described also vary considerably as to the duration of the branding iron application. In part this depends on whether the brand is applied through the hair or with the hair removed. It is not a problem with the bare flippers of otariid seals or the relatively hair-free elephant seals. S.S. Anderson (pers. comm.) and her co-workers achieve suitable brands on grey seals by applying the brand directly to wet or dry pelage for 3–7 seconds with a firm even pressure. The resulting fresh brand should appear tan-coloured. Hoek (1979) working with grey seals and Arctic phocids reports that the appropriate branding time on a dry pelt using a red-hot rod is approximately two seconds.

Other workers stress the importance of removing the hair in order to achieve uniform brands (Gentry & Holt, 1982). This is particularly important on heavily furred animals because of the tendency of the fur to char, forming a protective layer over the skin. Fur (or hair) removal can be accomplished by clipping, or by several light applications of the branding iron, followed by brushing away the singed fur before branding.

On northern fur seals, three or four such treatments were required before the actual branding was achieved by a light quick application of the brand to the skin (Gentry & Holt, 1982). Of 120 fur seals observed one year after branding, none had inflamed wounds, but several had incomplete marks due to inadequate branding.

Although there has been considerable success achieved in marking seals with hot brands, research workers have not implemented standardized marking schemes. A number of branding programmes have used brands simply to mark cohorts. Other workers have devised individual identification schemes.

In the Australian studies of southern elephant seals (Chittleborough & Ealey, 1951; Carrick & Ingham, 1960; Ingham, 1967), seals were marked with an ingenious four-letter code system permitting the immediate identification of the locality and year the seal was marked, as well as the identity of individual seals. The first letters H or M denoted Heard or Macquarie Islands, the second letter A through R denoted years, and the last two digits identified individuals. To avoid possible confusing brands, certain letters, G, O, Q, etc, were not used. The number of combinations could be doubled and tripled by simply varying the position of the brands – for example, right and left side or middle of back.

A difficulty associated with branding individual seals is the need for sufficient branding irons to apply sequential markings. Obviously, a major marking programme would require a large number of branding irons offering a variety of numbers or letters which could be used in combination. Two or even three brands might need to be used to mark a single animal. To reduce the number of branding irons but not the number of applications, some workers have devised branding schemes utilizing one or more symbols which could be used variously sequenced to achieve individual markings (R.B. Warneke, pers. comm.). Examples include:

One difficulty with the use of such symbols is the need to orient the marks on the animal's body so as to prevent the symbols being read upside down or in reverse order. As regards the best placement of brands to maximize subsequent observability, most workers favour the upper saddle, middle back, or upper shoulder (Rand, 1950; Gentry & Holt, 1982). The majority of workers administered no wound dressing or antibiotics following branding. However, Anderson (1985) and her co-workers apply a burn ointment to alleviate pain and infection and administer tetracycline as a prophylactic, which has the added merit of serving as a vital stain mark.

Freeze branding

In recent years there has been a trend to marking animals with cryothermic brands in preference to hot brands, presumably because it is more humane. However, seal researchers have experienced mixed success and many investigators have abandoned the procedure, because it is difficult to use and generally far less reliable than other marking techniques (Keyes & Farrell, 1979; S.S. Anderson pers. comm.; J.K. Ling pers. comm.). Conversely, Warneke (1979), Cornell, Antrim & Asper (1979), and L.H. Cornell (pers. comm.) have had greater success.

When correctly employed, cryothermic branding results in the selective killing of the melanocyte or pigment-producing cells, thus leaving a non-pigmented pelage mark (Farrell, 1979). However, there is a tendency for the melanocytes to be replaced by ones peripheral to the frozen site, thus causing eventual loss of the mark. In the case of seals, the duration of brands is generally only a year or two (Cornell *et al.*, 1979; L.H. Cornell pers. comm.). The technique is relatively painless both at the time of the freezing and after the skin has thawed. However, S.S. Anderson (pers. comm.) reports veterinary advice that tissue damage is essentially the same in hot and freeze branding.

In discussing the generally poor success of cryothermic brands on seals, Farrell (1979), the inventor of the procedure, expressed the view that most of the work done to date on pinnipeds suggests that the branding has been excessive, causing scars similar to hot brands edged with depigmented hair rather than a depigmented area of pelage. To avoid this he recommends that research workers attempt to achieve milder freezes at lower temperatures followed by faster thaws.

A major difficulty associated with freeze branding of seals is the necessity of removing the hair before application (Cornell *et al.*, 1979; Warneke, 1979; L.H. Cornell pers. comm.; J.K. Ling pers. comm.).

Cold branding apparatus In many respects, cold branding is similar to hot branding, the major difference being the substitution of a super-cooled branding iron for a heated iron.

The branding heads used by most workers are apparently the same as used for hot branding, but Warneke (1979) found that copper alloy heads specifically designed for cryothermic branding were superior to cast-iron heads. These heads were 2 inches (*c.* 50 mm) high with a contact surface between 5/16 inches (*c.* 8 mm) and 3/8 inches (*c.* 10 mm) across and were of sufficient mass to be an effective 'heat sink'.

The cooling mixture used by R. Warneke (pers. comm.) was methanol and dry ice which cooled irons to about $-95°$ F ($-70°$ C). Cornell *et al.* (1979), reported that alcohol and dry ice coolant medium was the most economical, requiring less preparation and equipment for its use. However, liquid nitrogen produced more rapid skin cooling in pinnipeds. Also, Anderson (1985) reports that liquid nitrogen is not permitted to be carried in helicopters (and presumably other aircraft) in the UK. Other workers have used Freon 12 and 22 as a coolant (Keyes & Farrell, 1979; J.H.M. David pers. comm.).

Another drawback associated with cryothermic marking, even when it is initially successful, is the short duration of the brands. In the closely controlled studies of Cornell and co-workers (pers. comm.), brands were applied to pinnipeds for 20 sec at 20 to 30 lbs of pressure. In phocids, including elephant seals, the resulting brands were dark, remained readable up to a year, and were discernible for three years. The brands on California sea lions varied from pink to dark, were readable for about 18 months, and discernible up to four years. In the walrus, the brands were pink, and it appears that they may remain readable for years.

Warneke (1979) also reported brands of fairly long duration on Australian sea lions. Before applying the iron, he reports that the brand site is wetted with alcohol to improve the branding contact. Brands applied to the flippers were clearly legible after seven years and flank brands were still quite good after four years.

Explosive branding

Both hot iron and freeze branding involve prior immobilization or restraint of the animal. Another method has been described that is claimed to be portable and instantaneous and so can be used on unrestrained wild animals (Homestead, Beck & Sergeant, 1971). A lead-coated instantaneous fuse is set into a rubber template of the desired shape and detonated by a percussion detonator triggered by a modified spear gun. This equipment has been tested on hooded seal, harp seal, harbour seal and northern elephant seal. Animals observed up to 12 months after branding had clearly visible brands. Subsequently, however, Hoek (1979) reported that several hundred harp and hooded seals were marked with explosive brands but the hair grew back after several years. For this reason, and as yet unresolved patent problems, the use of the method has been discontinued. Pitcher (1979) also marked 100 harbour seal pups with explosive brands but a subsequent evaluation of the procedure was never made.

Tagging

Of all the methods of marking seals, tagging is the most widely used. Over the years, a variety of tags has been employed to mark seals (Sivertsen, 1941; Scheffer, 1950; Laws, 1952; Hewer, 1955; Rasmussen & Øritsland, 1964), but the principal tags today are metal band tags or plastic disc tags.

Metal tags

The principal metal tags currently used to mark seals are monel or stainless steel tags of the type used to earmark livestock (Fig. 4.5). They are

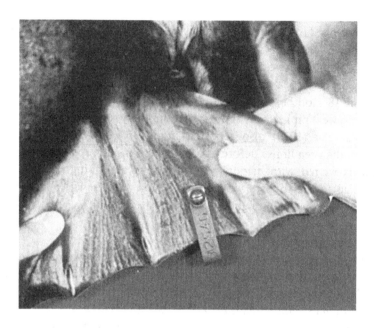

Fig. 4.5 Metal strap tag applied to interdigital web of a phocid seal.

self-piercing and can be quickly applied using special pliers. Essentially, they consist of a flat strip of metal, pointed at one end and with a slot at the other. When the strip is inserted in the pliers and applied, the point pierces the skin and is channelled and locked to the other end of the tag. The tags carry stamped serial numbers and a 'Return to organizer' legend. The size varies from large cattle size to small poultry wing clips. For Antarctic seals, most workers now use the monel cattle ear tags (National Band and Tag Company, Newport, Kentucky, USA) Type 1005, size 49 for the phocid seals and size 681 for fur seals. Sharp edges produced in stamping out the tag should be smoothed to minimize tag loss.

The tags are inserted in the hind edge of the fore flipper of fur seals and the hind interdigital web of phocid seals (Fig. 4.5). Although the retention of these tags applied to the ears of livestock, deer and other wild animals is highly reliable, the loss rate of metal band tags on seals is often quite high. At South Georgia, Dickinson (1967) had only 17 recoveries after eight years from a total of 15 000 tagged southern elephant seals. In the same species, Condy (1977) reported losses of seals tagged as pups to be 8% at six months, 14.9% at 12 months, 23.9% at

18 months, 24.9% at two years. At Kerguelen and Heard Island, Burton (1985) reported resightings of only 4% for monel tags as opposed to an 8% resighting rate for plastic roto-tags.

Although reliable figures for loss rates of metal tags do not appear to have been reported for other phocid species, the general experience with these tags has been unsatisfactory. Stirling (1979) and Siniff & DeMaster (1979a) reported high but unstated losses on Weddell seals. Similar results were reported for harbour seals (S. Jefferies pers. comm.) and for grey seals (S.S. Anderson pers. comm.); both workers report that the metal tags caused post-attachment tears and cuts which became septic.

Experience with metal strap tags on otariid seals is similar, except that occasionally an animal will retain a tag for many years (Payne, 1977; Roppel, 1979; J.H.M. David pers. comm.). On Australian sea lions tagged as pups, J.K. Ling (pers. comm.) reports loss rates of 16% at four months, 40% at six months, and 27% at 13–14 months. J.H.M. David (pers. comm.) reports that the loss of metal tags on 50 000 Cape fur seals tagged at six weeks of age and harvested at nine months since 1971 has varied between 3% and 15%, the rate being clearly influenced by the skill of the individual taggers. He did not specify the reasons for high or low success rates.

Other researchers reporting high but unspecified losses of metal strap tags on otariid seals include Roppel (1979), who reported that over 650 000 northern fur seals were marked on the Pribilof Islands, and Warneke (1979), working with Australian fur seals.

Gentry & Holt (1982) report the tendency of metal tags to rotate, thus causing injuries to seals. They report that the tag can be improved by moulding that portion of the tag which penetrates the flipper into a rounded post.

In addition to the generally low retention rate of metal tags, several workers reported that tags caused mortalities, apparently due to infections (Scheffer, 1950; Johnson, 1971; Roppel, 1979; M.W. Cawthorn pers. comm.).

Plastic tags

In recent years, most seal biologists have switched to self-piercing plastic ear tags, widely used on domestic stock. These tags consist of two disc elements joined by a round post which interlocks with the female element of the opposing disc. They are applied with special pliers and although self-piercing they are more easily applied if a small hole is made first with a leather punch, scalpel, or narrow-bladed pocket knife. The last two methods are preferred, as a punch tends to bruise the flesh.

Correspondence established that the two makes of tags currently used most frequently are the Dalton 'Rototag' made by Dalton of Nettlebed, Henley-on-Thames, Oxfordshire, U.K. (Fig. 4.2), and the 'Allflex' tag made by Delta Plastics, Palmerston North, New Zealand (Fig. 4.6). The main difference between Allflex and Rototag is that the male portion of the Allflex tag is enclosed when it is locked, thus lessening abrasion wear and potential loss of the tag. To lessen tag wear, the peg of the tags should face upward from the top of the flipper.

A desirable feature of plastic tags is the use of combinations of colours for the visual identification of cohorts or a limited number of individual seals. The Dalton tags, for example, are available in 12 bright colours, which could give 300 colour combinations. However, their quality varies and some tag types become brittle and break (Pitcher, 1979; Mate, in Pitcher, 1979; Hoek, 1979). When ordering tags it is advisable to specify little pigment, because tags with pigment tend to crack more easily. To avoid this, some researchers have switched to tags made of flexible plastic, others to tags of heavier materials.

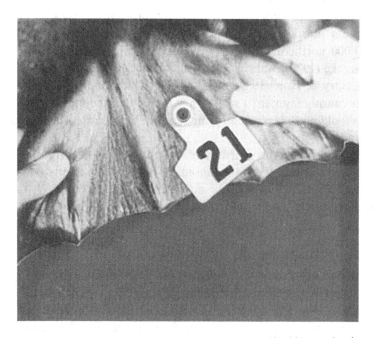

Fig. 4.6 Delta Plastics' 'Allflex Tag'. Note protected locking mechanism.

The retention rate of plastic tags on seals, with minor exceptions, appears in general to be higher than for metal strap tags. On Weddell seals at McMurdo Sound, Siniff & DeMaster (1979b) calculated the annual losses of Allflex tags over a period of several years as ranging between 5% and 10% as opposed to a higher loss rate estimated at 15% for Dalton Rototags (D.P. DeMaster pers. comm.). On northern elephant seals, Reiter (1984) experienced an 11% loss rate of Dalton Junior Rototags at age two for animals tagged as pups and a 6% annual loss rate thereafter. In the UK, Anderson (1984) reported a 9.5% annual loss rate on grey seals marked when pups which indicates, although not statistically significant, an increase in loss rates with an increase in tag age. In contrast, Dalton Jumbo Rototags applied to successive southern elephant seal pup cohorts at Marion Island since 1983 yielded loss rates which are much lower than reported elsewhere. Loss rates reported in years 1 to 6 were 0.9%, 1.8%, 2.1%, 2.4%, 4.4% and 1.5%, respectively (Wilkinson, 1991), the increase in loss rate with age being in accordance with the findings of Anderson (1984).

Plastic tags have also been used extensively to tag otariid seals but relatively few data on retention have been reported. M.W. Cawthorn (pers. comm.) reported an estimated 5–10% annual loss rate of Allflex tags on Hooker's sea lion, and J.H.M. David (pers. comm.) reported low but unspecified losses of Dalton Reise (soft plastic) tags applied annually to 600 Cape fur seal pups since 1982. However, the marked pups are not subject to harvest so specific tag retention data are not available.

There are at least two reports of poor retention rates of plastic tags on seals. Stirling (1979) marked Weddell seals with New Zealand plastic sheep tags between 1966 and 1970 but tag retention was poor and tags lasted only about three years. He also tagged New Zealand fur seals in South Australia with the same tags but they were lost after only one year. The lower retention rate of tags applied to fur seals was attributed to the dirty and rocky substrate associated with the fur seal colony. However, the tagged seals did not develop infections as the result of tagging and Stirling did not identify the cause of the poor retention of tags in either species.

A low retention rate has also been reported for Allflex plastic cattle ear tags used on Australian sea lions with retention rates generally exceeding those for monel tags, at least on a short-term basis (J.K. Ling pers. comm.). Loss rates for three groups of pups totalling 67 animals marked in 1976 and 1977 were 31% and 57%, respectively, after three months and 61% after five months. The tag loss rate for 16 seals tagged as yearlings was 19% after five months and 38% after 7-8 months. Thus, the loss of tags was spread across all sex and age groups; its cause was not stated.

General comment on seal tagging

In general, tagging experiences reported to date indicate that seal workers have not yet devised truly reliable procedures for tagging seals. The major variability in results between studies even on closely related species suggests that individual procedures may vary widely. Unfortunately, there has been virtually no specific research on seal tagging and so the specifics of appropriate procedures have not been defined. At this time, it appears that plastic disc tags are generally more reliable than metal strap tags, but minimal tag losses will approximate 10% annually. Furthermore, the long-term durability of plastic tags is inferior to that of metal tags (DeLong, 1979).

There is little information on the causes of tag losses. For metal tags, septic wounds and some infection have been reported, which cause tag losses, but this does not appear to be a problem with the plastic post or temple tags. Nonetheless, it would seem to be appropriate to treat tagging sites with an antiseptic and to minister an antibiotic, preferably tetracycline. M.W. Cawthorn (pers. comm.) reports that before application, tags are soaked overnight in a disinfectant and the tag stem is smeared with a broad-spectrum antibiotic before application.

Another suggested cause for the loss of plastic post tags is that the attachment holes become enlarged, permitting the tags to pull through. Interestingly, one worker addressed this problem by reducing the size of the plastic discs to reduce drag, while another increased the size of the discs. Others reported that tag wear and breakage were possible causes of tag losses, but specific data on these factors are unavailable.

A possible cause of tag loss in male southern elephant seals is the pronounced increase in the thickness of the interdigital webbing of the hind flippers with growth, approximating or exceeding the height of the post of the tag. This results in breakage of the skin and subsequent infection, enlargement of the hole, and loss of the tag (Wilkinson, 1991). Tags should be positioned so that approximately one third of the tag clears the trailing edge of the interdigital web. If more of the tag protrudes from the webbing, then it becomes prone to pulling through (tearing out), while insertion of the tag too deeply in the webbing prevents it from nestling between the toes. This results in chafing which promotes infection and loss of the tag (Wilkinson, 1991).

DeLong (1979) reports that a variety of plastics have been used for tags including polyethylene, polystyrene, polyurethane and nylon. Although easily moulded, plastic tags have little resistance to abrasion and they have little stability when exposed to ultraviolet radiation. Consequently the tag inscriptions wear away, the colours fade and the breakage strength lessens. Of the several materials tried, polyurethane appears to have the required abrasion resistance and UV stability, as a tag for pinnipeds (DeLong, 1979).

However, polyurethane varies in hardness (Jennings, in DeLong, 1979), and there is still uncertainty as to the best type of plastic for a seal tag. Jennings conjectured that a Teflon sleeve surrounding the tag post would reduce tissue trauma since Teflon is biocompatible. Thus, the Teflon sleeve would presumably fuse by growth with epidermal tissue at the tag site and allow the post of the tag to rotate freely within the Teflon sleeve.

It is obvious that more directed research on seal tags and tagging procedures is urgently needed. A first step in this direction is to amass information on tag retention.

Tag loss analysis Tag loss is an ever-present problem when estimating population parameters and allowance must be made for this error. This requires a reliable means of identifying when and where individual animals were marked. Preferably, a permanent marking procedure is used in conjunction with the tagging programme such that future identification of a tagged seal can be assured. Suitable permanent marks might include tattoos, hot brands and toe clipping.

Alternatively, but preferably concurrently, tag losses can be estimated by analysis of data based on recovery or observation of multi-tagged animals (see for example, Siniff & DeMaster, 1979b; Seber, 1973; Eberhardt, Chapman & Gilbert, 1979). Most analyses based on double tagged animals, from which a binomial model of tag loss is constructed, assume that the probability of losing both tags is the square of the probability of losing one tag. The model further assumes that the loss of each tag is an individual event. Several equations have been proposed for calculating tag losses as follows (Fairley, in Siniff & DeMaster, 1979b):

$$P(\text{lose 1 tag}) = \frac{N_2 + N_3}{2(N_1 + N_2 + N_3)} \tag{1}$$

where

N_1 = number of animals with 2 tags retained
N_2 = number of animals with 1 tag retained, 1 tag lost
N_3 = number of animals with 2 tags lost

Shaughnessy (in Siniff & DeMaster, 1979b) suggests reducing this formula to

$$P(\text{lose 1 tag}) = \frac{N_2}{2N_1 + N_2} \qquad (2)$$

$$P(\text{lose 2 tags}) = \frac{N_2^2}{N_1} \qquad (3)$$

Alternatively,

$$P(\text{lose 2 tags}) = P(\text{lose 1 tag})^2 \qquad (4)$$

DeMaster (in Siniff & DeMaster, 1979b) proposed the following formula for identifying animals lost from the population:

$$P = p(1)P(\text{lose 1 tag}) + p(2)P(\text{lose 2 tags}) \qquad (5)$$

where

$p(1)$ = proportion of animals with 1 tag
$p(2)$ = proportion of animals with 2 tags

The probability exists, of course, that individual tag losses are not independent events and estimated tag loss rates may vary for example with tag age or attachment location (Eberhardt *et al.*, 1979). Recently, Testa (1986) examined this relationship for Weddell seals tagged at McMurdo Sound and estimated that year-to-year tag loss rates of new tags and tags attached one or more years earlier were 0.2478 and 0.1172, respectively. Conversely, S.S. Anderson (pers. comm.) calculated tag retention rates on grey seals on an instantaneous rate basis and found no significant differences in tag loss rates with time.

Standardization of marking

Almost without exception, seal marking programmes to date have concentrated more on marking animals and less on recovery efforts and analysis of results. A particular fault of most marking programmes has been failure to assess the efficacy of various marking techniques.

To surmount these difficulties standardization of marking methods and reporting should be part of the Antarctic seal research effort.

Tagging recommendations

Metal tags The metal tag recommended for seals is the monel metal cattle ear tag manufactured by National Band and Tag Company, Newport, Kentucky, USA. The tag recommended for Weddell, crabeater and Ross seals is tag type 1005, size 49, applied to the interdigital web of the hind flipper. For fur seals, tag type 1005, size 681, applied to the axilla web of the fore flipper, is recommended.

Plastic tags The plastic tags recommended for both phocid and otariid seals are the Dalton Jumbo Rototag manufactured by Dalton of Nettlebed, Henley-on-Thames, Oxfordshire, UK, and the Delta Allflex plastic tag manufactured by Delta Plastics, Palmerston North, New Zealand. The attachment sites for these tags are the same as for the metal tags.

In all tagging, both right and left flippers should be tagged, and a permanent marker should be applied to ensure the future identification of marked seals.

Identifying and numbering tags and tag reports

One difficulty associated with seal marking programmes in the Antarctic has been that there is no uniform procedure which enables the marks and tags of individual research projects to be identified and reported. The problem is compounded by the fact that the marking efforts of individual workers have not been faithfully reported to the appropriate National SCAR Committee. To be fully useful this information needs to be collated by the SCAR Group of Specialists on Seals. Such a procedure would serve three important purposes: (1) facilitate communication between persons initially placing tags on seals and those resighting the seals at a later date; (2) assist in identifying the source of unfamiliar tags encountered in the field; and (3) help prevent different projects placing identical tags on different seals.

In order to construct an Antarctic pinniped master tag record, the Secretary of the SCAR Group of Specialists on Seals (currently

Dr J.L. Bengtson, National Marine Mammal Laboratory, Northwest and Alaska Fisheries Center, National Marine Fisheries Service, NOAA, 7600 Sand Point Way N.E., Seattle, Washington 98115, USA.) is compiling information on tagging activities and records, and will communicate the information to the National SCAR committees and to appropriate research workers.

To simplify the identification of tags, particularly worn tags with illegible return to sender instructions, the Group of Specialists on Seals has proposed that a prefix letter be used on each tag to denote the National SCAR Committee under whose auspices the tagging programme was performed. The prefix letters which have been proposed are as follows:

A	Australia	J	Japan
B	United Kingdom	N	Norway
C	Chile	P	Poland
D	Argentina	R	Russia
F	France	S	South Africa
G	Germany	U	USA
I	India	Z	New Zealand

This scheme will accommodate the addition of new National Committees using either remaining alphabet letters or symbols.

As regards the identification of worn and illegible tags it should be noted that US Fish and Wildlife Service, Migratory Waterfowl Banding Office, Patuxent, Maryland, has devised sophisticated techniques for identifying illegible tags. Accordingly, illegible seal tags should be directed to Dr Bengtson for transfer and deciphering by that office.

References

Anderson, S.A. 1985. *Rationale behind branding work.* Submitted to A.S.A.B. Ethical Committee, March 1985, 2 pp. (Unpublished.)
Anderson, S.S. 1984. *Interactions Between Grey Seals and UK Fisheries.* Natural Environment Research Council.
Balcomb, K.C., Boran, J.R., Osborne, R.W. & Haenel, N.J. 1980. *Observations of Killer Whales* (Orcinus orca) *in Greater Puget Sound, State of Washington.* U.S. Dept. Commerce, Springfield, VA, NTIS PB80-224728.
Baumgartner, L.L. 1940. Trapping, handling, and marking fox squirrels. *J. Wildl. Mgmt.* 4, 444-50.
Bengtson, J.L. 1988. Long-term trends in foraging patterns of female Antarctic fur seals at South Georgia. In *Antarctic Ocean and Resources Variability*, ed. D. Sahrhage, pp. 286-91. Springer-Verlag, Berlin.

Best, P.B. 1976. Tetracycline marking and the rate of growth layer formation in the teeth of a dolphin (*Lagenorhynchus obscurus*). *S. Afr. J. Sci.* **72**, 216-18.

Bigg, M.A., MacAskie, I.B. & Ellis, G. 1976. *Abundance and Movements of Killer Whales off Eastern and Southern Vancouver Island with Comments on Management.* Prelim. rep., Arctic Biological Station, Quebec, Canada.

Blair, W.F. 1941. Techniques for the study of mammal populations. *J. Mammal.* **22**, 148-57.

Bonner, W.N. 1968. *The fur seal of South Georgia.* British Ant. Survey Sci. Rep. 56, 81 pp.

Boyd J.M. & Campbell, R.N. 1971. The grey seal (*Halichoerus grypus*) at North Rona, 1959 to 1968. *J. Zool.* **164**, 469-512.

Burton, H.R. 1985. Tagging studies of male southern elephant seals (*Mirounga leonina*) in the Vestfold Hills area, Antarctica, and some aspects of their behaviour. In *Studies of Sea Mammals in the South Latitudes*, ed. J.K. Ling & M.M. Bryden, pp. 2-29. South Australia Museum.

Carrick, R. & Ingham, S.E. 1960. Ecological studies of the southern elephant seal, *Mirounga leonina* (L), at Macquarie Island and Heard Island. *Mammalia* **24**, 325-42.

Chittleborough, R.G. & Ealey, E.H.M. 1951. *Seal Marking at Heard Island, 1949.* Aust. Natl. Antarct. Res. Exp. Interim Rep. 1.

Condy, P. 1977. *The Ecology of the Southern Elephant Seal*, Mirounga leonina *(Linnaeus 1758), at Marion Island.* D.Sc. Thesis, Univ. Pretoria, S.A.

Condy, P. & Bester, M.N. 1975. Notes on the tagging of seals at Marion and Gough Islands. *S. Afr. J. Antarctic Res.*, **5**, 45-7.

Cornell, L.H., Antrim, J.E., Jr. & Asper, E.D. 1979. Cryogenic marking of pinnipeds and California sea otters. In *Report on the Pinniped and Sea Otter Tagging Workshop*, ed. L. Hobbs & P. Russell, pp. 19-20. Amer. Inst. Biol. Sciences, Arlington, VA.

Craighead, J.J., Hornocher, M., Woodgerd, W. & Craighead, F.C., Jr. 1960. Tracking, immobilizing, and color-marking grizzly bears. *Trans. N.A. Wildl. Conf.* **25**, 347-63.

David, J.H.M., Meyer, M.A. & Best, P.B. 1990. The capture, handling and marking of free-ranging adult South African (Cape) fur seals. *S. Afr. J. Wildl. Res.* **20**, 5-8.

DeLong, B. 1979. Tag materials. In *Report on the Pinniped and Sea Otter Tagging Workshop*, ed. L. Hobbs & P. Russell, Amer. Inst. of Biol. Sciences, Arlington, VA.

Dickinson, R.B. 1967. Tagging elephant seals for life history studies. *Polar Rec.* **13**, 443-6.

Eberhardt, L.L., Chapman, D.G. & Gilbert, J.R. 1979. A review of marine mammal census methods. *Wildl. Monog.* **63**, 1-46.

Erickson, A.W. 1957. Techniques for live-trapping and handling black bears. *Trans. N.A. Wildlife Conf.* **22**, 520-43.

Farrell, R.K. 1979. Comments on freeze marking and lasers. In *Report on the Pinniped and Sea Otter Tagging Workshop*, ed. L. Hobbs & P. Russell, Append. C, pp. 44-5. Amer. Inst. Biol. Sciences, Arlington, VA.

Fogden, S.C.L. 1961. Mother-young behaviour at grey seal breeding beaches. *J. Zool.* **164**, 61-2.

Gallucci, V.F. & Gribben, M.R. 1979. Comment. In *Report on the Pinniped and Sea Otter Tagging Workshop*, ed. H.L. Hobbs & P. Russell. p. 16. Amer. Inst. Biol. Sciences, Arlington, VA.

Gentry, R.L. 1979. Adventitious and temporary marks in pinniped studies. In *Report on the Pinniped and Sea Otter Tagging Workshop*, ed. H.L. Hobbs & P. Russell. Append. B. pp. 39–43. Amer. Inst. Biol. Sciences, Arlington, VA.

Gentry, R.L. & Holt, J.R. 1982. *Equipment and Techniques for Handling Northern Fur Seals*. NOAA Tech. Rept. NMFS-758.

Gribben, M.R. 1979. A study of the intermixture of subadult male fur seals *Callorhinus ursinus* (Linnaeus 1758) between the Pribilof islands of St. George and St. Paul, Alaska, M.S. Thesis, Univ. Washington, Seattle, 191 pp.

Gribben, M.R., Johnson, H.R., Gallucci, B.B. & Gallucci, V.F. 1984. A new method to mark pinnipeds as applied to the northern fur seal. *J. Wildl. Mgmt.* **48**, 945–9.

Hewer, H.R. 1955. Notes on marking of Atlantic seals in Pembrokeshire. *Proc. Zool. Soc. London* **125**, 87–95.

Hoek, W. 1979. Methods and techniques for long-term marking of pinnipeds. In *Report on the Pinniped and Sea Otter Tagging Workshop*. H.L. Hobbs & P. Russell. Append. A. Amer. Inst. Biol. Sciences, Arlington, VA.

Homestead, R., Beck, B. & Sergeant, D.E. 1971. A portable instantaneous branding device for permanent identification of wildlife. In *Symposium: Biological Sonar and Diving Mammals*, pp. 1–12. Proc. 8th Ann. Conf. Biol. Sonar Diving Mamm., Stanford Res. Inst., Menlo Park, CA.

Ingham, S.E. 1967. Branding elephant seals for life-history studies. *Polar Rec.* **13**, 447–9.

Johnson, A.M. 1971. Recoveries of marked seals. In *Marine Mammal Biological Laboratory, Fur Seal Investigations, 1970*, pp. 26–31. Natl. Mar. Mammal Lab., Northwest and Alaska Fish. Center, Natl. Mar. Fish. Serv., NOAA, Seattle, WA.

Katona, S., Baxter, B., Brazer, O., Kraus, S., Perkins, J. & Whitehead, H. 1979. Identification of humpback whales by fluke photographs. In *Behavior of Marine Mammals – Current Perspectives in Research*, vol. 3. *Cetaceans*, ed. H.E. Winn & B. Olla, pp. 33–44. New York, Plenum Press.

Kaufman, G.W., Siniff, D.B. & Reichle, R. 1975. Colony behavior of Weddell seals, *Leptonychotes weddelli at Hutton Cliffs, Antarctica. Rap. P.-v. Reun. Cons. Inter. Explor. Mer* **169**, 228–46.

Keyes, M.C. & Farrell, R.K. 1979. Freeze marking the northern fur seal. In *Report on the Pinniped and Sea Otter Tagging Workshop*, ed. H.L. Hobbs & P. Russell. Amer. Inst. Biol. Sciences, Arlington, VA.

Laws, R.M. 1952. Seal marking methods. *Polar Rec.* **6**, 359–61.

Laws, R.M. 1956. The elephant seal (*Mirounga leonina* Linn.) II. General, social and reproductive behaviour. *Falkland Isl. Depend. Surv., Sci. Rep.*, no. 13, 1–88.

LeBoeuf, B.J. 1971. The aggression of the breeding bulls. *Nat. Hist.* **136**, 82–94.

LeBoeuf, B.J. 1972. Sexual behavior in the northern elephant seal, *Mirounga angustirostris. Behaviour* **41**, 1–26.

LeBoeuf, B.J. & Peterson, R.S. 1969. Social status and mating activity in elephant seals. *Science* (New York) **163**, 91–3.

Lindsey, A.A. 1937. The Weddell seal in the Bay of Whales. *J. Mammal.* **18**, 126–44.

Matthews, L.H. 1929. The natural history of the elephant seal with notes on other seals found at South Georgia. *Discov. Rep.* **1**, 233–56.

Melchior, H.R. & Iwen, F.A. 1965. Trapping, restraining and marking Arctic ground squirrels for behavior observations. *J. Wildl. Mgmt.* **29**, 671–8.

Neal, B.J. 1959. Techniques for trapping and tagging the collared peccary. *J. Wildl. Mgmt.* **23**, 11–16.

Osgood, W.H., Preble, E.A. & Parker, G.H. 1915. The fur seals and other life of the Pribilof Islands, Alaska, in 1914. *U.S. Fish & Wildlife Service, Bul.* **34**, 1–172.

Payne, R., Brazier, O., Dorsey, E., Perkins, J., Rowntree, V. & Titus, A. 1981. *External Features in Southern Right Whales* (Eubalaena australis) *and their use in Identifying Individuals.* U.S. Dept. Commerce, Springfield, VA, NTIS PB81-161093.

Payne, M.R. 1977. Growth of a fur seal population. *Phil. Trans. Roy. Soc. Lond.* **279**, 67–79.

Pitcher, K. 1979. Pinniped tagging in Alaska. In *Report on the Pinniped and Sea Otter Tagging Workshop.* ed. L. Hobbs & P. Russell, p. 3. Amer. Inst. Biol. Sciences, Arlington, VA.

Rand, R.W. 1950. Branding in field work on seals. *J. Wildl. Mgmt.* **14**, 128–33.

Rasmussen, B. & Øritsland, T. 1964. Norwegian tagging of harp seals and hooded seals in North Atlantic waters. *Fiskerdir. Skr.* **13**, 43–55.

Reiter, J. 1984. Studies of female competition and reproductive success in the northern elephant seal. Ph.D. dissertation, Univ. California, Santa Cruz.

Roppel, A.Y. 1979. Northern fur seal – Pribilof Islands, Alaska. In *Report on the Pinniped and Sea Otter Tagging Workshop.* ed. L. Hobbs & P. Russell, Amer. Inst. Biol. Sciences, Arlington, VA.

Scheffer, V.B. 1950. *Experiments in the Marking of Seals and Sea Lions.* U.S. Fish & Wildlife Service, Spec. Sci. Rep. No. 4. Washington, DC.

Seber, G.A.F. 1973. *Estimation of Animal Abundance.* Hofner Press, NY.

Siniff, D.B. & DeMaster D.P. 1979a. Antarctic seal tagging. In *Report on the Pinniped and Sea Otter Tagging Workshop.* ed. L. Hobbs and P. Russell, pp. 9–10. Amer. Inst. Biol. Sciences, Arlington, VA.

Siniff, D.B. & DeMaster, D.P. 1979b. Tag loss correction. In *Report on the Pinniped and Sea Otter Tagging Workshop.* L. Hobbs and P. Russell, pp. 28–9. Amer. Inst. Biol. Sciences, Arlington, VA.

Sivertsen, E. 1941. On the biology of the harp seal. *Hvalradets-Skrifter* **26**, 1–166.

Spinage, C.A. 1973. A review of the age determination of mammals by means of the teeth, with special reference to Africa. *E. Afr. Wildl. J.* **11**, 165–87.

Stirling, I. 1966. A technique for handling live seals. *J. Mammal.* **47**, 543–4.

Stirling, I. 1979. Tagging Weddell and fur seals and some general comments on long-term marking studies. In *Report on the Pinniped and Sea Otter Tagging Workshop.* ed. L. Hobbs & P. Russell, pp. 13–14. Amer. Inst. Biol. Sciences, Arlington, VA.

Taber, R.D. & Cowan, I. McT. 1969. Capturing and marking wild animals. In *Wildlife Techniques*, 3rd edn, rev. R.H. Giles, Jr., pp. 277–317. The Wildlife Society, Washington, DC.

Testa, J.W. 1986. Long-term patterns in life history characteristics and population dynamics of Weddell seals (*Leptonychotes weddelli*) in McMurdo Sound, Antarctica. Ph.D. Dissertation, Univ. Minnesota, Minneapolis, 165 pp.

Warneke, R.B. 1979. Marking of Australian fur seals, 1966–1977. In *Report on the Pinniped and Sea Otter Tagging Workshop*. ed. L. Hobbs and P. Russell, Amer. Inst. Biol. Sciences, Arlington, VA.

Wilkinson, I.S. 1991. Reproductive success of southern elephant seals (*Mirounga leonina*) at Marion Island, Ph.D. dissertation, Univ. Pretoria, Pretoria, S. Africa.

Yagi, T., Nishiwaki, M. & Nakajima, M. 1963. A preliminary study on the method of time marking with lead salt and tetracycline on the teeth of fur seals. *Whales Res. Inst., Sci. Rep.* **17**, 191–5.

5

Telemetry and electronic technology

J.L. BENGTSON

Introduction

Rapid advances in the field of electronics over the past several years have made powerful research tools available to pinniped investigators. In particular, the remarkable miniaturization of solid-state circuitry through semiconductor technology has made it feasible to deploy highly sophisticated instruments on free-ranging individuals. These instruments have become important assets in the study of pinniped behaviour, ecology and physiology.

This chapter briefly outlines some of the types of instruments that have relevance to studies of free-ranging Antarctic seals. No attempt has been made to include a description of instruments and devices principally used in captive or laboratory studies. The main categories of instruments described are radio tracking and telemetry, sonic devices, archival recorders (behavioural, ecological and physiological) and satellite-linked hardware. A section on attachment methods is also included. A list of the addresses of research groups or manufacturers is provided in Appendix 5.1, at the end of this chapter.

Radio tracking and telemetry

Radio transmitters have been utilized in wildlife studies since the early 1960s (Cochran & Lord, 1963), and have assisted research on Antarctic pinnipeds since 1968 (Siniff, Tester & Kuechle, 1969, 1971). Transmitters can be monitored for two basic types of information: (1) the presence or absence of a signal (e.g. haul-out patterns, location) and (2) the relay of various types of data (e.g. heart rate, ambient temperature). Most studies using radio transmitters attached to Antarctic seals have involved monitoring activity patterns and movements of individuals (e.g. Siniff *et al.*, 1975). In addition to the brief description given below, several other sources offer a helpful introduction to using radio transmitters in wildlife studies (Cochran, 1980; Amlaner & Macdonald, 1980; Cheeseman & Mitson, 1982; Kenward, 1987).

Transmitters

Transmitter frequencies typically range from 30 to 220 MHz, with 150 to 166 MHz being frequencies commonly used for Antarctic seal studies. A popular design for such units is one in which a pulsed-tone transmitter and batteries are cast in epoxy resin with an attached whip antenna (plastic-coated, stainless steel cable, approximately 30 cm long and 2–3 mm in diameter) (Fig. 5.1). The operating life of the transmitter depends on various combinations of the voltage, pulse rate and battery capacity. For example, a 3-volt transmitter (available from Advanced

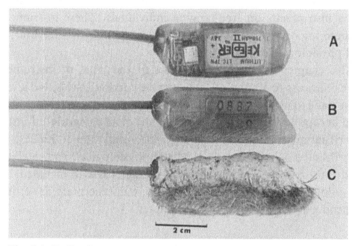

Fig. 5.1 Radio frequency transmitters (Advanced Telemetry Systems, Inc.) used for tracking Antarctic fur seals (mounted on the back). The leading edge of each transmitter is shown to the right, with the trailing whip antennae extending out of the photograph to the left. A = Bottom view, with the electronic circuitry (left side) and lithium batteries (right side) visible through the epoxy resin. B = Side view, with the epoxy resin (on the bottom and lower sides) roughed up in preparation for attachment to the seal's pelage using quick-setting epoxy resin. C = Side view showing a transmitter that was deployed and recovered from a fur seal. Fragments of the seal's fur are visible where they emerge from the white quick-setting epoxy resin used to attach the instrument to the seal (Photo by M.M. Muto).

Telemetry Systems, Inc.)[1] operating at 1 pulse per second will last approximately one year if powered by two lithium batteries, each with a capacity of 2–3 amp hours. Such a transmitter weighs approximately 65 g and measures 5 × 4 × 2 cm. Transmitters with higher voltage requirements or pulse rates would have a reduced operating life unless the battery capacity were increased. Depending on the specific requirements of a study, the antenna configuration for transmitters will vary: (1) to monitor haul-out or attendance ashore, a low profile (trailing against the seal's body) antenna can be used and (2) for tracking at sea, an upright antenna is needed to ensure that it extends above the sea surface when the seal surfaces to breathe.

Receivers

Radio receivers are required to monitor the transmitters deployed on seals. The receivers can be tuned to the specific frequencies of the transmitters, each of which can be built with a frequency unique to

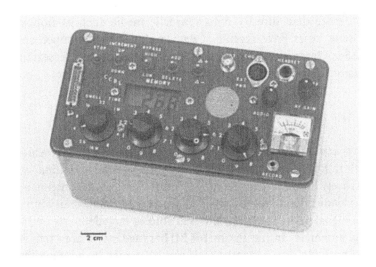

Fig. 5.2 Scanning radio receiver (Cedar Creek Bioelectronics Laboratory). Frequencies can be tuned manually (using the three dials in the lower centre of the face plate) or can be programmed to be scanned automatically using the four programming switches (upper left) and the scan dial (lower left) (Photo by M.M. Muto).

[1] Reference to trade names does not imply endorsement by the National Marine Fisheries Service, NOAA, Washington, USA.

each seal in a particular study. By tuning in to the appropriate frequency, it is possible to monitor or locate the transmitter attached to an individual seal. A useful feature available on some receivers is programmable scanning of frequencies (Fig. 5.2). This option allows the receiver to repeatedly scan a series of frequencies pre-set by the investigator. Such an arrangement is particularly useful when automated signal monitoring is desired. Haul-out patterns can be monitored in this manner; reception of radio signals is blocked when seals are submerged in salt water.

Recorders

When automated signal monitoring is needed (e.g. to record the haul-out or attendance patterns of seals on a beach or rookery), some type of data recorder must be interfaced with the receiver. Chart recorders (e.g. Esterline Angus event recorders) have been used for this purpose in several studies of Antarctic pinnipeds. However, the data from such analog charts must be summarized by visual inspection, a method involving many hours of manual transcription. Digital data recorders that store data directly onto magnetic media such as floppy disks or cassette tapes have recently been developed for pinniped research by the Sea Mammal Research Unit (Natural Environment Research Council, Cambridge, UK) and Advanced Telemetry Systems, Inc.

Tracking

A variety of receiving antennae can be employed to locate seals equipped with radio transmitters. Whip antennae are omnidirectional and have a relatively limited reception range. Multi-element yagi antennae provide a good indication of direction, and have a longer range than whip antennae. Popular designs for Antarctic seal studies include 2-, 3- and 4-element yagi antennae. In the 164 to 166 MHz range, these are easily hand-held and portable. Instrumented seals can be located by panning the antenna across the horizon; the radio signal is received with the greatest strength when the antenna is pointed directly at the transmitter.

Single yagi antennae can be mounted to the struts of aircraft for aerial tracking (Gilmer *et al.*, 1981). An antenna should be mounted pointing outward and down at about 45°, one to each side of the aircraft, with coaxial cable running from the antennae to a switchbox inside the cabin. By switching the receiver input between the left and right antennae, it is possible to locate instrumented seals from the air. This technique works well from both fixed-wing aircraft and helicopters. The principal advantages of using aircraft for locating instrumented seals are: (1) wider and faster coverage of the survey area, and (2) greater radio reception range at higher altitudes. Because VHF radio reception is essentially determined by line-of-sight, increased altitude allows reception at greater distances. Workable radio reception range at sea surface for the type of transmitters described above is often only about 5 km. However, reception range from an aircraft at 1300 m altitude may be 45 km or greater.

To track radio-tagged seals at sea from a ship, a system similar to the one used for aerial tracking can be utilized. Several yagi antennae (two to four) should be mounted as high as possible on the ship's mast, with each antenna oriented towards the horizon in an opposing direction. The investigator can either manually switch between antennae to find the strongest signal or use an automatic direction finding system which electronically switches between antennae (available from Cedar Creek Bioelectronics Laboratory). Such systems have been used for at-sea tracking of northern fur seals (*Callorhinus ursinus*) (Loughlin, Bengtson & Merrick, 1987) and Antarctic fur seals (*Arctocephalus gazella*) (Bengtson, Boveng & Jansen, 1991).

The use of a double-yagi antenna system is helpful if seals are to be tracked at sea from land-based sites. A null-peak reception system using tandem four-element yagi antennae will provide a more precise indication of the transmitter's direction than would otherwise be obtainable from a single hand-held yagi (Kenward, 1987). By simultaneously determining the direction of a transmitter from two fixed positions on land, it is possible to estimate the location of instrumented seals by triangulation. This system has worked successfully in tracking Antarctic fur seals up to 20 km offshore at Bird Island, South Georgia (Bengtson & Schneider, 1983).

Sonic Instruments

Sonic devices have been used to investigate a wide variety of Antarctic pinniped topics ranging from vocalizations to underwater movements and behaviour. In general, such devices can be used for three purposes: (1) to monitor sounds produced by seals, (2) to locate the positions of seals instrumented with transducers, and (3) to receive data transmitted sonically from transponders mounted on seals.

Vocalizations

It is beyond the scope of this chapter to review the methodology associated with the study of pinniped vocalizations. For more information on this topic, the reader is referred to studies on the vocalizations of Weddell seals (*Leptonychotes weddellii*) (Schevill & Watkins, 1965, 1971; Ray, 1967; Kooyman, 1981; Thomas & Kuechle, 1982; Thomas & Stirling, 1983; Thomas, Zinnel & Ferm, 1983, Thomas *et al.*, 1984), crabeater (*Lobodon carcinophagus*) and leopard (*Hydrurga leptonyx*) seals (Poulter, 1968; Ray, 1970; Stirling & Siniff, 1979; Thomas *et al.*, 1982; Thomas & DeMaster, 1923), Ross seals (*Ommatophoca rossii*) (Ray, 1970), southern elephant seals (*Mirounga leonina*) (Laws, 1956; LeBoeuf & Petrinovich, 1974) and southern fur seals *Arctocephalus* spp. (Bonner, 1968; Stirling, 1971; Stirling & Warneke, 1971; Miller, 1974).

Sonic transducers

Siniff *et al.* (1977) mounted sonic transducers on male Weddell seals to monitor their underwater movements and territories during the breeding season. These sonic tags broadcast at approximately 60 kHz. A triangulation array of underwater hydrophones was used to determine positions of instrumented seals by electronically comparing differences in arrival times of acoustic pulses from the transducers to the various hydrophones. This system only measured the relative horizontal position of seals; correcting for depth was not feasible. The results indicated that this method worked well in a fast ice situation where an array of hydrophones could be positioned.

The Sea Mammal Research Unit has developed a short range tracking system for seals using sonic transmitters operating at frequencies above 75 kHz. A suitable hydrophone can pick up the signal at least one kilometer away. Initially a yacht was fitted with a single directional hydrophone, but attachment of VHF transmitters also enabled seals to be tracked for up to nine days and 270 km. Various sensors were incorporated in the sonic pingers to enable monitoring of the animals' behaviour and physiology at the same time; these include swimming speed, swimming depth and heart rate (Sea Mammal Research Unit, 1988).

Sonic transponders

Although they have not been used commonly in studies of Antarctic pinnipeds, sonic transponders have been employed in field research on seals but results have been disappointing. These units differ from simple transducers in that they respond to an acoustic interrogation signal broadcast by the investigator. In responding to the query, they are able to provide data on both direction and range from the investigator. It is also possible to encode information on the depth of the transponder in its response to the interrogator.

Archival recorders
Archival recorders are self-contained instruments that collect and archive data in some fashion suitable for recovery at a later time. The development of such recorders to be carried by free-ranging seals represented a major step forward in pinniped field studies. These units were initially developed to monitor the diving behaviour of Weddell seals (Kooyman, 1965). Through the 1970s the early dive recorder design was modified and improved, and new units were developed to collect other behavioural, ecological and physiological data.

Dive recorders
Time-depth recorders (TDRs) have undergone dramatic changes over the past 20 years. The initial models (for Weddell seals) used a rotating smoked glass disk which was scored by a needle fastened to a pressure transducer (Kooyman, 1965). As described below, these units were further developed utilizing photographic film, and more recently, solid-state circuitry.

Photographic film TDRs
A miniature strip chart recorder, designed initially to use spooled paper (Kooyman, Billups & Farwell, 1983) and later photographic film (Kooyman, Gentry & Urquhart, 1976; Kooyman *et al.*, 1983), was developed to record the diving behaviour of fur seals. The photographic film TDRs incorporate a light emitting diode (LED) attached to a pressure transducer. As the film is advanced by a small electric motor, the LED traces a path over the film, indicating the instrument's relative depth in the water column. The electronics are encased in a cylindrical, aluminum pressure housing measuring 5 × 21 cm. The total instrument (excluding harness for attachment) weighs approximately 500 g (Fig. 5.3).

Photographic film TDRs (available from Meer Instruments, Inc.) can record the dive profile of a seal up to 20 days (running at 3 volts);

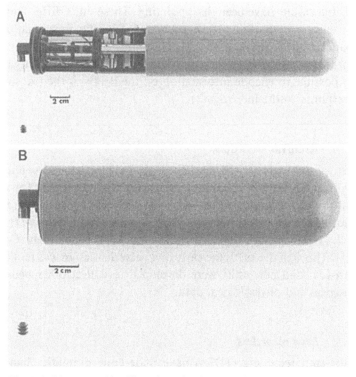

Fig. 5.3 Photographic film time-depth recorder (Meer Instruments, Inc.). A = The internal mechanism has been partially exposed by retracting the pressure housing. B = The instrument encased in its pressure housing and almost ready for deployment. During actual deployment, the oil-filled tube (attached to the pressure sensor) at the left end of the unit is retracted inside a protective cover (Photo by M.M. Muto).

with this configuration, events of 30 seconds or greater duration can be detected. Operating at 6 volts, the resolution of events increases to 8 sec or more, but the total record lasts only seven days (Gentry & Kooyman, 1986). The film record from recovered instruments is processed and transferred to paper using copy flow-xerography. The analog record on the xerox copy must then be manually digitized for computer analysis.

Coated paper TDRs

The Japanese National Institute of Polar Research has developed a coated paper TDR somewhat similar to the early units designed by Kooyman (1965). The TDR is composed of three main parts: a depth sensor, a quartz motor and a recording unit. A thin coat of carbon applied to paper is scratched off by a diamond needle linked to the depth sensor. To save battery power, the motor was geared to the very slow speed of 0.04 mm/minute in turning the coated paper. This TDR is reportedly small enough to attach to newly born pups[2]. The TDR measures 25 × 80 mm, weighs 80 g, and has a life span of approximately 20–30 days of continuous operation using 3-volt lithium batteries. The instrument can withstand pressure at depths up to 150 m.

Solid state dive recorders

The recent advances in microchip technology have allowed the development of solid state and microprocessor-controlled recorders. Several models of this type of unit are currently available, and many new changes and improvements are occurring rapidly. For example, maximum

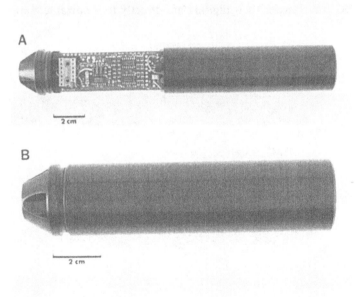

Fig. 5.4 Maximum depth recorder (Maret Consulting Services, Inc.). A = The pressure housing has been drawn back to reveal some of the internal circuits. B = The unit is nearly seated in the pressure housing (note the O-ring seal at the left end) and ready for deployment (Photo by M.M. Muto).

[2] Y. Naito, National Institute of Polar Research, 9–10, Kaga 1-Chome, Itabashi-ku, Tokyo 173, Japan. Pers. comm., October 1987.

depth recorders (MDRs) (available from Maret Consulting Services) measure the maximum depth for each dive and tally those depths in pre-assigned depth categories (Fig. 5.4). These units weigh 100 g and measure 12.5 × 3 cm. The data output from recovered units provides a depth frequency histogram of maximum depth of dives.

Microprocessor-controlled TDRs have been developed and successfully deployed on crabeater, Ross, Weddell, southern elephant and Antarctic fur seals (available from Wildlife Computers). These units digitally measure depth at pre-programmed time intervals. The time and depth data are stored in random access memory (RAM) chips with a capacity of up to 512 kilobytes. The sampling rate can be selected at any interval from 0.1 sec or greater. At a sampling rate of every 10 sec (6/min), the memory capacity of a 386 kilobyte unit is filled at approximately 45 days. Once the memory is filled, the unit switches to a low power mode, which can continue to maintain the data in RAM for up to one year. Recovered TDRs can transfer their digital data directly to a portable computer via

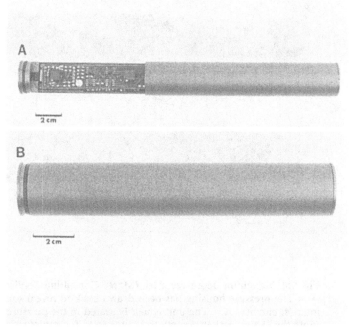

Fig. 5.5 Mark III microprocessor time-depth recorder (Wildlife Computers). A = The pressure housing has been drawn back to reveal some of the internal circuits. B = The left-hand endcap is almost seated in the pressure housing (note the O-ring seal near the endcap) and ready for deployment (Photo by M.M. Muto).

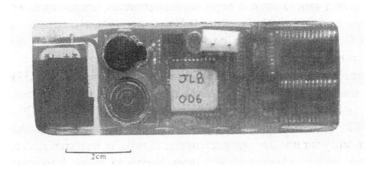

Fig. 5.6 Mark IV microprocessor time-depth recorder (Wildlife Computers). Through the transparent epoxy resin in which this unit is potted, one can see (from left to right) the batteries, the computer connector (upper) and pressure transducer (lower), the unit number, the conductivity switch, and the RAM chips (Photo by M.M. Muto).

a serial interface. The Mark III units are encased in an aluminium pressure housing, weigh approximately 120 g, and measure approximately 2.5 × 15 cm (Fig. 5.5). The Mark IV model is potted in epoxy resin, allowing a low profile for reduced drag in the water. Like the Mark III, it is equipped with a conductivity switch which monitors whether the seal is in the water or hauled-out (Fig. 5.6).

A recent modification of the Mark III time-depth recorder (Hill, 1993) is the geolocation option, which stores light-level and sea water temperature when the animal surfaces. Since the change in light-level at sunrise and sunset is quite rapid, dawn and dusk can be measured quite reliably and these data used to calculate day length. Estimation of longitude is based on local apparent noon and latitude upon day length. Standard navigational formulae are used for calculating dawn and dusk based upon location and these formulae are incorporated into algorithms that estimate position based upon dusk and dawn times. The temperature data can be used in conjunction with surface seawater temperature maps to supplement latitude estimates. Locations can generally be plotted with an accuracy of ± 111 km (60 nautical miles).

Gastro-thermo recorders

To study the feeding habits of Antarctic seals, the Japanese National Institute of Polar Research developed the gastro-thermo recorder (GTR). This unit is designed to evaluate feeding behaviour by monitoring fluctuations in stomach temperature. After food is swallowed, its arrival in the stomach should cause a depression of stomach temperature, which will then gradually again increase. It is assumed that when greater amounts of food are consumed, a longer time will be required for the stomach to return to its former temperature. The GTR is given to the seal to swallow and is supposed to remain in the stomach until retrieved. The GTR design is similar to the Japanese coated paper TDR except for its sensor, which is smaller and has a response time estimated at less than 30 sec. The GTR is encased in a metal housing for extra weight (to reduce the tendency for it to be regurgitated) and to assist in its retrieval. A magnetic 'cow sucker' (developed for farm livestock) is used to retrieve the GTR from the stomach. These units measure 25 × 90 mm and weigh 210 g. Using 3-volt lithium batteries, the instrument has a life span of approximately 20–30 days.

Other microprocessor-controlled recorders

Multi-channel microprocessor-controlled recorders have the capacity to monitor a variety of parameters. For example, microcomputer packages developed for Weddell seals simultaneously measure and record depth of dive, heart rate and body temperature (Hill *et al.*, 1983, 1984, 1987). These instruments have four channels of analog to digital conversion for data acquisition, 8 kilobytes of read only memory (ROM) for program storage, and 56 kilobytes of RAM for data storage. The circuitry was assembled on three boards measuring 10 × 10 cm each and stacked on top of each other. The circuit boards and batteries (two lithium 'C' cells) were cast in resin epoxy. Data stored in the unit's RAM were transferred to a computer via a fibre-optic lead which was connected for several seconds when the seals surfaced to breathe (at a controlled-access breathing hole drilled in the sea ice for these experiments).

Further developments in microprocessor-controlled instruments include a device designed to sample arterial blood at depth from free-diving Weddell seals (Hill *et al.*, 1984; Hill, 1986). A microprocessor backpack was used to control an aortic blood sampling pump, which allowed initiation of blood sampling at flexible combinations of depth and time during either the descending or ascending phases of the dive. A serial sampler was used to obtain up to eight sequential blood samples during a single dive. Concurrently with the blood sampling, the microprocessor unit recorded swimming depth, heart rate and body temperature at selected intervals.

Another microprocessor-controlled instrument was developed and is being used by research workers at the Scripps Institution of Oceanography[3]. In its current configuration, this device is capable of simultaneously measuring up to eight different variables (e.g. diving pattern, heart rate, swimming velocity) which are selected or changed according to the needs of particular experiments. Also with this device, alterations in circuitry and interfaces between external sensors and the microprocessor may be made to increase the number of variables that could be monitored. Data are stored digitally in RAM until they are recovered via computer link once the unit has been retrieved.

Satellite-linked instruments

Despite the exciting possibilities opened up by the development of the remarkable archival recorders described above, a major drawback remains: the instruments must be recovered to retrieve their data. It is reasonable, then, that investigators should focus their attention on developing instruments that will relay their precious data via satellite. However, there are still several difficulties that must be overcome, such as perfecting attachment methods or reducing the size of the instrument package. Another problem concerns the limitations to the quantity of data that can be relayed through a satellite link. The Argos system (the most practical polar-orbiting satellite system available for Antarctic studies) only allows 32 bytes of data per channel per satellite pass. To circumvent this limitation, researchers have generally used several (up to eight) channels sequentially to expand their data transmission capability, but data must still be summarized before transmission to get results through this data bottleneck. Additionally, data transmission is performed 'blind', meaning that the researcher does not know if the transmitted data have been received correctly by the satellite.

Several wildlife research groups are actively developing satellite-linked instruments, and there have been several successes with a variety of terrestrial, marine and avian species (Schweinsburg & Lee, 1982; Stoneburner, 1982; Timko & Kolz, 1982; Fuller *et al.*, 1984; Strikwerda *et al.*, 1985, 1986). Progress in utilizing this technology with marine mammals has been slower, due in large part to problems related to attachment, protection from pressure at depth (while keeping the package small) and timing satellite communications when the animal is at the

---

[3] G.L. Kooyman, Physiological Research Laboratory, Scripps Institution of Oceanography, La Jolla, California, USA. Pers. comm., March 1988.

ocean surface. Efforts to place satellite-linked instruments on great whales have been partially successful in recent years (Mate, 1987), and applying similar technology to seals is finally within reach. Two independent programmes relating to satellite monitoring of Antarctic seals are outlined below.

The SCAR Group of Specialists on Seals, supported by the United Nations Environmental Programme, commissioned the Sea Mammal Research Unit, Cambridge, UK, to develop a data-logger and satellite transmitter to monitor the behaviour of Antarctic seals. By May, 1986, the transmitter had been successfully built and tested, and the data-logger was in the final stages of testing. The unit has sensors to monitor haul-out, depth of dive, and heartbeat. It is encased in a plastic pressure housing measuring approximately 10 × 7 cm. Two types of data are transmitted via satellite: (1) summary data collected over two 6 hour periods concerning heartbeat and activity pattern, and (2) detailed data on the characteristics of recent dives. The satellite uplink transmitter is manufactured by Mariner Radar.

Additional satellite packages for Antarctic seals were designed, built and successfully deployed in the field by R.D. Hill and W.M. Zapol (Department of Anesthesiology, Massachusetts General Hospital, Boston, MA 02114, USA. Pers. comm., March 1988). These instrument packages are an extension of the microprocessor-controlled recorders developed for Weddell seals and described above. Those packages were modified to include a platform terminal transmitter (PTT) built by Telonics, Inc. The PTT provides the uplink for the recorders to relay their data via the Argos satellite data retrieval system. The units are encased in an aluminium pressure housing measuring 18 × 12 × 4 cm, and weighing approximately 4 kg (Fig. 5.7). Two models of instruments were built and deployed: (1) ecological units (time, depth, haul-out, geographical location for up to six months), and (2) physiological units (ECG, body temperature, time and depth data for up to 28 h). By June, 1987, several of the ecological units had successfully provided geographical locations of crabeater seals for up to five months in the Antarctic Peninsula area. Some of the physiological units were partially successful in that dive pattern data were successfully recovered via satellite.

Attachment methods

Several methods of attaching electronic instruments have been used during studies of Antarctic pinnipeds. Some of the methods tested in

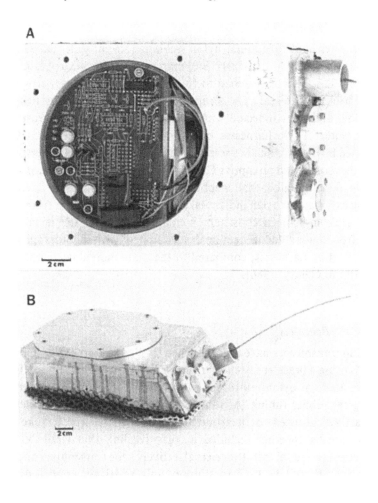

Fig. 5.7 Satellite-linked instrument package (Massachusetts General Hospital). Both the ecological and physiological units are housed in the same type of aluminium pressure housing illustrated. A = The pressure housing cover has been removed to show some of the internal circuits. Features at the right end of the unit are (top to bottom) the uplink antenna (partial view), the conductivity switch, and the pressure transducer. B = The pressure housing has been sealed and the unit is ready for deployment. The entire uplink antenna is pictured in this photograph (Photo by M.M. Muto).

early projects (e.g. sutures and tail mounts; Siniff *et al.*, 1971) were found to be unsatisfactory and are no longer used. Three methods that are currently employed are harnesses, bracelets and glues or epoxy resins.

Harnesses

Harnesses made of tubular nylon webbing (2.5 cm wide) have been used to attach TDRs to fur seals (Kooyman *et al.*, 1983; Gentry & Kooyman, 1986). The TDR is fastened to two straps on the harness crossing over the back of the seal. The straps continue underneath the belly of the animal and are connected with steel rivets that eventually rust away in salt water. These harnesses hold TDRs in position well, but the drag caused by the webbing has some effect on a seal's swimming performance. Studies conducted on captive California sea lions (*Zalophus californianus*) have demonstrated that webbing harnesses significantly increase the drag experienced by an individual (Feldkamp, 1985). Data from northern fur seals indicate a consistent increase of 20% in the metabolic rate of free-ranging lactating females equipped with photographic TDRs attached by harnesses, compared to the same individuals without TDRs (Costa & Gentry, 1986).

Bracelets

Radio transmitters have been attached to Weddell, crabeater and leopard seals using a bracelet fastened around the ankle of a rear flipper. Bracelets were made of nylon-reinforced rubber straps (1.5 cm wide) inserted into surgical rubber tubing. A transmitter was bolted to the bracelet, which was then adjusted to fit around the seal's ankle. The linkage holding the bracelet together included a corrosive link that would eventually deteriorate and fall off. Bracelets effectively keep transmitters on phocids, but they should be used to support only relatively small instruments. Long-term deployment of larger instruments would be likely to impede a seal's ability to swim and might be harmful to the ankle.

Care must be taken in adjusting the tightness of bracelets because improper tightness or placement can cause abrasion and inflammation of the ankle. Bracelets should be fastened tight enough so that an index finger can be slid between the bracelet and the seal's ankle. Young animals or emaciated adults may encounter problems with bracelets because of growth and weight gain after bracelet adjustment.

Glues and epoxy resins

The use of glues and epoxy resins for both otariids and phocids has become increasingly widespread in recent years (Fedak, Anderson &

Curry, 1982; Testa *et al.*, 1983; Harvey, 1987; Loughlin, Bengtson & Merrick, 1987). The technique involves fastening the instrument or a base-plate to the pelage of the seal using a quick-setting adhesive. Cyano-acrylic adhesives ('super glue') such as Loctite 422 form a very strong bond with hair in a few seconds. However, it degrades in water over time. Quick-setting epoxy resins such as Devcon EK-40 (no longer available) or Devcon 5-min epoxy harden within 5–20 min depending on the ambient temperature. These products have the advantage of being durable and resistant to degradation in salt water. Devcon 5-min epoxy should be applied in layers no deeper than 1 cm because of the heat generated during curing. Thicker layers give rise to possibility of generating sufficient heat to burn the seal's epidermis. In cases requiring more than 1 cm of epoxy resin to attach an instrument, several layers of resin should be applied in turn, with each being applied after the previous layer has cooled.

Instruments are variously mounted depending on the species and the size of the package. Radio transmitters have been successfully fixed with epoxy resin to the top of the heads or the backs of Antarctic fur seals (up to three months retention) and on the backs of Weddell and crabeater seals (up to 10 months retention). Special care must be taken when mounting a transmitter on the head of a fur seal to ensure that the instrument and epoxy attachment are properly centered. Improper orientation of the unit, or application of too much epoxy, may cause the instrument to fall off sooner than desired because of the hydrostatic pressure when the seal is swimming. Microprocessor TDRs have been fastened to the backs of crabeater seals (without the need for drug immobilization during handling), using a combination of super glue (to hold the unit in place while the epoxy is setting) and epoxy (for long-term attachment). An advantage of using glues and epoxy resins for attaching instruments is that if it is not possible for an investigator to recapture the seal to remove the instrument, the unit will fall off by itself when the pelage is shed during the next moult period.

References

Amlaner, C.J. & Macdonald, D.W. 1980. *A Handbook on Biotelemetry and Radio Tracking*. Pergammon Press, Oxford.

Bengtson, J.L. & Schneider, D.J. 1983. Fur seal research at Bird Island, South Georgia. *Antarct. J. U.S.* **18**, 212.

Bengtson, J.L., Boveng, P.L. & Jansen, J.K. 1991. AMLR program: Foraging areas of krill-consuming penguins and fur seals near Seal Island, Antarctica. *Antarct. J. U.S.*, **26**, 217–18.

Bonner, W.N. 1968. The fur seal of South Georgia. *Br. Antarct. Surv. Sci. Rep.* **56**, 1-81.

Cheeseman, C.L. & Mitson, R.B. 1982. *Telemetric Studies of Vertebrates.* Symposia of the Zoological Society of London, 49. Academic Press, London.

Cochran, W.W. 1980. Wildlife telemetry. In *Wildlife Management Techniques Manual*, 4th ed, ed. S.D. Schemnitz, pp. 507-20. Wildlife Society: Washington, D.C.

Cochran, W.W. & Lord, R.D. 1963. A radio-tracking system for wild animals. *J. Wildl. Mgmt.* **27**, 9-24.

Costa, D.P. & Gentry, R.L. 1986. Free-ranging energetics of northern fur seals. In *Fur Seals: Maternal Strategies on Land and at sea*, R.L. Gentry & G.L. Kooyman, pp. 79-101. Princeton University Press, Princeton, N.J.

Fedak, M.A., Anderson, S.S. & Curry, M.G. 1982. Attachment of a radio tag to the fur of seals. *Notes Mammal. Soc.* **49**, 298-300.

Feldkamp, S.D. 1985. Swimming and diving in the California sea lion, *Zalophus californianus*. Ph.D. thesis, University of California, San Diego. 176 pp.

Fuller, M.R., Levanon, N., Strikwerda, T.E., Seegar, W.S., Wall, A., Black, H.D., Ward, F.P., Howey, P.W. & Partelow, J. 1984. Feasibility of a bird-borne transmitter for tracking via satellite. *Biotelemetry* **8**, 375-8.

Gentry, R.L. & Kooyman, G.L. 1986. Methods of dive analysis. In (eds.) *Fur Seals: Maternal Strategies on Land and at Sea*, ed. R.L. Gentry & G.L. Kooyman pp. 28-40 Princeton University Press, Princeton, N.J.

Gilmer, D.S., Cowardin, L.M. Duval, R.L. Mechlin, L.M. Shaiffer, C.W. & Kuechle, V.B. 1981. Procedures for the use of aircraft in wildlife biotelemetry studies. *US Fish Wildl. Serv. Resour. Publ.* **140**, 1-19.

Harvey, J.T. 1987. Population dynamics, annual food consumption, movements, and dive behaviors of harbor seals, *Phoca vitulina richardsi*, in Oregon. Ph.D. thesis, Oregon State University. 171 pp.

Hill, R.D. 1986. Microcomputer monitor and blood sampler for free-diving Weddell seals. *J. Appl. Physiol.* **61**, 1570-6.

Hill, R.D. (1993). Theory of geolocation by light-levels. In *Population Ecology of Elephant Seals*, ed. B.J. LeBoeuf & R.M. Laws, Proceedings of an international conference, Santa Cruz, California, 19-20 May, 1991. Univ. California Press. (In press).

Hill, R.D., Schneider, R.C., Liggins, G.C., Shuette, A.H., Elliott, R.L., Guppy, M., Hochachka, P.W., Qvist, J., Falke, K.J. & Zapol, W.M. 1987. Heart rate and body temperature during free diving of Weddell seals. *Am. J. Physiol.* **253**, R344-51.

Hill, R.D., Schneider, R.C., Schuette, A.H., Zapol, W.M., Liggins, G.C. & Hochachka P.W. 1983. Microprocessor-controlled monitoring of bradycardia in free-diving Weddell seals. *Antarct. J. U.S.* **18**, 213-14.

Hill, R.D., Schneider, R.C., Zapol, W.M., Liggins, G.C., Qvist, J., Falke, K., Guppy, M & Elliott, R. 1984. Microprocessor-controlled monitoring of free-diving Weddell seals. *Antarct. J. U.S.* **19**, 150-1.

Kenward, R. 1987. *Wildlife Radio Tagging*. Academic Press, London.

Kooyman, G.L. 1965. Techniques used in measuring diving capacities of Weddell seals. *Polar Rec.* **12**, 391-4.

Kooyman, G.L. 1981. *Weddell Seal: Consumate Diver*. Cambridge University Press, Cambridge.

Kooyman, G.L., Billups, J.O. & Farwell, W.D. 1983. Two recently developed recorders for monitoring diving activity of marine birds and mammals. In *Experimental Biology at Sea* ed. A.G. Macdonald & I.G. Priede, pp. 197-214 Academic Press, New York.

Kooyman, G.L., Gentry, R.L. & Urquhart, D.L. 1976. Northern fur seal diving behavior: a new approach to its study. *Science* 193, 411-12.

Laws, R.M. 1956. The elephant seal (*Mirounga leonina* Linn.). II. General, social and reproductive behaviour. *Falkl. Isl. Depend. Surv., Sci. Rep.* 13.

LeBouef, B.J. & Petrinovich, L.F. 1974. Elephant seals: interspecific comparisons of vocal and reproductive behaviour. *Mammalia* 33, 16-32.

Loughlin, T.R., Bengtson, J.R. & Merrick, R.L. 1987. Characteristics of feeding trips of female northern fur seals. *Can. J. Zool.* 65, 2079-84.

Mate, B.R. 1987. *Development of Satellite-Linked Methods of Large Cetacean Tagging and Tracking in OCS Lease Areas - Final Report.* U.S. Department of the Interior, Minerals Management Service, Contract AA-730-79-4120-0109.

Miller, E.H. 1974. Social behaviours between adult male and female New Zealand fur seals *Arctocephalus forsteri* (Lesson) during the breeding season. *Aust. J. Zool.* 22, 155-73.

Poulter, J.C. 1968. Underwater vocalization and behavior of pinnipeds. In *The Behavior and Physiology of Pinnipeds.* ed. R.J. Harrison, R.C. Hubbard, R.S. Peterson, C.E. Rice & R.J. Schusterman, pp. 69-84. Appleton-Century-Crofts, New York.

Ray, G.C. 1967. Social behavior and acoustics of the Weddell seal. *Antarct. J. U.S.* 2, 105-6.

Ray, G.C. 1970. Population ecology of Antarctic seals. In (ed.) *Antarctic Ecology*, vol. I., ed. M.W. Holdgate, Academic Press, London.

Schevill, W.E. & Watkins, W.A. 1965. Underwater calls of *Leptonychotes* (Weddell seal). *Zoologica* 50, 45-6.

Schevill, W.E. & Watkins, W.A. 1971. Directionality of the sound beam in *Leptonychotes weddelli* (Mammalia: Pinnipedia). *Antarct. Res. Ser.* 18, 163-8.

Schweinsburg, R.E. & Lee, L.J. 1982. Movement of four satellite-monitored polar bears in Lancaster Sound, Northwest Territories. *Arctic* 35, 504-11.

Sea Mammal Research Unit. 1988. *Report for 1986/87.* Natural Environment Research Council, Swindon, England.

Siniff, D.B., Tester, J.R. & Kuechle, V.B. 1969. Population studies of Weddell seals at McMurdo Station. *Antarct. J. U.S.* 4, 120-121.

Siniff, D.B., Tester, J.R. & Kuechle, V.B. 1971. Some observations on the activity patterns of Weddell seals as recorded by telemetry. *Antarct. Res. Ser.* 18, 173-80.

Siniff, D.B., Reichle, R.A., Hofman, R.J. and Kuehn, D. 1975. Movements of Weddell seals at McMurdo Sound, Antarctica, as monitored by telemetry. In *Symposium on the Biology of the Seal (Guelph, Ontario, Canada, August 1972).* ed. K. Ronald & A.W. Mansfield, pp. 387-393. Council for the International Exploration of the Sea, Charlottenlund Slot, Denmark.

Siniff, D.B., DeMaster, D.P., Hofman, R.J. & Eberhardt, L.L. 1977. An analysis of the dynamics of a Weddell seal population. *Ecol. Monogr.* 47, 319-35.

Stirling, I. 1971, Studies on the behaviour of the South Australian fur seal,

Arctocephalus forsteri (Lesson). I. Annual cycle, posture and calls, and adult males during the breeding season. *Aust. J. Zool.* **19**, 243–66.

Stirling, I. & Siniff, D.B. 1979. Underwater vocalizations of leopard seals (*Hydrurga leptonyx*) and crabeater seals (*Lobodon carcinophagus*) near the South Shetland Islands, Antarctica. *Can. J. Zool.* **57**, 1244–8.

Stirling, I. & Warneke, R.M. 1971. Implications of a comparison of the airborne vocalizations and some aspects of the behaviour of the two Australian fur seals *Arctocephalus* spp., on the evolution and present taxonomy of the genus. *Aust. J. Zool.* **19**, 227-241.

Stoneburner, D.L. 1982. Satellite telemetry of loggerhead sea turtle movement in the Georgia Bight. *Copeia* 1982, 400–8.

Strikwerda, T.E., Black, H.D., Levanon, N. & Howey, P.W. 1985. The bird-borne transmitter. *Johns Hopkins APL (Appl. Phys. Lab.) Tech. Dig.* **6**, 60-7.

Strikwerda, T.E., Fuller, M.R., Seegar, W.S., Howey, P.W. & Black, H.D. 1986. Bird-borne satellite transmitter and location program. *Johns Hopkins APL (Appl. Phys. Lab.) Tech. Dig.* **7**, 203-8.

Testa, J.W., Braun, S.E., Siniff, D.B., Reichle, R., Ferm, L. & Winter, J.D. 1983. Ecology of Weddell seals in McMurdo Sound 1982-1983. *Antarct. J. U.S.* **18**, 214-15.

Thomas, J.A. & DeMaster, D.P. 1983. An acoustic technique for determining the haulout of leopard (*Hydrurga leptonyx*) and crabeater (*Lobodon carcinophagus*) seals. *Can. J. Zool.* **60**, 2028-31.

Thomas, J.A., Fisher, S.R., Evans, W.E. & Awbrey, F.T. 1982. Ultrasonic vocalizations of leopard seals (*Hydrurga leptonyx*). *Antarct. J. U.S.* **17**, 186.

Thomas, J.A. & Kuechle, V.B. 1982. Quantitative analysis of the underwater repertoire of the Weddell seal (*Leptonychotes weddelli*). *J. Acoust. Soc. Am.* **72**, 1730-8.

Thomas, J.A., Puddicombe, R., George, M. & Lewis, D. 1984. Investigations on geographic variation of Weddell seal vocalizations around the Antarctic. *Antarct. J. U.S.* **19**, 154 - 5.

Thomas, J.A. & Stirling, I. 1983. Geographic variation in Weddell seal (*Leptonychotes weddelli*) vocalizations between Palmer Peninsula and McMurdo Sound, Antarctica. *Can. J. Zool.* **61**, 2203-10.

Thomas, J.A., Zinnel, K.E. & Ferm, L.M. 1983. Investigation of Weddell seal (*Leptonychotes weddelli*) underwater calls using playback techniques. *Can. J. Zool.* **61**, 1448-56.

Timko, R.E. & Kolz, A.L. 1982. Satellite sea turtle tracking. *U.S. Natl. Mar. Fish. Serv. Mar. Fish. Rev.* **44**, 19-24.

APPENDIX 5.1

Addresses of companies or research centres referred to in this chapter.

Advanced Telemetry Systems, Inc.
470 First Avenue North Box 398 Isanti, Minnesota 55040 USA

Cedar Creek Bioelectronics Laboratory
University of Minnesota 2660 Fawn Lake Drive, N.E. Bethel, Minnesota 55005 USA

Devcon Corporation
Danvers, Massachusetts 01923 USA

Esterline Angus Instrument Corporation
P.O. Box 24000 Indianapolis, Indiana 46224 USA

Maret Consulting Services
9865 Mozelle Lane La Mesa, California 92041 USA

Mariner Radar Ltd
Bridle Way Camps Heath Lowestoft, Suffolk United Kingdom

Massachusetts General Hospital
Department of Anesthesiology, Boston, Massachusetts 02114 USA

Meer Instruments
Box 591 del Mar, California 92014 USA

National Institute of Polar Research
9–10, Kaga 1-Chome Itabashi-ku Tokyo 173 Japan

Scripps Institution of Oceanography
University of California at San Diego, P.O. Box 109, La Jolla, California 92037 USA

Sea Mammal Research Unit
c/o British Antarctic Survey, High Cross, Madingley Road, Cambridge CB3 OET United Kingdom

Telonics, Inc.
932 E. Impala Avenue Mesa, Arizona 85204 USA

Wildlife Computers
20630 N.E. 150th Street Woodinville, Washington 98072 USA

6

Behaviour

I. STIRLING, R.L. GENTRY AND T.S. McCANN

Introduction

On a superficial level most seal behaviour seems straightforward, but recording behavioural observations for scientific use can be deceptively difficult. The object of this chapter is to provide some guidelines on the kinds of data that are important, and the methods that can be used. Of necessity, this discussion is orientated towards recording the terrestrial behaviour of seals; recording behaviour at sea requires special instrumentation and is discussed in chapter 5. Renouf (1991) provides a detailed review of pinniped behaviour.

We will begin by describing briefly how to quantify behaviour and then discuss in more detail some of the categories of behavioural observation that are applicable to population ecology, the biology of species, and to defining taxonomic affinities. For each category the types of observations to be made are described, along with the species or situations to which they apply. Specific applications of behavioural knowledge are discussed, and references for further study are given.

Background information

Behavioural questions

Whenever possible, we recommend that overall objectives and specific behavioural questions be posed before the study begins. It is especially true of behavioural studies, that researchers are tempted to collect data in the field and then distil the questions from the mass of data accumulated. For those going to Antarctica or the sub-Antarctic islands specifically to study seals the questions should be based on extensive prior reading about the behaviour of the target species and of related species. These questions may be posed as hypotheses that can be tested by the results of field research. For those who find themselves in a position to record useful observations on an opportunistic basis, it may be more difficult to assess what would be most valuable to document. For both these types of observer we suggest that the following list of questions be consulted before behavioural work is begun on any Antarctic species. These are not the only questions that can be asked, but they address existing gaps in our knowledge of some species. Even more importantly, the answers can vary within the same species depending upon variation in factors such as geographical latitude and type of substrate (e.g. rocks, pebbles, sand, ice) at the study site. Quantitative comparison of such variation within a species in different parts of its range can provide valuable insight into its evolution. Some of these questions are discussed in greater detail in the sections on intraspecific and interspecific behaviour.

1. What is the duration of the reproductive season (dates of observed copulations and/or births)?
2. What is the duration of neonate dependency?
3. What is the pattern of daily movement (if any) in and out of the water of all animals, especially females with young? How is it affected by weather?
4. Is there evidence that animals feed during the breeding season (from scats or regurgitations)? Collections of these can be valuable.
5. Does geographic segregation occur? Is it based on sex or age?
6. What is the adult sex ratio on breeding areas (e.g. polygyny, monogamy, serial monogamy); what is the ratio when non-breeding areas are considered? Does it change through the breeding season?
7. What is the duration of tenure of adult males in breeding aggregations?
8. Is male–male aggression associated with specific locations (territoriality), with body size (dominance relations) or other factors?
9. Is the substrate the same throughout the breeding area and if not, are there any behavioural differences associated with different substrates (e.g. degree of territoriality in relation to specific locations)?

10. What is the nature of male–female relations (dominance/ subordinance, male 'control' of females, mutual consort)?
11. What is the size and composition of groups in which breeding occurs?
12. How do females associate (gregarious, solitary, aggressive, positively or negatively thigmotactic)?
13. What is the duration of sexual receptivity in females; how many times do they copulate?
14. Does mating occur on land or ice or aquatically?
15. Where do the virgin females mate?
16. What kinds of associations do juveniles form? Describe the forms of play, if any.
17. List, describe and make tape recordings of all calls, postures, displays, or signals that are used repeatedly (this list is called an 'ethogram').
18. What are the group sizes outside the breeding season?

In addition to these questions Gentry (1975) has listed several relatively stereotyped behavioural patterns that should be recorded for otariids. Some of his categories apply to phocids as well. All these questions can be addressed by observation of undisturbed animals. However, behaviour of some animals can be manipulated experimentally in the field to answer specific questions (Gentry & Holt, 1982; 1986).

Basic data

No matter what observations are being recorded the following should be noted; date, time, location, weather (preferably temperature, wind speed, cloud cover, precipitation), and solar radiation if the equipment is available. The duration of observation should always be recorded so that behavioural results can be expressed as rates.

Physical aspects of the habitat can significantly influence the behaviour and distribution of seals. Consequently the study area should be described and photographed with the date, time and other relevant information documented for reference photographs. For comparison, nearby areas that are not used by seals should also be described. Ice may be separated into pack ice and land fast ice. The following variables of ice should be noted: percentage of ice cover, size of floes in general and of those being used by seals, description of floe and the distribution of leads and single holes (especially important in fast ice) (Gilbert & Erickson, 1977).

Terrestrial habitat may be smooth sand or gravel beaches, tussock grass, boulder beaches, rugged shorelines broken up by rocky ridges or large boulders, bays and isolated coves. Shorelines may be open to the sea or have ice against them. The differential uses of habitat by males, females and juveniles should be indicated. Note also whether seals are using the only available habitat, such as a single floe in open water, or the only beach on a cliff-bound island.

Sampling, recording and analysing behavioural data

In more detailed studies the essence of quantifying behaviour is to break it into units that are recognisable, relevant to the questions being asked and small enough to manage. These units can be viewed as a series of pigeonholes into which given behavioural events are sorted. Obviously the units are arbitrary because behavioural patterns occur from variable to stereotyped, and from simple to complex along a continuum having no firm boundaries between types. No universal set of behavioural units exists for seals; they must be devised on the spot, specific to the questions being asked. This should be done by first observing the animals closely and describing all the relevant behavioural patterns. These patterns are then named and listed on paper, and sorted into units. The breadth, or inclusiveness, of each unit depends on the questions being asked.

Three major techniques exist for quantifying these behavioural units; scan sampling, event recording and focal animal studies. All these techniques list the behavioural units on a form called a protocol and follow various sampling regimes to record the frequency of repetition and to correlate one type of behaviour with others.

Scan sampling

This uses the broadest categories of behaviour (for example, 'head up' and 'head down' may be used to quantify activity levels). This sampling technique is usually applied to large numbers of anonymous animals at frequent intervals. A common format is for the behavioural units to be listed in rows and some time unit (e.g. five minutes) is listed in columns. Using a clock the researcher scans the population as a whole or a number of preselected individuals at each predetermined time interval and records the number of animals exhibiting each category of behaviour at that moment.

Event recording

This is used to gather information about parts of complex behavioural patterns, variation in a pattern over time, or the frequency of repetition. Here a protocol is constructed having the categories of information as column heads and blank spaces for each repetition of the event as rows. For example to quantify copulatory behaviour the column heads might be date, time, identity of the male and female (if known), number of mounts, duration, and ano-genital contact, etc. The event might also be recorded on a map of the study site to show its location. This technique lends itself to recording many types of detailed behaviour and to making correlations between different behavioural categories. It is the main technique used in many field studies of behaviour, often in conjunction with the technique described next.

Focal animal studies

These are useful for quantifying sequences of events, or behavioural events that are subtle and easily missed by the previous two techniques. Here a single, usually known, animal is observed continuously for a predetermined period of time and the sequence in which all its behavioural patterns occur is recorded. The behavioural units used here are often simple movements, like a head lunge or one kind of bite. The behavioural units are treated as if they do not change over time, so the focus is on the sequence in which the units occur. This technique lends itself to analysing action and counteraction when animals interact in pairs, or dyads, although the analysis of such records can be complex. It can also be used to measure spacing between animals over time by recording nearest neighbour differences of focal animals at regular intervals.

Data collection

Often several different versions of protocol must be tried before one is devised that is neither too detailed nor too vague, and that is simple enough that the recording of one event does not prevent the observation of others. Pads of paper printed with 10 or 12 vertical columns and 25 to 30 rows are very convenient for establishing protocols and collecting quantitative data. Protocols should be filled out in the field as the behaviour occurs. Data should not be collected in field notebooks and then transcribed onto protocols because of the time costs and the introduction of errors. Alternatively, the data can be collected directly on a portable computer. This system requires that the protocol format be converted to computer form, but data collection and analysis are faster than with the paper method. However, tape recorders and computers have a tendency to break down, a problem not encountered when using paper and pencil. (See Chapter 15.)

Protocol formats may be changed or abandoned at will if they do not adequately cover the behaviour being observed, or if the questions change as a result of experience in the field. No data collection scheme has ever been devised that was perfect from the outset.

If the study population comprises less than five animals, behavioural data can probably be recorded on all of them continuously. However, the periods of observation must still be selected to represent the full cycle of diurnal activity. In larger colonies the behaviour of individually recognizable animals may be focused on as representative of behaviour at large. Alternatively, a subgroup of animals representing each age and sex class can be selected for observation. The degree of emphasis to be placed on each age and sex group should be clearly established at the outset depending on the priorities of the research.

Before starting any behavioural research one should consult Altmann's (1974) review of observation and data collection methods, and Colgan's (1978) book on statistical analysis for behavioural data. In the event that time does not permit these preparations, motion picture film, or video and audio tape recordings should be used to the greatest extent possible. These methods record events exactly and completely, and can be analysed in the laboratory.

Intraspecific behaviour

General

This section discusses in more detail some of the questions listed at the beginning of this chapter.

Grouping

The size of breeding aggregations, as well as the criteria used to define these groups, are as important to note as the total numbers. Seals are often distributed in a series of small groups. For example, the mean group size of crabeater seals in the pack ice in January and February was 2.3

(Siniff, Cline & Erickson, 1970) while aggregations of sub-adults at the edge of the fast ice in October could number several hundred (Siniff *et al.*, 1979). Elephant seals spread along a beach may form a continuum of groups from small to large, or a series of small groups (Van Aarde, 1980; McCann, 1980a). The absence of clearly separable social units should also be noted.

Spacing

Spacing within a group is often indicative of reproductive associations; attention to the details of spacing can be valuable in confirming the identity of species. About the only constant in spacing is that females and their young tend to lie in close body contact. Aside from that trend, spacing in a particular area can vary by species, by season, by substrate and by reproductive status. For example, Weddell seals and male elephant seals are more widely spaced during the breeding season (spring) than they are later in the summer (Stirling, 1969a; McCann, 1983). Adult male and female crabeater seals may lie in body contact from after the female has weaned her pup to sometime after mating, but not outside the breeding season (Siniff *et al.*, 1979). Female sea lions usually lie in close body contact but female fur seals do not. Spacing in some seals may be strongly influenced by the distribution, movements and age of the pups.

The *sex ratio* of seals hauled out on land or ice during and outside the breeding season is important but poorly documented for most Antarctic seals. The number of adult females with each adult male should be noted (and preferably their positions sketched) and their numbers kept separate from sub-adults and non-breeding adults in total censuses. The presence or absence of sub-adult males or females in breeding colonies should be noted. The ratio of females to males may also vary with the type of habitat or the status of the population (Van Aarde, 1980; McCann, 1980a). It is also important to note whether males consort serially with more than one female.

Diel rhythm

The diel pattern of haul-out is well documented only for Weddell seals on the fast ice in the austral spring and summer (Muller-Schwarze, 1965; Smith, 1965; Tedman & Bryden, 1979; Thomas & DeMaster, 1983). Only very limited data are available for any of the other Antarctic seals (Erickson *et al.*, 1970; Condy, 1977; Gilbert & Erickson, 1977; Thomas & DeMaster, 1983). When possible, data on different age or sex groups should be segregated. Detailed information on the diel rhythm of haul-out should be collected for all species as the present data are still preliminary, for the most part, and limited in quantity of sampling and geographical representation.

Mother–pup relations

Parturition and post-partum behaviour have been described for the Weddell seal (Mansfield, 1958; Stirling, 1969b; Tedman & Bryden, 1979), elephant seal (Laws, 1956; Carrick, Csordas & Ingham, 1962; McCann, 1982) and Antarctic fur seal (Bonner, 1968), but not for the other Antarctic species. The details to note include the time of day observations began, interval between all the major events (contractions, first appearance of membranes and pup, full delivery) whether presentation is caudal or cephalic, behavioural interactions between mother and young, first responses of the pup, time until placenta appears, presence of scavengers and time to first suckling.

The duration of suckling bouts, length of time between bouts and time to weaning may vary between mothers of different ages, between populations, or according to latitude or habitat. For otariids, the duration of feeding trips to sea by a suckling female are important to record. In Antarctic fur seals these have been shown to be sensitive to annual variations in prey abundance (Doidge, McCann & Croxall, 1986; Croxall *el al.*, 1988). For some pack ice phocids neither the length of the lactation period nor whether the female leaves the pup to feed are known (studies which require animals to be individually marked). Information should be obtained on these subjects whenever possible.

Territorial and reproductive behaviour

Despite the fact that territorial behaviour has been documented often, especially in land breeding species, untrained observers regularly overlook key points. In territorial systems, such as that of Antarctic fur seals, it is important to note the date on which adult males begin to defend territories or become spaced apart from each other. If the boundaries are fixed the size of initial territories should be calculated. If individual males can be recognized by natural marks or scars, the length of time each is present should be noted, as well as changes in the size of their defended area through the season. Displays at territorial boundaries, and the events precipitating these displays, should be described, as should the proportion of encounters that are resolved by means other than physical contact (McCann, 1980b, 1981). The absence of any territorial defence is also noteworthy and in phocids is probably the more common situation. Resource defence may occur but is likely to be defence of a female or group of females rather than defence of a geographical area (Corner, 1971; Siniff *et al.*, 1979; McCann, 1981).

Since the function of territoriality and dominance hierarchies is to improve the male's reproductive advantage, male–male behaviour must be correlated with copulation frequency whenever possible. Because the copulation rate in a colony varies through the season, it is important to record which individual males are present at the site, and their tenure dates relative to the peak of reproductive activity (Kaufman, Siniff & Reichle, 1975; McCann, 1980b, 1981, 1983).

The behaviour of males and females towards each other in different circumstances is also important to document. For example changes in the responses of females to males through the lactation period can indicate when mating will occur and can also relate to the interpretation of different breeding strategies (Siniff *et al.*, 1979; McCann, 1981, 1983). The degree to which males can control the movements of females, and the degree to which control depends on the presence of a pup, or varies with habitat and species should be noted. The relevant information to collect for copulations are the animal that initiates the encounter, sequence of components, total duration and the outcome.

The extent to which copulation occurs on land and in the water is also important. In species which usually copulate either in water or on land or ice, the circumstances leading up to copulation in the alternative situation are important. For example, if access to a beach is precluded by ice, do the seals mate on the ice or in the water? Are the seals that are behaving differently full adults, primiparous females or sub-adults? What percent of the matings are either aquatic or terrestrial? These questions are relevant to theoretical arguments on the evolution of mating systems (Bartholomew, 1970; Stirling, 1975, 1983; Le Boeuf, 1986).

Interspecific behaviour

Reactions to humans

The effects of man and his machines on Arctic and Antarctic seals have not been examined very extensively or intensively. Opportunities to record distribution, abundance or behaviour, of seals before and after large or small disturbances are particularly valuable. In general, the potential for detrimental effects may be greater in otariids because they appear to be more frightened of people and of sudden disturbances than are southern phocids. Similarly, human activities that do not appear to disturb seals are worth recording. For example, responses of fur seals to overflights by aircraft may vary greatly with the type of aircraft, the altitude, or the time of the year. The cost of designing experiments to test such aspects is prohibitive so that the gathering of such data on an opportunistic basis is particularly important. Disruption or displacement of seal colonies by shore parties or conversely the lack of visible changes are noteworthy.

Sometimes it is possible to document the distribution, numbers, movements or behaviours of seal colonies before they are disturbed. This is particularly valuable because it provides a baseline against which to evaluate whether or not a particular disturbance causes detectable effects. At places where longer term disturbances are expected, such as the sites of bases, places visited by tourists, or by scientists themselves, a regular monitoring programme should be established in which specific data are collected on distribution, movements or reproductive behaviour.

Interactions with other animals

There are few accounts of predation on seals or of feeding behaviour of seals. For example, in shallow water, or near the ice edge it is occasionally possible to watch seals feeding on invertebrates or fish yet few descriptions of such behaviour exist in the literature. Of particular interest are observations of predation by one seal on another (e.g. leopard seals/crabeater seals; leopard seals/fur seals), of seals on penguins (Muller-Schwarze & Muller-Schwarze 1975; Siniff & Bengtson, 1977; Bonner & Hunter, 1982) or of killer whales on seals (Condy, Van Aarde & Bester, 1978, Smith et al., 1981). Descriptions of these aspects would be extremely interesting and, if possible, should include reference to the situation, the age and sex of the prey, how the seal or prey was approached, how it was attacked, how it was killed, how long it took, what was eaten, in what order and whether other predators or scavengers shared the kill.

Where species occur together, observations on competition for space, separation by substrate preferences, differential timing of peak numbers (for breeding or moult) should be recorded. Examples are elephant and fur seals (separated by preference for sandy or rocky beaches; different breeding season chronology), crabeater and Ross seals (prefer different types of ice floe). See the section on basic data.

Vocalizations

The most valuable vocalizations are those related to reproductive behaviour, that is, male threat calls, calls between males and females in relation to mating and mother–young recognition calls.

Tape recordings of airborne vocalizations are of value in assessing taxonomic affinities. For example, LeBoeuf & Peterson (1969) and LeBoeuf & Petrinovich (1974, 1975) have shown significant variation in vocalizations between different populations of northern elephant seals, and between northern and southern elephant seals. Stirling & Warneke (1971) demonstrated interspecific variation between the vocalizations of different species of *Arctocephalus*. The absence of certain calls may also be significant in comparative studies. For example, Gentry (1975) noted that while *Arctocephalus* may yelp like a dog during submission, sea lions remain silent. Playback to seals of recorded sounds, and recording of their responses, may be critical in determining the function of individual vocalizations (Watkins & Schevill, 1968; Trillmich, 1981; Thomas & Keuchle, 1982).

A good quality tape recorder, such as a Uher 4000L, 4200 stereo or Nagra III, running at a tape speed of 7½ ips, with a recording level that reduces distortion, and an M516 (or equivalent) microphone that is sheltered from wind will produce the best recordings. However, recordings should be made with whatever equipment is at hand if the opportunity presents itself. Recordings should be made from as close to the seal as possible without endangering oneself or disturbing the seal. It is often possible to predict from watching seals when they will vocalize. If recordings are made only at those times, rather than continuously, tape will be spared.

Adequate research on vocalizations of ice breeding seals can only be undertaken using underwater recordings, which require the use of hydrophones. Such sounds are more difficult to interpret than airborne vocalizations because the vocal animal usually cannot be seen. In addition to species identification it is possible that sub-ice distribution of species and relative abundance, if not actual numbers, can be derived from underwater recordings (Stirling & Siniff, 1979; Thomas & DeMaster, 1982; Stirling, Calvert & Cleator, 1983). Detailed work on underwater vocalizations has only been carried out on Weddell seals (Thomas &

Kuechle, 1982; Thomas & Stirling, 1983; Thomas, Zinnel & Ferm, 1983) and these references provide details of recording techniques and equipment. Schusterman & Balliet (1969) showed that the structure and function of barking by sea lions (*Zalophus californianus*) in air and in water were very similar. Comparisons of the air and underwater repertoires of the southern phocids and otariids may give new insights into their functions.

The field of hydroacoustics is a specialized area and an expert should be consulted before hydrophones are purchased. Several good models of hydrophones and amplifiers are available (e.g. International Transducer Corporation, Goleta, California, model 6050C). To reduce background noise, amplifiers should be attached side by side with the hydrophone before waterproofing, rather than attached further back toward the recorder. The date, time and location of the recordings should be spoken into the tape recorder at the beginning of each recording session. Because the seals are not usually visible, underwater recording should be continuous; at least 10 min per session is advisable.

References

Altmann, J. 1974. Observational study of behaviour; sampling methods. *Behaviour* **49**, 227-67.

Bartholomew, G.A. 1970. A model for the evolution of pinniped polygyny. *Evolution*, **24**, 546-59.

Bonner, W.N. 1968. The fur seal of South Georgia. *Sci. Rep. Br. Antarc. Surv.* **56**, 1-81.

Bonner, W.N. & Hunter, S. 1982. Predatory interactions between Antarctic fur seals, macaroni penguins and giant petrels. *Br. Antarct. Surv. Bull.* **56**, 75-9.

Carrick, R., Csordas, S.E. & Ingham, S.E. 1962. Studies on the southern elephant seal, *Mirounga leonina* (L.) IV. Breeding and Development. *CSIRO Wildl. Res.* **7**, 161-97.

Colgan, P.W. (Ed.) 1978. *Quantitative Ethology.* John Wiley and Sons, London.

Condy, P.R. 1977. Results of the 4th Seal survey in the King Haakon VII Sea, Antarctica. *S. Afr. J. Antarct. Res.*, **7**, 10-13.

Condy, P.R., Van Aarde, R.J. & Bester, M.N. 1978. The seasonal occurrence of killer whales (*Orcinus orca*) at Marion Island. *J. Zool., Lond.*, **184**, 449-64.

Corner, R.W.M. 1971. Observations on a small crabeater seal breeding group. *Br. Antarc. Surv. Bull.* **30**, 104-106.

Croxall, J.P., McCann, T.S., Prince, P.A. & Rothery, P. 1988. Reproductive performance of seabirds and seals at South Georgia and Signy Island, South Orkney Islands, 1976-1987; implications for Southern Ocean monitoring studies. In *Antarctic Ocean and Resources Variability*, ed. D. Sahrhage, pp. 261-85, Springer-Verlag, Berlin.

Doidge, D.W., McCann, T.S. & Croxall, J.P. 1986. Attendance behaviour of Antarctic fur seal *Arctocephalus gazella*. In *Fur Seals: Maternal Strategies on Land and at Sea*, ed. R.L. Gentry, & G.L. Kooyman, pp. 102–14. Princeton University Press, Princeton.

Erickson, A.W., Siniff, D.B., Cline, D.R. & Hofman, R.J. 1970. Distributional ecology of Antarctic seals. In *Symposium on Antarctic Ice and Water Masses*, pp. 55–76. SCAR, Cambridge.

Gentry, R.L. 1975. Comparative social behaviour of eared seals. *Rapp. p.-v. Reun. Cons. Int. Explor. Mer.* **169**, 189–94.

Gentry, R.L. & Holt, R.J. 1982. *Equipment and Techniques for Handling Northern Fur Seals*. NOAA Tech. Rep. NMFS-SSFR-758.

Gentry, R.L. & Holt, R.J. 1986. Attendance behaviour of northern fur seals. In *Fur Seals: Maternal Strategies on Land and at Sea*, ed. R.L. Gentry & G.L. Kooyman, pp. 41–60. Princeton University Press, Princeton.

Gilbert, J.R. & Erickson, A.W. 1977. Distribution and abundance of seals in the pack ice of the Pacific Sector of the Southern Ocean. In *Adaptations within Antarctic Ecosystems*, ed. G.A. Llano, pp. 703–48. Smithsonian Institution, Washington DC.

Kaufman, G.W., Siniff, D.B. & Reichle, R. 1975. Colony behaviour of Weddell seals *Leptonychotes weddelli*, at Hutton Cliffs, Antarctica. *Rapp. P.-v. Reun. Cons. Int. Explor. Mer.* **169**, 228–46.

Laws, R.M. 1956. The elephant seal (*Mirounga leonina* Linn.) II. General, social and reproductive behaviour. *Sci. Rep. Falkl. Is. Dep. Surv.* **13**, 1–88.

LeBoeuf, B.J. 1986. Sexual strategies of seals and walruses. *New Scientist*. No. 1491, 36–9.

LeBoeuf, B.J. & Peterson, R.S. 1969. Dialects in elephant seals. *Science*, **166**, 1654–6.

LeBoeuf, B.J. & Petrinovich, F. 1974. Elephant seals: interspecific comparisons of vocal and reproductive behaviour. *Mammalia* **38**, 16–32.

LeBoeuf, B.J. & Petrinovich, F. 1975. Elephant seal dialects: are they reliable? *Rapp. P.-v. Reun. Cons. Int. Explor. Mer.* **169**, 213–18.

McCann, T.S. 1980a. Population structure and social organization of southern elephant seals, *Mirounga leonina* (L.) *Biol. J. Linn. Soc.* **14**, 133–50.

McCann, T.S. 1980b. Territoriality and breeding behaviour of adult male Antarctic fur seals, *Arctocephalus gazella. J. Zool., Lond.* **192**, 295–310.

McCann, T.S. 1981. Aggression and sexual activity of male southern elephant seals, *Mirounga leonina. J. Zool., Lond.* **195**, 295–310.

McCann, T.S. 1982. Aggressive and maternal activities of female southern elephant seals (*Mirounga leonina*). *Anim. Behav.* **30**, 268–76.

McCann, T.S. 1983. Activity budgets of southern elephant seals, *Mirounga leonina* during the breeding season. *Z. Tierpsychol.* **61**, 111–26.

Mansfield, A.W. 1958. The breeding behaviour and reproductive cycle of the Weddell seal (*Leptonychotes weddelli*, Lesson). *Sci. Rep. Falkl. Isl. Dep. Surv.* **18**, 1–41.

Muller-Schwarze, D. 1965. Zur Tagesperiodlik der allgemeinen aktivitat der Weddell robbe (*Leptonychotes weddelli*) in Hallett, Antarktika. *Z. Morph. Oekol. Tierre*, **55**, 796–803.

Muller-Schwarze, D. & Muller-Schwarze, C. 1975. Relations between leopard seals and Adelie penguins. *Rapp. P.-v. Reun. Cons. Int. Explor. Mer.* **169**, 394–404.

Renouf, D. (Ed.) 1991. *Behaviour of Pinnipeds.* Chapman & Hall, London.
Schusterman, R.J. & Balliet, R.F. 1969. Underwater barking by male sea lions (*Zalophus californianus*). *Nature*, 222, 1179-81.
Siniff, D.B. & Bengtson, J.L. 1977. Observations and hypotheses concerning the interactions among crabeater seals, leopard seals and killer whales. *J. Mammal.* 58, 414-16.
Siniff, D.B., Cline D.R. & Erickson, A.W. 1970. Population densities of seals in the Weddell Sea, Antarctica, 1968. In *Antarctic Ecology*, vol. I, ed. M.W. Holdgate, pp. 377-94. Academic Press, London.
Siniff, D.B., Stirling, I., Bengtson, J.L. & Reichle, R.A. 1979. Social and reproductive behaviour of crabeater seals (*Lobodon carcinophagus*) during the austral spring. *Can. J. Zool.* 57, 2243-55.
Smith, M.S.R. 1965. Seasonal movements of the Weddell seal in McMurdo Sound, Antarctica. *J. Wildl. Mgmt.* 29, 464-70.
Smith, T.G., Siniff, D.B., Reichle, R. & Stone, S. 1981. Co-ordinated behaviour of killer whales, *Orcinus orca*, hunting a crabeater seal, *Lobodon carcinophagus. Can. J. Zool.* 59, 1185-9.
Stirling, I. 1969a. Ecology of the Weddell seal in McMurdo Sound, Antarctica. *Ecology,* 50, 573-86.
Stirling, I. 1969b. Birth of a Weddell seal pup. *J. Mammal.* 50, 155-6.
Stirling, I. 1975. Factors affecting the evolution of social behaviour in the pinnipedia. *Rapp. P.-v. Reun. Cons. Int. Explor. Mer.* 169, 205-12.
Stirling, I. 1983. The evolution of mating systems in pinnipeds. In ed. J.F. Eisenberg & D.G. Kleiman, *Recent Advances in the Study of Mammalian Behaviour*, pp. 489-527. Spec. Publ. No. 7. American Society of Mammalogists.
Stirling, I. & Siniff, D.B. 1979. Underwater vocalization of leopard seals (*Hydrurga leptonyx*) and crabeater seals (*Lobodon carcinophagus*) near the South Shetland Islands, Antarctica. *Can. J. Zool.* 57, 1244-8.
Stirling, I. & Warneke, R.M. 1971. Implications of a comparison of the airborne vocalizations and some aspects of the behaviour of the two Australian fur seals, *Arctocephalus* spp., on the evolution and present taxonomy of the genus. *Aust. J. Zool.* 19, 227-41.
Stirling, I., Calvert, W. & Cleator, H. 1983. Underwater vocalizations as a tool for studying the distribution and abundance of wintering pinnipeds in the High Arctic. *Arctic* 36, 262-74.
Tedman, R.A. & Bryden, M.M. 1979. Cow-pup behaviour of the Weddell seal, *Leptonychotes weddelli* (Pinnipedia), in McMurdo Sound, Antarctica. *Aust. Wildl. Res.* 6, 19-37.
Thomas, J.A. & DeMaster, D.P. 1982. An acoustic technique for determining diurnal activities in leopard (*Hydrurga leptonyx*) and crabeater (*Lobodon carcinophagus*) seal. *Can. J. Zool.* 60, 2028-31.
Thomas, J.A. & DeMaster, D.P. 1983. Diel haulout patterns of Weddell seal (*Leptonychotes weddelli*) females and their pups. *Can. J. Zool.* 61, 2084-6.
Thomas, J.A. & Kuechle, V.B. 1982. Quantitative analysis of Weddell seal (*Leptonychotes weddelli*) underwater vocalizations at McMurdo Sound, Antarctica. *J. Acoust. Soc. Am.* 72, 1730-8.
Thomas, J.A. & Stirling, I. 1983. Geographic variation in the underwater vocalizations of Weddell seals (*Leptonychotes weddelli*) from Palmer Peninsula and McMurdo Sound, Antarctica. *Can. J. Zool.* 61, 2203-12.

Thomas, J.A., Zinnel, K.C. & Ferm, L.M. 1983. Analysis of Weddell seal vocalizations using underwater playbacks. *Can. J. Zool.* **61**, 1448-56.
Trillmich, F. 1981. Mutual mother-pup recognition in Galapagos fur seals and sea lions: cues used and functional significance. *Behaviour.* **78**, 21-42.
Van Aarde, R.J. 1980. Harem structure of the southern elephant seal *Mirounga leonina* at Kerguelen Island. *Rev. Ecol. Terre. Vie.* **34**, 31-44.
Watkins, W.A. & Schevill, W.E. 1968. Underwater playback of their own sounds to *Leptonychotes* (Weddell seals). *J. Mammal.* **49**, 287-96.

7

Killing methods

W.N. BONNER

Destruction comes most swiftly upon seals when they are smitten on the head.   Oppian, *Halieutica* v.1.391.

Introduction

Killing of seals has attracted more attention than any other group of animals and several official publications on approved methods have been issued, perhaps largely on account of public concern about inhumane methods alleged to be used in commercial seal hunting. In the past much of the information on seal biology was obtained from post-mortems carried out on animals taken commercially, killed for dog food or specifically for the purpose of biological research. However, few species are now commercially exploited – certainly no Antarctic species – and very few seals are taken for dog food; also there has been a change in public attitudes on killing animals for scientific purposes and non-destructive methods can now be used to obtain much of the data formerly obtained in that way (e.g. capture, drug immobilization, blood sampling and telemetry). Nonetheless, some information and material is still only obtainable from dead animals. Animals should not be killed unless absolutely necessary and humane methods must be used to avoid unnecessary suffering. (Appendix 16.7.)

It is likely that a scientist collecting seal material will in most cases choose a firearm to kill his specimens, though special methods may be required for particular purposes and chemical euthanasia will often be chosen when the animals are already restrained.

Shooting

Rifles

Rifles have the advantage in that they offer a reliable means of dispatching even the largest animals that will be encountered, and they can be used without coming into close contact with the animals. Their disadvantages are that their use may disturb other seals, the bullets may cause damage or contamination to tissues or other specimens that are to be collected, and perhaps most importantly, they present a risk from ricochet or over-carry to other people in the field.

Official regulations generally stipulate the use of a powerful weapon, probably to avoid charges of allowing the use of inadequate weapons that might cause suffering. Norwegian regulations for a weapon to shoot adult seals specify it must have a rifled barrel of at least 5.6 mm (.22 ins) calibre and fire expanding ammunition developing at least 200 kgm energy at 100 m. For the shooting of pup seals the requisite energy is reduced to 100 kgm at 100 m (Øritsland, 1976). Canadian regulations stipulate a rifle firing centre-fire cartridges not made with a metal-cased hard point bullet and with a muzzle velocity of not less than 1800 ft/sec (549 m/sec) and developing a muzzle energy of 1100 ft lbs (152 kgm). Shot guns of not less than 20 gauge firing rifled or 'Poly-Kor' slugs are also permitted (Environment Canada, 1976). Republic of South Africa regulations allow rifles firing soft-nosed or hollow-point bullets of not less than 45 grains (2.9 g) and with a muzzle velocity of not less than 762 m/sec (this corresponds to an energy of 86 kgm); or with a calibre of 5.6 mm (.22 ins) and use 'Hornet' type ammunition. (Hornet ammunition has bullets weighing 2.9 g, a muzzle velocity of 820 m/sec and develops 99.5 kgm) (Republic of South Africa, 1976). British regulations do not set a minimum calibre but demand a minimum muzzle energy of 600 ft lbs (83 kgm) and bullets weighing at least 45 grains (2.9 g) (Conservation of Seals Act, 1970). These limits effectively permit the use of nearly all types of centre-fire cartridge except the .32/20 Winchester.

Much smaller ammunition can be safely used for shooting seals, particularly when the animals can be approached closely, as in the Antarctic. Laws (1953) used a rifle firing .22 ins (5.6 mm) Long Rifle 'High Velocity' ammunition (bullet weight 2.6 g, muzzle velocity 419 m/sec, muzzle energy 20 kgm) to kill all but the very largest elephant seals. Bonner (1968) reduced even these light loads for collecting Antarctic fur seals. For general work weapons firing .222 Remmington type ammunition (bullet weight 3.2 g, muzzle velocity 975 m/sec, muzzle energy 157 kgm) offer a very good compromise. They combine range and flat trajectory with killing power and not too much noise.

Shotguns

Rifles should never be used from small boats or large vessels moving in the open sea to shoot at seals in the water. If collection is *absolutely* necessary for scientific purposes in this situation a smooth-bore shotgun may be needed. A British Special SG cartridge, 12-gauge 2¾ inch contains 16 pellets, each weighing 2.6 g and developing 12.7 kgm at 20 yards (18.3 m). With buck-shot there is much less spread than with figure shot and normally all the pellets will be contained within a 30 in (76.2 cm) circle at 40 yards (36.6 m) (Bonner, 1970). There is thus a good chance at short ranges that several pellets will strike the target simultaneously developing energy of the same order as a .22 Long Rifle 'High Velocity' bullet. The method is unlikely to be effective for adult Antarctic phocids and should not be used.

Pistols

The use of pistols for killing seals, except at pointblank ranges, is not recommended. Captive-bolt pistols, as used in slaughter houses, offer no risk of ricochet or over-carry, but must be used in actual contact with the seal's skull. This will usually only be possible with restrained specimens. Captive-bolt pistols were not found satisfactory for killing grey seal pups (Nature Conservancy, 1963) or harp and hooded seal pups (Rasmussen, 1954). This was attributed to the inability to secure the head of a pup against a rigid surface.

Point of aim

Clearly, the killing power of a bullet depends more on where it strikes the target than the energy it develops, so skill in marksmanship is essential. As quick killing is desirable on practical as well as on ethical grounds (wounded seals may escape into the water) head shots are indicated. The heart shot should never be attempted, as it does not kill a seal instantaneously, an effect related to the seal's physiological adaptation to diving. The point of aim chosen will be dictated by what part of the animal is exposed; when seals are shot on land a shot from the side aimed at a point 2.5 cm behind and at the same level as the opening of the ear will kill the seal instantly (Fig. 7.1) (Bonner, 1970). South African regulations stipulate a point of aim 2.5 cm *above* the ear; the difference

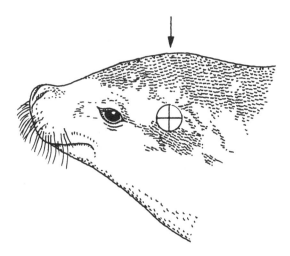

Fig. 7.1 Recommended point of aim for shooting a seal. The arrow indicates the position of the rifle muzzle when shooting at point-blank range.

is negligible. If an animal must be shot from in front, the point of aim should be between and slightly above the level of the eyes.

It is important not to aim too low when shooting at seals in the water. A bullet that strikes the water in front of a seal will lose much of its energy and be deflected upwards and may inflict a non-fatal wound in the muzzle or facial part of the head. Shooting at seals in the sea is made very much easier if the marksman can be stationed a few metres above the water level. It should not be practised in the Antarctic other than in exceptional circumstances.

If neck shots have to be used (as, for example, when perfect skulls are to be collected) it is preferable to shoot as nearly as possible in the sagittal plane (i.e. a vertical plane dividing the body longitudinally). This is not difficult with seals that rear up (fur seals and elephant seals) but much more so with those that lie prone, as it involves standing over the seal. The strong and variable curvature of the spine in the cervical region makes neck shots from the side very difficult. The use of very powerful ammunition (e.g. Norwegian 6.5 mm × 55 sealing ammunition which develops 297 kgm at 100 m) is an advantage here.

Post-shooting procedure

If there is any doubt that the first bullet has not completely smashed the skull and destroyed the brain (and it is not specially desired to save the head from damage), or the spinal cord in the cervical region, another shot should be fired into the head at point-blank range (Fig. 7.1). The muzzle of the weapon should be held about 1 cm from the top of the skull (and not against it, lest the barrel explode) and the shot directed downwards into the base of the skull. The object is to destroy the basal areas of the brain which control the vital respiration centres. When firing in this way, a rifle should be held in the hand in the manner of a pistol, and not held to the shoulder, as the latter makes it difficult to see if the muzzle is in the right place. With a rifle held in a near vertical position the recoil is not great and does not result in loss of aim.

When a seal has been shot it is often advisable to bleed the animal as soon as possible. Either severing the arteries to the fore flipper (in the 'arm-pit'), or stabbing into the heart, dorsally or ventrally, will be effective. In the case of a large elephant seal, a knife with a blade at least 30 cm long will be required for the latter technique.

Chemical euthanasia

This method may be chosen for killing a seal when it is already restrained, when an intact skeleton is required, or where contamination with heavy metals must be avoided. Any anaesthetic drug that results in loss of sensation can be used to render the seal unconscious and it is then killed by severing the major blood vessels and allowing it to bleed. Pentobarbitone sodium is often used in veterinary practice for euthanasia. It is commonly available as a 200 mg/ml solution or a more concentrated solution containing 25.9% of the drug. Dose rates for seals are not accurately known, but 200 mg/ml per kg has been found effective on common seals and young grey seals. There is, of course, no danger of overdose.

Although most drugs can be administered by the intraperitoneal or intrapleural routes, the effect is much speedier when administered intravenously. Superficial veins are not easy to find on seals because of the amount of blubber. However, a suitable vein can often be found in the interdigital web of the hind flipper. Using a long needle a skilled anatomist may be able to inject into the extradural vein within the spinal canal, dorsal to the spinal cord. This is best done about 40–60 cm anterior to the origin of the hind flippers, where the blubber is thin, feeling for the vertebral spines and injecting between a pair. For further details on injection sites, the use of dart-guns, etc., see Chapter 3.

Clubbing

There may be occasions when it is necessary to collect material that has not been contaminated either by metals or chemicals. The use of a heavy club on the head of a young seal of any Antarctic species will produce loss of consciousness which should be followed by complete destruction of the brain by crushing the skull and severing of major blood vessels, as described above. There is no reason to suppose that clubbing is less humane than shooting but the method should not be attempted with large active seals.

References

Bonner, W.N. 1968. *The Fur Seal of South Georgia*. British Antarctic Survey Scientific Reports No. 56.

Bonner, W.N. 1970. *Humane killing of seals*. Seals Research Unit, Natural Environment Research Council. Occas. pubs. No. 1.

Conservation of Seals Act 1970. Chapter 30 (UK Parliamentary Act). HMSO, London.

Environment Canada 1976. Seal protection regulations made under the Fisheries Act, Amendment List, April 5, 1976.

Laws, R.M. 1953. *The Elephant Seal* (Mirounga leonina Linn) *1. Growth and Age*. Falkland Islands Dependencies Survey Scientific Reports No. 8. 62 pp.

Nature Conservancy 1963. *Grey seals and fisheries: report of the consultative committee on Grey seals and fisheries*. HMSO, London.

Øritsland, T. 1976. *Fangst av sel. Orientering for selfangere. Fangstsesongen 1976*. Fiskeridirektoratets Havforskningsinstitutt, Bergen.

Rasmussen, B. 1954. Rapport vedrørende avliving av sel med Cash boltpistol. *Dyrenes ven*, **64** (9-10), 56.

Republic of South Africa 1976. Sea Birds and Seals Protection Act, 1973 (Act 46 of 1973). Sealing regulations, pp. 81-3. Staatskverant (Government Gazette) 22 October 1976. No. 5317.

8

Morphometrics, specimen collection and preservation

W.N. BONNER AND R.M. LAWS

Introduction

In the past much of the information on seal biology was obtained from post-mortems carried out on animals taken commercially, killed for dog food or specifically for the purpose of biological research. However, few species are now commercially exploited – certainly no Antarctic species – and very few seals are taken for dog food; also there has been a change in public attitudes on killing animals for scientific purposes and non-destructive methods can now be used to obtain much of the data formerly obtained in that way (e.g. capture, drug immobilization, blood sampling and telemetry). Nonetheless, some information and material is still only obtainable from post-mortem examination and so standard methods for collection of data, samples and specimens from dead animals are described here. Animals should not be killed unless absolutely necessary.

Historically, many measurements and specimens and samples were taken from all animals collected or captured. However, this is time consuming and often the material is not used. In this chapter we have largely restricted ourselves to the collection of data and specimens from the three areas most relevant to research and management needs in population ecology:

1. fundamental cataloguing of the original specimens, samples and essential measurements;
2. age determination;
3. reproductive status.

The collection of skeletal material is also described here. Studies of food habits are also important; the collection of complete stomachs (and intestines) from killed seals is described in chapter 13. The collection of tissue specimens for organochlorine, heavy metal or electrophoretic studies is covered in chapters 9 and 10. This does not preclude the collection of additional measurements or specimens for more specialized purposes, for example physiology, parasitology or pathology, but specialized text books should be consulted as necessary. The collection of additional specialized reproductive material is described in chapter 12. There is a need for great care in recording, as carelessness may result in a significant reduction, or even the total loss of value, of a data set that may have been gathered only after considerable cost and effort.

Fundamental cataloguing of specimens

The following data should be recorded to catalogue each specimen collected or captured:

> species: Latin name essential, common name as well if the recorder desires.
> sex: This should always be determined by checking the genital openings (Fig. 12.4) and never by accepting apparent sexual differences, such as those of size or colour dimorphism.
> date: Record in the order: day/month/year, e.g. 27 Oct 94; time of day should be recorded for fresh food specimens.
> location: A map location should be given if the specimen is collected on shore, or latitude and longitude in degrees and minutes if collected at sea.
> habitat: Note the general habitat type such as beach, fast ice, pack ice, etc. (For ice type and cover use codes given in chapter 2.)

Additional notes or photographs may be taken of such features as moult, external wounds or scar patterns and external parasites, depending on the interests or time available to the observer. The least known species, for which additional notes would be more valuable, are the leopard and Ross seals.

Basic measurements and weights

The following measurements should be recorded:

Standard length

Measure to the nearest cm the straight-line distance from snout to tip of tail flesh on the unskinned body, belly up, ideally with the head and body in a straight line (see Fig. 8.1). It approximates the length of the axial skeleton. Use a metal or other non-stretch tape. If in doubt check the

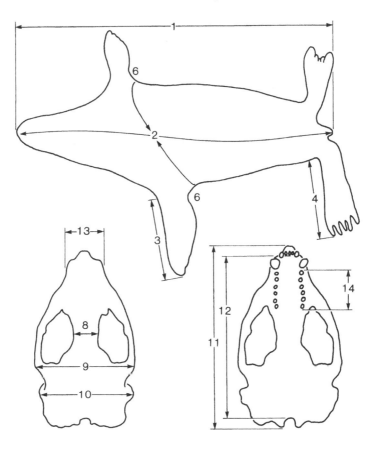

Fig. 8.1 Standard measurements of a seal. 1. Standard length; 2. Curvilinear length; 3. Anterior length of front flipper; 4. Anterior length of hind flipper; 5. Thickness of blubber (not shown); 6. Axillary girth; 7. Weight (not shown); 8. Interorbital width; 9. Zygomatic width; 10. Cranial width; 11. Condylobasal length; 12. Basilar length of Hensel; 13. Rostral width; 14. Length of upper post-canine series. (After Scheffer, 1967.)

accuracy of the tape against another; fabric tapes stretch considerably. In commercial operations, it may be necessary to make allowances for some curvature of the neck after bleeding or if rigor has occurred. The principal use of length measurements is the calculation and comparison of age-specific growth curves within and between populations. In the Antarctic, probably most length measurements collected to date have been taken from seals (either alive or dead) lying on their belly. Measurements taken of seals lying on their backs are probably longer because of less flexion in the spinal column, but to date no research has been done to check this point. Another aspect worth comparing would be to assess the potential variability in length measurements taken from living undrugged seals, immobilized seals and dead seals.

Axillary girth

Measure around the surface of the body underneath the fore flippers to the nearest cm (Fig. 8.1). An easy way to do this without having to roll over heavy seals is to pull a cord underneath the body from the head end, using a sawing motion. When the cord is in place, mark the girth with a metal clip, or simply by holding the spot, pull out the cord and measure it.

Blubber thickness

Measure in mm the combined thickness of the blubber and skin over the sternum on a line between the axillae, to the nearest mm (Fig. 8.1). The principal value of this measurement is as an index of physical condition, which can be compared within or between populations in different states or densities, seasons, and geographic locations (e.g. Stirling, 1971). Blubber thickness can also be measured using a portable ultrasonic probe (see chapter 14).

Total weight

This should be recorded when practical, especially for newborn pups, because their weight can vary significantly in different populations or in the same population in different stages of population growth or intensity of commercial exploitation (Bryden, 1968 a,b). If seals are weighed in pieces, Scheffer (1967) recommends adding a standard correction of 10% to allow for blood loss. A comparison of the weights of immobilized and killed seals of comparable sizes, or preferably the same individuals, would be valuable for refining this correction factor. Weights should be recorded to the greatest accuracy permitted with the equipment available, but it is essential to ensure that the scales, of whatever kind, are calibrated regularly.

In studies of Weddell seals near Antarctic bases, an electronic platform scale, commonly used for domestic livestock, has been used to obtain weights (Braun-Hill, 1987). In this case the platform scale is attached to a sledge and moved from place to place with a tractor or snow-mobile. Weddell seals were guided across the scale by biologists, and weights were recorded. Such a device should be considered where obtaining a large number of weights on live or immobilized seals, on fast ice or land, is an objective of the research.

Total weights can also be estimated quite accurately from measurements of total length and axillary girth (Hofman, 1975) once a sample of weighed individuals is available for calibration of the values. Age-specific growth curves and weights are of particular value in comparing the status of different populations or the same population at different times. The following simple regression equations may be used for the estimation of body weights of adult Antarctic phocid seals from length and girth measurements (derived from expressions in Hofman, 1975):

Species	Equation 1 (lb/ins)	Equation 2 (kg/cm)
Crabeater seal	$W = 11.5 + LG^2/592$	$W = 5.2 + LG^2/21, 834$
Ross seal	$W = 85.8 + LG^2/761$	$W = 38.9 + LG^2/27, 489$
Leopard seal	$W = 169 + LG^2/822$	$W = 76.7 + LG^2/29, 692$
Weddell seal	$W = 346 + LG^2/924$	$W = 156.9 + LG^2/33, 376$

(W = weight; L = standard length; G = axillary girth)

For leopard and Weddell seals the large value of the intercept means that younger animals do not fall on the same line as adults and hence their weights cannot be estimated from these equations.

Historically most measurements of Antarctic seals have been made in pounds and inches. However, metric measurements are recommended for future studies, because most publications now use only metric values and field recording in metric units reduces the chance of errors in converting from English values.

Smirnov (1934) and Scheffer (1967) list additional measurements that other workers have found useful in the USSR and USA respectively.

Collection and preservation of specimens

Specimens for age determination, assessment of reproductive status and determination of food habits are desirable (chapters 11, 12, 13). Additional specimens may be collected depending on the needs or interests of the collector. Parasites should be collected when possible, cataloguing them by their location in the body. Record any notable abnormalities of the organs. However, any additional use that can be made of specimens from killed animals is encouraged.

The best all purpose preservative is 10% buffered formal saline. In the field it is not always possible to prepare this solution and in this event, most specimens can be preserved for macroscopic examination in a solution of one part 40% formaldehyde (formalin) to nine parts of seawater (if fresh water is not available). Alcohol (70%) may also be used in the absence of an alternative; the volume of fixative should preferably be about ten times that of the tissue to be preserved.

It is essential that each specimen be labelled with a permanent plastic, metal or high wet-strength paper label. The length of time that specimens are to be preserved is important in choosing a label because some plastics or metals may dissolve or corrode in certain solutions over time. Only India ink or soft pencil should be used on paper labels; paper labels should be enclosed with the specimen inside the container.

Risk of infection

After using bare hands to autopsy seals it is possible to develop a variety of infections, particularly through minor cuts on the fingers: the worst such condition is seal finger. The main symptom is acute swelling of the finger, redness of the tissue and severe pain. If untreated and left to run its course it can cause permanent disability.

The causative organism has not been identified and apparently it cannot be treated successfully with penicillin. Achromycin, a tetracycline preparation, has been found to be effective and is recommended for the first aid kits of field parties expecting to collect from or autopsy seals (Beck & Smith, 1976). Wearing rubber gloves reduces the risk of coming into contact with infectious organisms, but is inconvenient in the field. Regular washing of the hands with hot water and soap is recommended.

Specimens for age determination

Teeth

Background information and further details are given in chapter 11. Teeth from the lower jaw of Antarctic seals are easier to collect, but upper teeth are of equal value for age determination. The lower jaw can be collected by using a cable-cutter, axe or butcher's bone saw and severing the mandible on each side, half-way between the canine and the articulation with the skull. This ensures the availability of canines for age determination by laminae in the dentine, as well as molar roots for laminae in the cementum.

For elephant seals it is easier, and adequate, simply to remove the canine tooth at the gum line with a hacksaw, but extracted whole teeth are better (chapter 11). For crabeater seals, the best tooth for age determination is the third post-canine tooth. Comparisons of the relative suitability of different teeth have not been made in other Antarctic seals. Incisor and post-canine teeth can be extracted from living immobilized seals by using dental forceps (chapter 11); the technique should first be perfected by practice on dead animals.

Teeth are robust specimens and any method, even dry storage, will suffice to store labelled mandibles, but storage in dry salt, brine, freezing or 10% buffered formalin is preferable. Individually extracted teeth may be stored in glycerol alcohol. The suitability of the method chosen will vary with the methods available, convenience (such as dry salt), and the length of time the specimens are stored before examination. Dry storage, however, should only be used as a temporary measure, because teeth may dry out and crack, often in only a few months. Some treatments may affect the readability of teeth and this should be experimentally determined for each species. Low & Cowan (1963) suggested that boiling may decrease the stainability of some tooth tissues and this needs to be checked.

Fore flipper toe nails

The nail from the third digit on either fore flipper should be pulled if they are in good condition. Otherwise collect the best available. This can be done very easily using dental forceps, grasping the claw as close to the base as possible, and giving a twist as the nail is pulled. Nails should be dried and carefully labelled before storage (Hofman, 1975). Claws from bearded seals have been successfully preserved in brine prior to thin-sectioning but no other methods of preservation have yet been tried for the nails of Antarctic seals. In general, jaws only are required from dead seals, while single teeth or nails would be collected from living seals.

Reproductive material

In general, priority for collection of reproductive material should be given to females because of the importance of evaluating such parameters as pregnancy rates and age of first reproduction.

Ovaries

These should be carefully removed (chapter 12, Fig. 12.7) and preserved in 10% formal saline, preferably buffered. Ideally fluid volume should be about ten times that of the tissues to be preserved. Do not section prior to preserving, because this causes distortion of structures (follicles, corpora albicantia) to be measured later in the laboratory. Specimens can, if necessary, be transferred with some preserving fluid to a plastic bag after being fixed for 48 hours, but preferably a week should be allowed for adequate fixation.

The reproductive tract (chapter 12, Fig. 12.7) should be examined for an embryo or foetus and its presence or absence recorded. If a foetus is present and less than 15 cm in length, it should be collected. If it is greater than 15 cm, a standard straight line measurement should be taken if possible or a crown–rump length taken (chapter 12, Fig. 12.9) if the body cannot be properly straightened. Foetuses should be weighed as accurately as possible, without extraneous material and after severing the umbilical cord where it enters the abdomen. A balance should be used for weighing smaller specimens to the nearest gram when possible.

See chapter 12 for interpretation of reproductive material in the field and recording of data when specimens are not collected. Also a fuller account is given on the collection of material additional to the basic requirements, including preservation for electron microscopy.

Testes

Usually it is not necessary to collect whole testes, but some portions of tissue may be necessary to determine whether or not spermatogenesis is occurring. Epididymal smears can be made by cutting open the epididymis (which is attached to the testis) at the mid-point, removing a drop of fluid, drawing it along a slide as for a blood film, fixing in alcohol, staining with nigrosin and eosin (Campbell, Dott & Glover, 1956) and storing dry (see chapter 12). Sperm can be identified in stained or unstained smears in the field using a compact microscope such as the Macarthur.

Weights of testes are valuable in assessing the development of reproductive maturity. Preferably, both testes should be weighed on a sensitive spring balance or on a platform balance, but it should be noted whether the weight is of one or both organs. Collection of a slice, no more than 1 cm thick, from the centres of the testes and preserved in 10% buffered formal saline is adequate for histology. In this case a piece of epididymis and the prostate should also be preserved.

Other organs

Most other soft organs can be preserved in 10% buffered formal saline or Bouin's solution; tissues for histological examination should be no more than 1 cm thick. Spleen and bone marrow should be preserved in Helly's fluid and parasites (internal and external) should be preserved in 70% alcohol.

Skeletal material

Skull

The highest priority should be given to the skull. Of special value are the skulls of Ross, leopard and fur seals, but most museums will still welcome material from any Antarctic seal. Although skulls seem robust, they should be prepared and handled with care, because minor topographic features may be important in taxonomic studies. Skulls may be preserved whole in the field in salt or brine, or cleaned and dried (see below). To conserve space and reduce the weight, the skull should be defleshed and the eyes, tongue and brain removed.

Skeleton

If time and facilities permit, whole skeletons may be collected. These are of particular value to museums because there are few in existence for any of the Antarctic seals and especially the less accessible species. For palaeontological and evolutionary studies, the limb bones, carpals and tarsals are of particular importance. If time or space is short, note that the ribs are the least useful skeletal component in most studies, followed by the vertebrae (although more useful than ribs). The limbs are bilaterally symmetrical, and collecting only one front limb, with shoulder blade, and one hind limb, with pelvis, is just as useful in study specimens (as opposed to display specimens), as collecting all four limbs. The minimal amount collected should be the skull and two limbs.

The following procedure may be followed for collecting a skeleton. Again, the extent to which this procedure is followed will vary with time and facilities, but full details are given for those wishing to do a complete job:

1. Skin seal, down to base of flippers (do not remove skin from flippers at this stage).
2. Remove blubber, open abdominal and thoracic cavities (avoid cutting the ribs), and remove all organs.
3. Remove eyes and tongue (but be careful not to damage the hyoid bones, so cut trachea well below the glottis), and cut away as much flesh as possible from the head, body and limbs down to the base of the flippers. Care should be taken not to damage cartilaginous parts of the ribs or sternum, or to remove the patella from the rear limbs. Be sure to leave the distal phalanges and cartilaginous extensions with the skeleton and not with the skin! Any sharp knife, with blade 15–20 cm long, is suitable for initial removal of flesh; knives with shorter blades should be used for the final cuts. Although defleshing could proceed until most bones are bare the actual amount of flesh left at this stage is not critical.
4. If required, for ease of storage and transport, the skeleton can be disarticulated at this stage as follows:
   (a) Skull (plus lower jaw and glottis, so as not to damage hyoid) – from the rest of body.
   (b) Rib cage plus attendant part of vertebral column, and sternum.
   (c) Rest of vertebral column.
   (d) Each front limb, including scapula.
   (e) Pelvic girdle (innominate bones).
   (f) Each hind limb.

Skulls and skeletons, sunk in the sea in a wire cage, will often be cleaned by amphipods, depending on the field location. The skeleton, with whatever flesh remains on it, should now be thoroughly wind-dried, unless the skeleton can be cleaned further in a reasonable time (e.g. at base camp). If this is done outside, some cover (such as wire mesh) should be used to discourage scavenging birds. Further processing should be carried out in a laboratory.

Baculum

Traditionally the baculum, or penis bone, has been collected from most mammals, including seals. The baculum no longer has a useful function for determining the age of seals, but its development may indicate the relative degree of sexual maturity. This can be valuable when collecting specimens from seals found dead, which precludes the collection of testis samples or epididymal smears. The baculum can simply be cut out and stored with the surrounding tissue in salt or brine, or boiled to remove the tissue and stored dry.

Acknowledgements

We thank Dr A.J. Nel and Dr C.A. Repenning for their contributions to this chapter.

References

Beck, B. and Smith, T.G. 1976. *Seal Finger – An Unsolved Medical Problem in Canada*. Canadian Fisheries Marine Service, Technical Report 625.

Braun-Hill, S.E. (1987). Reproductive ecology of Weddell seals (*Leptonychotes weddelli*) in McMurdo Sound, Antarctica. Ph.D. thesis, University of Minnesota, Minneapolis, MN. pp. xiv + 106.

Bryden, M.M. 1968a. Control of growth in two populations of elephant seals. *Nature*, **217**, 1106-7.

Bryden, M.M. 1968b. Lactation and suckling in relation to early growth of the southern elephant seal, *Mirounga leonina* (L.). *Aust. J. Zool.,* **16**, 739-48.

Campbell, R.C., Dott, H.M. & Glover, T.D. 1956. Nigrosin-eosin as a stain for differentiating live and dead spermatozoa. *J. Agric. Sci.,* **48**, 1.

Hofman, R.J. 1975. *Distribution Patterns and Population Structure of Antarctic Seals*. Ph.D. Thesis, University of Minnesota, Minneapolis.

Low, W.A. & Cowan, I.McT. 1963. Age determination of deer by annular structure of dental cementum. *J. Wildl. Mgmt*, **27**, 466-71.

Scheffer, V.B. (ed.) 1967. Standard measurements of seals. *J. Mammal.,* **48**, 459-62.

Smirnov, N.A. 1934. [Instructions on biological investigations of seals.] Leningrad, VSES. Arkt. Inst. Biull., 4.

Stirling, I. 1971. Population dynamics of the Weddell seal (*Leptonychotes weddelli*) in McMurdo Sound, Antarctica, 1966-1968. In *Antarctic Pinnipedia*, ed. W.H. Burt, *Antarctic Research Series*, vol. 18, 141-61. American Geophysical Union, Washington, DC.

9

Genetic-based studies for stock separation

P.D. SHAUGHNESSY, R.J. HOFMAN,
T.E. DOWLING AND W.M. BROWN

Introduction

Effective conservation of Antarctic seals requires knowledge of the relative discreteness of populations or stocks[1] in different geographic areas. Studies of Antarctic seals carried out to date are reviewed by Laws (1984), and suggest that the terrestrial and fast ice colonial breeding species – the Antarctic and sub-Antarctic fur seals, the southern elephant seal and the Weddell seal – may be composed of several or many relatively small, more or less discrete populations. The more solitary, pelagic species – the crabeater, leopard and Ross seals – may be composed of only one or at most a few relatively large, panmictic populations. Available data are insufficient to confirm these suggestions or to determine the number, sizes and geographic ranges of populations comprising various species.

Possible means for discrimination between populations

Determination of the degree of isolation between seals from different geographic areas can be approached directly, by mark–resighting and/or radio tagging and tracking to ascertain home ranges, dispersal distances and migration patterns, and indirectly by assessment of possible environmental-related and genetic-based variables. Direct methods for determining population discreteness are of limited use in the Antarctic because the area is remote, there is little air or ship traffic, and the seals are not being commercially exploited. Mark–resighting programmes, for example, are of little use, except in the vicinity of coastal scientific stations, because the lack of ship and air traffic makes tag resighting difficult. Similarly, while radio tagging and tracking have been and can be

---

[1] In this paper, the terms 'stock' and 'population' are used interchangeably.

used to study local movements and activity patterns, they cannot be used effectively to assess at-sea movements and movements in the vast pack ice ecosystem unless, and until, satellite-linked tracking capability is developed (see chapter 5).

Variables that may reflect environmental variability and indicate possible geographic isolation of populations include the types, quantities and ratios of environmental contaminants absorbed and stored in blubber and other tissues, the structure and usage of vocalizations and various behaviours, and the mean dates of pupping, mating, implantation and moulting. Some of these variables may also have genetic components. As examples, behaviour patterns and timing of pupping, mating, implantation and moulting no doubt are genetically based as well as being potentially affected by environmental variability. Other potentially useful genetic-based variables include skull-characteristics, body size and morphology, colour and marking patterns, chromosome number and morphology, protein structure, and both RNA and DNA sequences.

Techniques for marking and tracking seals are described in chapters 4 and 5. Techniques for assessing and comparing contaminant levels are described in chapter 10. Techniques for assessing and characterizing vocalizations and behaviour patterns are described in chapter 6. Basic techniques for electrophoretic analysis of proteins, assessment of chromosome number and morphology, and assessment of both nuclear and mitochondrial DNA are described below. These techniques have recently been discussed thoroughly for cetaceans (Hoelzel, 1991). Sherwin (1991) has reviewed the collecting of mammalian tissue and data for genetic studies.

Electrophoretic analysis of proteins

Protein structure is genetically determined and may be highly variable. Variation can be detected by a number of means including amino acid sequencing, immuno-chemical comparison and electrophoresis. Electrophoresis, a procedure for separating proteins or polypeptides with different net electrical charges and thus different migration rates in electric fields, is the simplest, most versatile, and least expensive, of the techniques for detecting variants.

Extracts for electrophoretic analysis can be prepared from various tissues such as blood, liver, muscle. They are applied to starch or polyacrylamide gels, an electric current is run through the gel under standard conditions (pH, buffer, time, temperature, etc.), and the gel is then treated with histochemical stains to mark and locate the protein(s) being

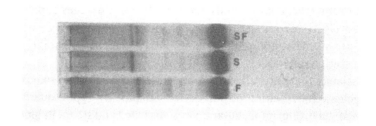

Fig. 9.1 Electrophoretic patterns of serum transferrins in crabeater seals. S is the slow migrating homozygote, F is the fast migrating homozygote, SF is the heterozygote. (From Hofman, 1975.)

evaluated. Differences in electrophoretic mobility and the presence of protein variants are indicated by different band positions for homozygotes and multiple bands for heterozygotes.

The stained gels, as illustrated in Fig. 9.1, are read and scored for homozygotes and heterozygotes of each allele from each individual being examined. The resulting data indicate gene frequencies at the loci being evaluated. Seals from different geographic areas are considered to be from discrete populations if samples from the two areas are composed entirely of protein variants having different electrophoretic mobilities, or if allelic frequencies of one or more polymorphic loci are significantly different. Lack of different protein variants and/or differences in allelic frequencies may not, however, be evidence that seals from the different geographic areas are part of the same population. Much of the material covered in this section is treated in detail by Richardson, Baverstock & Adams (1986).

Sample collection and storage

As noted earlier, electrophoretic analysis can be done with extracts from a variety of tissues including liver, skeletal muscle, whole blood, serum and red blood cells. Tissue samples should be collected from live animals, using standard biopsy techniques, or immediately after death of animals taken for other purposes. Whole blood, in acid-citrate-dextrose solution, should be kept cool but not frozen prior to analysis. Other tissues, including serum and red blood cells, should be frozen as soon as possible after collection and maintained at −20°C or lower until the samples are prepared for analysis. Liver and muscle samples (5–10 g from dead animals or 50–100 mg by biopsy) should be wrapped in aluminium foil or placed in air tight containers prior to freezing. All samples should be accurately labelled.

Blood samples can be collected from freshly killed seals by cardiac puncture. Blood samples can be collected from the pelvic flipper or the intravertebral extradural vein of live, phocid seals. In the former case, described by Geraci & Sweeney (1978), blood is extracted from the main vessel or the rete surrounding it by inserting a 1.5 inch (3.8 cm) 18 gauge needle into the ventral side of the flipper, at the proximal end of the inter-digital webbing, near the second or fourth digits. The latter technique is described by Hubbard (1968), illustrated by Ridgway (1972, Figure 10–43), and applied to Antarctic phocids by Cline, Siniff & Erickson (1969). Otariid seals do not have the large intravertebral extradural vein found in phocid seals and thus blood cannot be collected from the Antarctic or sub-Antarctic fur seals by lumbar puncture. Among otariids, blood has been collected by several techniques: from the anterior vena cavae by inserting a needle between the first and second ribs as near the mid-line as possible (Hubbard, 1968); from the hind flipper of northern fur seals and California sea lions (Chuba *et al.*, 1970); by jugular and carotid puncture of California sea lions (Palumbol *et al.*, 1971); from the caudal gluteal vein (Geraci & Sweeney, 1978); and by cardiac puncture of southern fur seals (Shaughnessy, 1970). The last technique is simplest, especially for pups, but is not without risk. Hubbard (1968) warns of the possibility of piercing one of the numerous large pericardial veins which may cause leaking into the pericardial sack resulting in death. 'The preferred location is the *caudal* gluteal vein, with the needle inserted just lateral to the vertebral column, about one-third the distance from the *palpable femoral trochanter* to *the base of* the tail'.

A single 15–20 ml blood sample is generally adequate. Larger samples can be taken with no ill effects and, as a general rule, one or more duplicate samples should be taken in case the initial sample is damaged or destroyed. In the field, samples should be kept cool but not frozen prior to processing. If freezing is a problem, the samples should be kept in an inner pocket close to the body, placed in a storage box heated with a hand-warmer, or placed in a vacuum flask.

Whole blood is usually separated into serum and red cells, or into plasma and red cells, and then stored at −20 °C or colder until analysis. In the field it should be stored in liquid nitrogen and in the laboratory at −70 °C, if possible. Plasma is simpler to prepare than serum, but has the disadvantage that fibrin clots may form. Samples should be kept frozen and not subjected to repeated thawing and freezing.

Preparation of serum

Whole blood (15–20 ml) is allowed to clot in the collecting bottle (e.g. a 1 oz McCartney bottle or 10 ml vacuum tube). Clot retraction can be enhanced by sharply tapping the bottom of the bottle on the open palm to free the clot from the walls of the bottle and then laying the bottle on its side. After a few hours, serum can be drawn off with a pasteur pipette or, preferably, decanted along with a few red cells into a centrifuge tube and then spun in a centrifuge at about 200 rpm for 10 min to remove any debris. The serum is then transferred with a pasteur pipette to a small (about 10 ml) vial for storage at −20 °C or colder. The serum will expand when frozen and the vials should not be filled to much more than half their volume to ensure that they do not split when frozen. Each vial should be labelled or marked to indicate the date, location, sample number, and the species, sex and relative age of the seal from which the sample was taken. Marking or labelling should be done in a way such that the labels cannot be separated from the samples and the writing cannot be blurred or obliterated by contact with blood, ice, water or other solutions. A master list of specimen numbers and related information should be maintained.

Preparation of plasma

If plasma is required, an anticoagulant such as heparin or EDTA must be added to the syringe or collection bottle. Cline *et al.* (1969) used sodium heparin in the ratio of 1000 units of heparin per 100 ml of blood. The container must be agitated gently during or immediately following collection to ensure thorough mixing of the blood and the anticoagulant. Samples should be kept cool, but not frozen, and centrifuged within four hours after collection to separate cellular components from the plasma. If the blood cannot be centrifuged, most of the cellular components will settle out if the samples are allowed to stand quietly for several hours. Separation by centrifuging is preferable because the activity of many enzymes decreases when they remain at room temperature.

As with serum, plasma is transferred with a pasteur pipette to a small vial for storage at −20 °C or colder. Storage vials should not be filled to much more than half of their volume and should be clearly labelled.

Preparation of red blood cells

If an anticoagulant was added to the blood, red cells and a buffy coat of white cells remain in the centrifuge tube after the plasma has been drawn off. Red cells should be washed two or three times with physiological saline solution (0.9 g sodium chloride per 100 ml of distilled water) by gently resuspending the cells in 10 - 15 ml of saline and then centrifuging and removing the supernatant. After being washed two or three times, the red cells are transferred to a small vial with a pasteur pipette and stored at −20°C.

Analysis

Electrophoretic analysis of samples should be done as soon as possible after collection. The objective is to identify as many polymorphic genetic loci as possible with gene frequencies of .05 or greater. Thus, as wide a spectrum of genetic markers as possible should be examined. Possibilities include but are not limited to:

adenosine deaminase	glyceraldehyde phosphate dehydrogenase
adenosine kinase	glycerophosphate dehydrogenase
alcohol dehydrogenase	hexokinase
alkaline phosphatase	isocitrate dehydrogenase
acid phosphatase	lactate dehydrogenase
catalase	malate dehydrogenase
creatine kinase	mannose isomerase
esterase-aryl	nucleoside phosphorylase
esterase-choline	peptidase
fumarase	phosphogluconate dehydrogenase
glutamate dehydrogenase	phosphoglucomutase
glukinase	superoxide dismutase
glucosephosphate isomerase	sorbital dehydrogenase
glutamate oxalacetate transaminase	transferrin
glucose-6-phosphate dehydrogenase	xanthine dehydrogenase

Techniques and equipment for electrophoresis are variable and are reviewed by Richardson *et al.* (1986). General background to the starch gel technique is provided by Brewer (1970), Shaw & Prasad (1970) and Smith (1976); the acrylamide gel technique is described by Tombs (1968). Reviews of the many buffer solutions used in the gels and electrode compartments for the electrophoresis of various proteins, along with the solutions used to stain them, are provided by Selander *et al.* (1971), Siciliano & Shaw (1976) and Harris & Hopkinson (1976).

Electrophoretic studies involving pinnipeds are described in Naevdal (1969), Shaughnessy (1969, 1970, 1974a,b), McDermid, Anathakrishnan & Agar (1972), Seal *et al.* (1971a,b) Bonnell & Selander (1974), Hofman (1975), McDermid & Bonner (1975), Lidicker, Sage & Calkins (1981), Simonsen *et al.* (1982a,b), Vergani (1985), Vergani, Spairani & Aguirre (1986), Testa (1986) and Gales, Adams & Burton (1989). Few of these studies address the matter of stock separation; a good example is provided by a study of the southern elephant seal (Gales *et al.*, 1989).

Sample sizes and selection of animals to be sampled

Before beginning field studies, consideration should be given to the age classes and number of animals to be sampled. To minimize possible sampling errors, sampling should ideally be done during the breeding season and only animals representative of the breeding population should be sampled – that is, breeding adults and pups.

Pups are good subjects for population identification studies because they are representative of the breeding population and easy to handle without killing or using immobilizing drugs. However, caution must be exercised when using pups because some proteins undergo ontogenetic changes which can be misinterpreted as polymorphism. This could be overcome by examining each protein in animals representative of each age class.

The most useful markers for population identification by electrophoresis are those for which the pattern of inheritance has been determined, for then it is known that observed variation is due to genetic, rather than ontogenetic or environmental variation. Mother-pup pairs generally can be identified prior to weaning and, when possible, collection of samples from such pairs should be included in the collection programme. If adults are being collected outside the breeding season, mother–foetus pairs should be included in the collecting programme to provide the basis for determining the pattern of inheritance of the proteins examined.

The numbers of samples required to discriminate between isolated breeding populations is dependent upon several factors. When dealing with paired comparisons from two different geographic areas, and involving single polymorphic loci, the sample size required to detect differences decreases as (1) the frequency of the most common allele approaches

Table 9.1. *Number of individuals that would have to be sampled from each location to permit differentiation of two populations (at the 0.05 confidence level) with a gene frequency difference of the magnitude (ΔP) for the average frequencies ($\bar{P}$) listed at the power level $1 - \beta = 0.5$ and with the non-centrality parameter $\lambda = 3.84$ (from Sharp, 1981)*

	$\bar{P}$				
ΔP	.95	.90	.80	.70	.55
.05	146	276	492	645	760
.10	50	69	123	162	190
.20	50	25	31	40	48
.50	50	25	13	9	6

unity and (2) the difference in allele frequency between the two sets of samples increases.

When the number of samples from both geographic areas is equal, the number (N_i) required to detect differences at the 0.05 level of significance can be determined from Table 9.1. Because of the manner in which the values in Table 9.1 are calculated, there is a lower limit, in some instances, to the size of the series required. These values are in italics. Examination of the Table indicates that, when the allelic frequencies differ by 0.1 or more and lambda is set equal to 3.84, a sample size of 200 individuals or less from each location should be sufficient to substantiate any differences.

Since substantial time and effort is required to collect samples, pilot studies should be carried out, when feasible, before initiating full scale field studies to identify potentially useful polymorphisms, estimate the relative frequencies of alleles in animals from the geographic areas in question, and determine the number of animals that should be sampled to determine whether observed differences are significant.

Data analysis

When few loci have been examined or the samples being compared are polymorphic for a small number of proteins, observed differences in allele frequencies can be evaluated one locus at a time using chi-square tests for homogeneity. When multiple loci have been examined and the samples being compared are polymorphic at a number of loci, the 'relatedness' of the sampled populations should be evaluated using one of the indices of genetic distance or genetic similarity such as those of Nei (1972) or Rogers

(1972). We recommend the latter method because it is simpler to understand and easier to apply; other studies which use this method can be used for comparison.

Using the method of Rogers (1972), the genetic distance (D) between two populations X and Y is:

$$D = \frac{1}{L} \sum_i \left[\frac{1}{2} \sum_j (P_{ijx} - P_{ijy})^2 \right]^{0.5}$$

where L is the number of gene loci controlling the proteins examined, and P_{ijx} and P_{ijy} refer to frequencies of the jth allele at the ith locus in the two populations. D will range from 0 when the two populations have the same alleles in identical frequencies to 1 when they have no alleles in common.

Alternatively, the genetic similarity of the two populations is simply $1 - D$. The estimate of genetic similarity can then be compared with values in the literature for conspecific populations, semispecies, sibling species and non-sibling species (e.g. see Selander & Johnson, 1973).

Karyotype analysis

The karyotypes of most pinniped species, including the Antarctic species, have been determined (Corfman & Richart, 1964; Hungerford & Snyder, 1964; Fay, Rausch & Feltz, 1967; Arnason, 1970, 1974; Seal *et al.* 1971b; Anbinder, 1971; Hofman, 1975). The karyotypes are very similar, both within and between species and species groups.

All otariids studied to date have 2n = 36 and nearly identical chromosome morphology. In the Phocidae, two chromosome numbers have been found, 2n = 34 and 2n = 32. Arnason (1970, 1972, 1974) demonstrated that the 32 chromosome phocid karyotype probably evolved from the 34 chromosome karyotype through a fusion of two chromosome pairs.

Only one study has been done to determine whether karyotype variability occurs and can be used to identify possible independent populations of Antarctic seals (Hofman, 1975). The results indicate that both the chromosome number and autosomal morphology are virtually identical in crabeater, leopard, Weddell and Ross seals. The only apparent differences are in the Y chromosomes. The study did not include assessment of G, C or other bands (see below).

Banding studies might identify polymorphisms which would be useful for determining the degree of isolation between seals from different geographic areas.

Materials and methods

Mitotic cells for chromosome analysis can be obtained by culturing samples of blood, skin, muscle and other tissues. Blood samples can be drawn from the extradural vein of immobilized seals using the procedures described in the previous section. Cultures can be initiated with 5-10 drops of whole blood in approximately 7 ml of culture medium, or by allowing the red blood cells to settle out of the blood sample and using approximately 1 ml of the buffy coat and plasma containing white blood cells to innoculate the cultures.

Tissues can be stored in culture medium for a week or more at 4 °C before beginning incubation. As a general rule, however, cultures should be initiated as soon as possible after sample collection. Occasionally, cultures can be successfully initiated from tissues collected from animals that have been dead for some time.

Cultures can be grown in a variety of media (e.g. Basal Medium for Suspension Cultures, TC 199 medium, McCoy's 5A medium). Antibiotics (e.g. 100 IU of penicillin and 0.2 mg of streptomycin per ml of medium) and, if necessary, fungicides are added to the medium to prevent the growth of bacteria and fungi. A mitotic stimulator, such as Bacto Phyto-hemaglutinin-P is added to the media to stimulate cell growth. Cultures are grown in sterile, 10-30 ml glass or plastic culture tubes or flasks. They are incubated at body temperature (37 °C) for 48 h or more and can be subcultured at approximately 72 h intervals by harvesting and innoculating cells into fresh culture medium.

Approximately two to four hours prior to harvesting cultures for analysis, a mitotic inhibitor such as colchicine or colcimid (1 ml per 10 ml of culture medium) is added to inhibit spindle formation and accumulate dividing cells during the last hours of incubation. Cells are harvested by transferring the cultures to centrifuge tubes and centrifuging for 5 min at 1000 rpm. The supernatant is discarded and the cell button is resuspended in a hypotonic solution (e.g. 0.75 M KCl) for 10 min at room temperature to swell the cells. The cells are again concentrated by centrifugation, the hypotonic solution is discarded, and the cells are fixed in several changes of methanol-acetic acid (3:1).

Slide preparation and staining

Slides can be prepared by dropping cell suspensions on clean slides and (1) allowing them to air dry, (2) passing them through a flame to expedite drying or (3) placing a cover slip on the cell suspension, using the thumbs to 'squash' the cells under the cover slip, and freezing the slides with freon, carbon dioxide or dry ice, so that the cover slips can be 'popped off' with a scalpel or razor blade.

Slides can be stained with Giemsa, carbol fuchsin, orcein or other standard stains to permit examination. Using the Giemsa stain (4 ml Giemsa, 4 ml pH 6.5 phosphate buffer and 92 ml distilled water), slides are immersed for 3–8 min in the stain and then rinsed in two changes of 95% ethanol and cleared in zylol. Cover slips are then mounted over the stained area with mounting media.

Since the early 1970s several techniques have been developed to permit more detailed analysis of chromosome structure. These include:

> Q-banding (staining with fluorescent derivatives of quinacrine and subsequent examination under UV light) (Caspersson, Zech & Johansson, 1970);
>
> R-banding (staining after heat denaturation) (Dutrillaux & Lejeune, 1971);
>
> G-banding (staining after enzyme or ionic treatment) (Seabright, 1971; Wang & Federov, 1972); and
>
> C-banding (differential staining of heterochromatin) (Pardue & Gall, 1971; Sumner, 1972).

G-banding with trypsin-Giemsa treatment is one of the simplest and most useful techniques. It involves treatment of air or flame dried slides in a trypsin solution for 10 s to 2 min (time is dependent upon the age of the slide), two rinses in a buffer solution, and staining in a buffered Giemsa solution for 2–5 min.

The buffer solution consists of:

NaCl	16.0 g
KCl	0.4 g
Na_2HPO_4	2.3 g
KH_2PO_4	0.4 g
H_2O	2.0 l

This solution can be stored at 4 °C.

The trypsin solution is made by adding 2 g of trypsin to 1 litre of the buffer solution. It can be subdivided into 50 ml units and frozen for storage.

The buffered Giemsa solution is:

Giemsa	1.5 ml
methanol	1.5 ml
citric acid, 0.1 M	2.0 ml
Na_2HPO_4, 0.2 M	4.0 ml
H_2O	50.0 ml

Photomicroscopy and analysis

Chromosome measurements and karyotype analysis are generally done from photomicrographs. These are obtained by searching prepared slides to locate metaphase spreads in which the chromosomes are well separated, flat and stained as desired, and then photographing these spreads under the oil immersion lens ($\times$ 100) of a high quality light microscope. The film is developed using standard techniques and prints, at enlargements of $\times$ 3-7, are made on high contrast copy film.

Idiograms are constructed by cutting the chromosomes from the photographs and arranging them according to relative size and centromere position (see the example in Fig. 9.2). A minimum of 3 to 10 idiograms should be prepared for each animal sampled.

Idiograms are compared to detect any observable differences in chromosome number, morphology or banding patterns. Individual chromosomes and chromosome arms can be measured and compared statistically. Measurements can be made with vernier calipers or a two-pointed compass by placing one of the caliper or compass points in the middle of a centromere and the other at the end of the chromosome arms. If the arms are bent, measurements are made serially by measuring straight line segments in two or more steps.

The absolute size of a given chromosome varies due to differential chromosome condensation as metaphase proceeds. Therefore, measurements must be normalized before chromosomes or chromosome arms can be compared meaningfully. Measurements can be normalized by dividing each independent measurement by the total length of the haploid autosomal complement plus the length of the X chromosome (i.e. $\Sigma AA/2 + X$).

Normalizing measurements will not avoid measurement errors or errors that may be caused by the inability always to identify homologous chromosomes with certainty. Also, the rate of individual chromosome condensation is not uniform throughout metaphase so that even normalized measurements may be quite variable. Therefore, large sample sizes may be needed for meaningful statistical analyses.

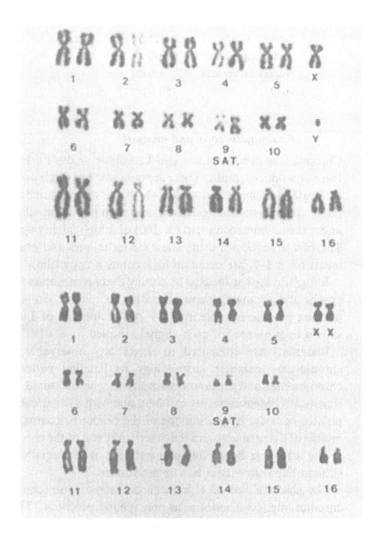

Fig. 9.2 Idiograms of a male and female Weddell seal. (From Hofman, 1975.)

Analysis of mitochondrial DNA

To date, most DNA studies of stock discreteness have used restriction endonucleases to characterize mitochondrial DNA (mtDNA) sequences. DNA variation may also be characterized by direct sequencing; however, because of the amount of time and expense required to directly sequence

DNA, it is less cost effective than restriction endonuclease analysis for the large numbers of animals typically surveyed in population level studies (Hillis & Moritz, 1990b). Technological developments (i.e polymerase chain reaction (PCR), automated sequencing) will soon make direct sequences more readily attainable. In this chapter, we will focus on restriction endonuclease analysis of mtDNA. For further information on sequencing or other analysis of other target sequences (i.e. single copy genes, rRNA, etc.), the reader is referred to Hillis & Moritz (1990a) and references therein.

Mitochondrial DNA (mtDNA) can be treated with restriction endonucleases – enzymes which recognize and cleave DNA at specific base sequences. The resulting fragments can be separated according to size by electrophoresis, providing characteristic DNA fragment patterns. These patterns may be used as genetic markers for the identification of specific individuals and for the estimation of genetic relatedness of individuals, populations or other goups. Many previous studies utilizing mtDNA have demonstrated its utility as a genetic marker at several levels, ranging from individuals to populations and species, including identification of maternal taxa involved in reciprocal and non-reciprocal hybridization events (reviewed by Avise & Lansman, 1983; Wilson *et al.*, 1985; Avise *et al.*, 1987; Moritz, Dowling & Brown, 1987; Dowling, Moritz & Palmer, 1990). The features which make this molecule particularly useful are its small size (reviewed by Brown, 1983, 1985), its relative ease of isolation, its rapid rate of evolution (Brown, George & Wilson, 1979; Vawter & Brown, 1986) and its maternal mode of inheritance (reviewed by Avise & Lansman, 1983).

Because of its relatively rapid rate of evolution in mammals (Brown *et al.*, 1979, 1982) polymorphisms are more likely to be detectable in mtDNA than in proteins or karyotypes. Therefore, smaller sample sizes are needed to differentiate isolated populations using mtDNA analysis. In addition, the strictly maternal inherited pattern of mtDNA makes it a more sensitive indicator of population structure (DeSalle *et al.*, 1987; Crease, Lynch & Spitze, 1990). (See also chapter 12.)

Materials and methods

Highly purified mtDNA can be prepared from either solid tissues (i.e heart, liver, gonads, kidney or muscle; a minimum of approximately 100 mg wet weight of fresh tissue is required) or from platelet and buffy coat fractions of large amounts (< 200 ml) of whole blood. When insufficient amounts of solid tissue or blood are available, total cellular DNA can be prepared from small amounts of frozen tissue or from small (< 20 ml) blood samples, and the mtDNA component detected by transfer hybridization and autoradiography (Dowling *et al.*, 1990), using mtDNA-specific probes.

Fresh tissue is preferred (but not essential) for mtDNA isolation as freezing of tissue results in greatly reduced yields. Fresh or frozen tissue works equally well for total DNA preparation. When fresh tissue cannot be processed immediately, freezing and storage at −20 °C minimizes this reduction. The tissues must be maintained in a continually frozen state and not subjected to cycles of freezing and thawing. For long-term storage (> 2 months) tissues should be kept at −80 °C or lower.

Preparation of purified mtDNA is performed as described in Dowling *et al.* (1990). The method consists of an initial step, in which an enriched mitochondrial fraction is prepared from a tissue homogenate by differential centrifugation, and a final step, in which the mtDNA is released from the mitochondria by detergent lysis and then separated from any remaining nuclear DNA by density equilibrium ultracentrifugation ('banding') in CsCl-intercalating dye gradients. Techniques which omit the CsCl-dye banding should not be used because even small amounts of contaminating nuclear DNA can result in serious artifacts with some types of analysis.

Preparation of total cellular DNA can be performed by a variety of methods (reviewed in Dowling *et al.*, 1990; Hillis *et al.*, 1990). For most tissues, total DNA can be isolated from homogenized tissues by detergent lysis of membranes, followed by a series of extractions with phenol and chloroform to remove protein contamination. When protein contamination is considerable (as in mammalian blood), it is best to purify DNA by density equilibrium centrifugation of the lysate.

Characterization of mtDNA is accomplished by digestion with restriction endonucleases which typically recognize and cleave at sequences of four to six DNA base pairs (bp) (Kessler, Neumaier & Wolf, 1985). Enzymes cleaving at six bp sites usually provide adequate resolving power for interspecific comparisons, because interspecific mtDNA sequence divergences are typically large. Enzymes cleaving at four bp sites are useful for comparisons between more closely related taxa, because four bp sites generally occur much more frequently than six bp sites and thus provide more extensive sampling of the mtDNA sequence, and hence finer resolution of differences between mtDNAs.

DNA fragments produced by restriction endonuclease digestion are separated according to size by electrophoresis in either agarose (0.7 to

1.5%) or polyacrylamide (3.5–6.0%) gels (discussed in Brown 1980, 1984). Several methods are available for the detection of mtDNA restriction fragments in electrophoretic gels or on membranes after their transfer from the gels. When large amounts ($\geq 10 \mu$g) of purified mtDNA are available, fragments may be detected directly by staining the DNA in the gel with an intercalating dye such as ethidium bromide (see Maniatis, Fritsch & Sambrook, 1982, p. 161 for details). When large amounts of purified mtDNA are not available (which is the more usual situation), restriction fragments are detected indirectly by attaching a radioactive group (usually ^{32}P) to the DNA and detecting the position of the radioactive DNA fragments in the electrophoretic gel autoradiographically (several methods for labelling DNA are given in Maniatis *et al.*, 1982; also see Brown, 1980). Equally sensitive non-radioactive methods are being developed (reviewed by Gardner, 1983).

When only total DNA preparations are available, transfer hybridization methods can be used (Southern, 1975; also see Dowling *et al.*, 1990). The alkaline buffer protocol described by Reed & Mann (1985) is used to transfer DNA from agarose gels to nylon membranes. Mitochondrial DNA sequences are detected by autoradiography after incubating the membranes under DNA hybridizing conditions with a radioactive, mtDNA-specific probe, made by nick-translation (see Rigby *et al.*, 1977) or by random priming (Feinberg & Vogelstein, 1983).

Data analysis

Analysis of mtDNA restriction enzyme data typically involves calculation of estimates of sequence relatedness or divergence. When enzymes that cleave at four bp sites are used, the amount of sequence divergence between mtDNA samples can be estimated by comparing fragment lengths using the formulae derived by Upholt (1977). When employing enzymes that cleave at six bp sites, it is preferable to map the cleavage sites by double digestion (Dowling *et al.*, 1990) and compare the resulting maps, because of the increased accuracy of divergence estimates obtained from cleavage site rather than from fragment length comparisons (Nei & Tajima, 1983). The cleavage map comparisons can be analyzed by treating individual restriction sites as characters and finding the most parsimonious among all possible branching orders (Swofford, 1985; Felsenstein, 1989; Swofford & Olsen, 1990). The accuracy of specific topologies can be tested using the modified signed-rank test (Templeton, 1983) or bootstrapping (Felsenstein, 1985). Alternatively, restriction fragment or site comparisons can be converted to estimates of distance by any of several methods (i.e., Nei & Li, 1979; Nei & Tajima, 1983; Nei, Stephens & Saitou, 1985). The distance estimates obtained can be clustered into phenograms using any of several clustering algorithms (i.e. Fitch & Margoliash, 1967; Farris, 1972, Sneath & Sokal, 1973) to depict the relationships among the taxa studied. Levels of variation (termed nucleon or haplotype diversity) within populations can be calculated as described in Nei & Tajima (1981). Estimates of among population variation can be quantified using parameters comparable to Fst (Takahata & Palumbi, 1985; Weir, 1990) and statistically evaluated for the existence of significant subdivision (DeSalle *et al.*, 1987; Roff & Bentzen, 1989). Because mtDNA is inherited as a single unit, restriction site variants are analogous to multiple allelic states of a single locus and should be treated accordingly.

References

Anbinder, E.M. 1971. Chromosomal sets of the Greenland, Caspian and Baikal seals and some problems of evolution of true seals (Phocidae). *Citologija* **13**, 341–7.

Arnason, U. 1970. The karyotype of the grey seal (*Halichoerus grypus*). *Hereditas* **62**, 237–42.

Arnason, U. 1972. The role of chromosomal rearrangement in mammalian speciation with special reference to Cetacea and Pinnipedia. *Hereditas* **70**, 113–18.

Arnason, U. 1974. Comparative chromosome studies in Pinnipedia. *Hereditas* **76**, 179–226.

Avise, J.C., Arnold, J., Ball, R.M., Bermingham, E., Lamb, T., Neigel, J.E., Reeb, C.A. & Saunders, N.C. 1987. Intraspecific phylogeography: The mitochondrial DNA bridge between population genetics and systematics. *Ann. Rev. Ecol. Syst.* **18**, 489–522.

Avise, J.C. & Lansman, R.A. 1983. Polymorphism of mitochondrial DNA in populations of higher animals. In *Evolution of Genes and Proteins*, ed. M. Nei & R.K. Koehn, pp. 165–90. Sinauer Associates, Sunderland, Massachusetts.

Bonnell, M.L. & Selander, R.K. 1974. Elephant seals: genetic variation and near extinction. *Science*, **184**, 908–9.

Brewer, G.J. 1970. *An Introduction to Isozyme Techniques*. Academic Press, New York.

Brown, W.M. 1980. Polymorphism in mitochondrial DNA of humans as revealed by restriction endonuclease analysis. *Proc. Nat. Acad. Sci. U.S.A.* **77**, 3605–9.

Brown, W.M. 1983. Evolution of mitochondrial DNA. In *Evolution of Genes and Proteins*, ed. M. Nei & R.K. Koehn, pp. 62–8. Sinauer Associates, Sunderland, Massachusetts.

Brown, W.M. 1984. The use of restriction endonuclease cleavage sites in mitochondrial DNA as genetic markers in vertebrate cell culture applications. In *Uses and Standardization of Vertebrate Cells in Culture*, ed. R.E. Stephenson, pp. 196-203. American Tissue Culture Association, Symposium Vol. 5.

Brown, W.M. 1985. The mitochondrial genome of animals. In *Molecular Evolutionary Genetics*, ed. R.J. MacIntyre, pp. 95-130. Plenum Publishing Co., New York.

Brown, W.M., George, M., Jr., & Wilson, A.C. 1979. Rapid evolution of animal mitochondrial DNA. *Proc. Nat. Acad. Sci. U.S.A.* **76**, 1967-71.

Brown, W.M., Prager, E.M., Wang, A. & Wilson, A.C. 1982. Mitochondrial DNA sequences of primates: tempo and mode of evolution. *J. Mol. Evol.* **18**, 225-39.

Caspersson, T., Zech, L. & Johansson, C. 1970. Differential binding of alkylating fluorochromes in human chromosomes. *Expt. Cell Res.* **60**, 315-19.

Chuba, J.V., Kuhns, W.J., Nigrelli R.F., Morris R.A. & Friese, U.E. 1970. B-like blood factor of fur seals and sea lions. *Haematologia* **4**, 85-96.

Cline, D.R., Siniff D.B. & Erickson A.W., 1969. Immobilizing and collecting blood from Antarctic seals. *J. Wildl. Mgmt.* **33**, 138-44.

Corfman, P.A. & Richart, R.M. 1964. Chromosomes of the ring seal. *Nature* **204**, 502-3.

Crease, T.J., Lynch, M. & Spitze, K. 1990. Hierarchical analysis of population genetic variation in mitochondrial and nuclear genes of *Daphnia pulex*. *Mol. Biol. Evol.* **7**, 444-58.

DeSalle, R., Templeton, A., Mori, I., Pletscher, S. & Johnston, J.S. 1987. Temporal and spatial heterogeneity of mtDNA polymorphisms in natural populations of *Drosophila mercatorum*. *Genetics* **116**, 215-23.

Dowling, T.E., Moritz, C. & Palmer, J.D. 1990. Nucleic acids II. Restriction site analysis. In *Molecular Systematics*, ed. D.M. Hillis & C. Moritz, pp. 250-317. Sinauer Associates, Sunderland, Massachusetts.

Dutrillaux B. & Lejeune, J. 1971. Sur une nouvelle technique d'analyse du caryotype humain. *C.R. Acad. Sci.* **272**, 2638-40.

Farris, J.S. 1972. Estimating phylogenetic trees from distance matrices. *Am. Nat.* **106**, 645-68.

Fay, F.H., Rausch V.R. & Feltz, E.T. 1967. Cytogenetic comparison of some pinnipeds (Mammalia: Eutheria). *Can. J. Zool.* **45**, 773-8.

Feinberg, A.P. & Vogelstein, B. 1983. A technique for radiolabelling DNA restriction endonuclease fragments to high specific activity. *Anal. Biochem.* **132**, 6-13.

Felsenstein, J. 1985. Confidence limits on phylogenies: an approach using the bootstrap. *Evolution* **39**, 783-91.

Felsenstein, J. 1989. *PHYLIP (Phylogenetic-inference package)*, version 3.2. University of Washington, Seattle.

Fitch, W.M. & Margoliash, E. 1967. Construction of phylogenetic trees. *Science* **155**, 279-84.

Gales, N.J., Adams, M. & Burton, H.R. 1989. Genetic relatedness of two populations of the southern elephant seal, *Mirounga leonina*. *Mar. Mammal Sci.* **5**, 57-67.

Gardner, L. 1983. Non-radioactive DNA labelling: detection of specific DNA and RNA sequences on nitrocellulose and *in situ* hybridizations. *Bio Techniques* **1**, 38-41.

Geraci, J.R. & Sweeney, J. 1978. Clinical techniques. In *Zoo and Wild Animal Medicine*, ed. M.E. Fowler, pp. 580-7. W.B. Saunders Co., Philadelphia.

Harris, H. & Hopkinson, D.A. 1976. *Handbook of Enzyme Electrophoresis in Human Genetics*. North Holland, Amsterdam.

Hillis, D.M., Larson, A., Davis, S.K. & Zimmer, E.A. 1990. Nucleic acids III. Sequencing. In *Molecular Systematics*, ed. D.M. Hillis, & C. Moritz, pp. 318-70. Sinauer Associates, Sunderland, Massachusetts.

Hillis, D.M. & Moritz, C. (Eds). 1990a. *Molecular Systematics*. Sinauer Associates, Sunderland, Massachusetts.

Hillis, D.M. & Moritz, C. 1990b. An overview of applications of molecular systematics. In *Molecular Systematics*. ed. Hillis D.M. & C. Moritz, 502-515. Sinauer Associates, Sunderland, Massachusetts.

Hoelzel, A.R. (Ed.). 1991. *Genetic Ecology of Whales and Dolphins: incorporating the Proceedings of the Workshop on Genetic Analysis of Cetacean Populations*. Rep. Int. Whaling Comm., Special Issue 13.

Hofman, R.J. 1975. *Distribution Patterns and Population Structure of Antarctic seals*. Ph.D. Thesis. Department of Ecology and Behavioral Biology. Univ. of Minnesota, Minneapolis.

Hubbard, R.C. 1968. Husbandry and laboratory care of pinnipeds. In *The Behavior and Physiology of Pinnipeds*, ed. R.J. Harrison, R.C. Hubbard, R.S. Peterson, C.E. Rice & R.J. Schusterman, pp. 299-358 Appleton-Century-Crofts, New York.

Hungerford, D.W. & Snyder, R.C. 1964. Karyotypes of two more mammals. *Am. Natur.* **98**, 125-7.

Kessler, C., Neumaier, D.S. & Wolf, W. 1985. Recognition sequences of restriction endonucleases and methylases – a review. *Gene* **33**, 1-102.

Laws, R.M. 1984. Seals. In *Antarctic Ecology*, vol. 2, ed. R.M. Laws, pp. 621-715. Academic Press, London.

Lidicker, W.Z. Jr. Sage, R.D. & Calkins, D.G. 1981. Biochemical variation in northern sea lions from Alaska. In *Mammalian Population Genetics* eds. M.H. Smith & J. Joule, pp. 231-41. Univ. Georgia Press, Athens.

Maniatis, T., Fritsch, E., & Sambrook, J. 1982. *Molecular Cloning: A Laboratory Manual*. Cold Spring Harbor Laboratory, New York.

McDermid, E.M. & Bonner, W.N. 1975. Red cell and serum protein systems of grey and harbour seals. *Comp. Biochem. Physiol.* **50B**, 97-101.

McDermid, E.M., Ananthakrishnan, R. & Agar, N.S. 1972. Electrophoretic investigation of plasma and red cell proteins and enzymes of Macquarie Island elephant seals. *Anim. Blood Groups Biochem. Genet.* **3**, 85-94.

Moritz, C., Dowling, T.E. & Brown W.M. 1987. Evolution of animal mitochondrial DNA: relevance for population biology and systematics. *Ann. Rev. Eco. Syst.* **18**, 269-92.

Naevdal, G. 1969. Blood protein polymorphism in harp seals off Eastern Canada. *J. Fish. Res. Bd., Canada*, **26**, 1397-9.

Nei, M. 1972. Genetic distance between populations. *Am. Nat.* **106**, 283-92.

Nei, M. & Li, W.H. 1979. Mathematical model for studying genetic variation in terms of restriction endonucleases. *Proc. Nat. Acad. Sci. U.S.A.* **76**, 5269-73.

Nei, M. Stephens, J.C. & Saitou, N. 1985. Methods for computing standard errors and branching points in an evolutionary tree and their application to molecular data from humans and apes. *Mol. Biol. Evol.* **2**, 66-85.

Nei, M. & Tajima, F. 1981. DNA polymorphism detectable by restriction endonuclease digests. *Genetics* **85**, 583–90.

Nei, M. & Tajima, F. 1983. Maximum likelihood estimation of the number of nucleotide substitutions from restriction site data. *Genetics* **97**, 145–63.

Palumbo, N.E., Allen, J., Whittow C. & Perri, S. 1971. Blood collection in the sea lion. *J. Wildlife Diseases* **7**, 290–1.

Pardue, M.L. & Gall, J.G. 1971. Chromosomal localization of mouse satellite DNA. *Science* **168**, 1356–8.

Reed, K.C. & Mann, D.A. 1985. Rapid transfer of DNA from agarose gels to nylon membranes. *Nucleic Acids Res.* **13**, 7207–21.

Richardson, B.J., Baverstock P.R. & Adams, M. 1986. *Allozyme Electrophoresis: A Handbook for Animal Systematics and Population Studies.* Academic Press, Sydney.

Ridgway, S.H. 1972. Homeostasis in the aquatic environment. In *Mammals of the Sea: Biology and Medicine*, ed. S.H. Ridgway, pp. 590–747. C.C. Thomas, Springfield.

Rigby, P.W.J., Dieckman, M. Rhodes, C. & Berg, P. 1977. Labelling deoxyribonucleic acid to high specific activity *in vitro* by nick translation with DNA polymerase. *I.J. Biol. Chem.* **239**, 222.

Roff, D.A. & Bentzen, P. 1989. The statistical analysis of mitochondrial DNA polymorphisms: X^2 and the problem of small samples. *Mol. Biol. Evol.* **6**, 539–45.

Rogers, J.S. 1972. Measures of genetic similarity and genetic distance. *Studies in Genetics, VII*. Univ. of Texas Publ., **7213**, 145–53.

Seabright, M. 1971. A rapid banding technique for human chromosomes. *Lancet* **2**, 971–2.

Seal, U.S., Erickson, A.W., Siniff, D.B. & Cline, D.R. 1971a. Blood chemistry and protein polymorphisms in three species of Antarctic seals (*Lobodon carcinophagus, Leptonychotes weddelli* and *Mirounga leonina*). In *Antarctic Pinnipedia*, ed. pp. 181–92. Am. Geophys. Union, Washington, D.C.

Seal, U.S., Erickson, A.W., Siniff, D.B. & Hofman, R.J. 1971b. Biochemical, population genetic, phylogenetic and cytological studies of Antarctic seal species. In *Symposium on Antarctic Ice and Water Masses.* ed. G. Deacon, pp. 77–95. SCAR, Cambridge, UK.

Selander, R.K. & Johnson, W.E. 1973. Genetic variation among vertebrate species. *Ann. Rev. of Ecol. and Syst.* **4**, 75–91.

Selander, R.K., Smith, M.H., Yang, S.Y., Johnson, W.E. & Gentry, J.B. 1971. Biochemical polymorphism and systematics in the genus *Peromyscus*. I. Variation in the old-field mouse (*Peromyscus polionotus*). *Studies in Genetics, VI*. Univ. of Texas Publ., **7103**, 49–90.

Sharp, G.D. 1981. Biochemical genetic studies, their value and limitations in stock identification and discrimination of pelagic mammal species. In *Mammals in the Seas*. FAO Fish. Series No. 5, III, pp. 131–136.

Shaughnessy, P.D. 1969. Transferrin polymorphism and population structure of the Weddell seal *Leptonychotes weddelli* (Lesson). *Aust. J. Biol. Sci.*, **22**, 1581–4.

Shaughnessy, P.D. 1970. Serum protein variation in southern fur seals, *Arctocephalus* spp., in relation to their taxonomy. *Aust. J. Zool.*, **18**, 331–43.

Shaughnessy, P.D. 1974a. An electrophoretic study of blood and milk proteins of the southern elephant seal, *Mirounga leonina. J. Mamm.*, **55**, 796–808.

Shaughnessy, P.D. 1974b. *Biochemical Identification of Populations of the Harbor Seal,* Phoca vitulina. Ph.D. Thesis. Univ. of Alaska, Fairbanks.

Shaw, C.R. & Prasad, R. 1970. Starch gel electrophoresis of enzymes - a compilation of recipes. *Biochem. Genet.,* **4,** 297-320.

Sherwin, W.B. 1991. Collecting mammalian tissue and data for genetic studies. *Mammal. Rev.* **21,** 21-30.

Siciliano, M.J. and C.R. Shaw. 1976. Separation and visualization of enzymes on gels. In *Chromatographic and Electrophoretic Techniques,* 4th edn, vol. 2, ed. I. Smith, pp. 185-209. Heinemann, London.

Simonsen, V., Allendorf, F.W. Eanes, W.F. & Kapel, F.O. 1982a. Electrophoretic variation in large mammals III. The ringed seal, *Pusa hispida,* the harp seal, *Pagophilus groenlandicus,* and the hooded seal, *Cystophora cristata. Hereditas,* **97,** 87-90.

Simonsen, V., Born, E.W. & Kristensen, T. 1982b. Electrophoretic variation in large mammals IV. The Atlantic walrus, *Odobenus rosmarus rosmarus* (L). *Hereditas* **97,** 91-4.

Smith, I. 1976. Techniques of starch gel electrophoresis. In *Chromatographic and Electrophoretic Techniques,* 4th edn, vol. 2, ed. I. Smith, pp. 153-76. Heineman, London.

Sneath, P.H.A. & Sokal R.R. 1973. *Numerical Taxonomy: The Principles and Practice of Numerical Classification.* W.H. Freeman Co., San Francisco.

Southern, E.M. 1975. Detection of specific sequences among DNA fragments separated by gel electrophoresis. *J. Mol. Biol.* **98,** 503-17.

Sumner, A.J. 1972. A simple technique for demonstrating centromeric heterochromatin. *Exp. Cell Res.* **75,** 304-6.

Swofford, D.L. 1985. *PAUP (Phylogenetic analysis using parsimony),* version 2.4. Ill. Nat. Hist. Survey, Champaign.

Swofford, D.L. & Olsen, G.J. 1990. Phylogeny reconstruction. In *Molecular Systematics,* ed. D.M. Hillis & C. Moritz, pp. 411-501. Sinauer Associates, Sunderland, Massachusetts.

Takahata, N. & Palumbi, S.R. 1985. Extranuclear differentiation and gene flow in the finite island model. *Genetics* **109,** 441-57.

Testa, J.W. 1986. Electromorph variation in Weddell seals (*Leptonychotes weddelli*). *J. Mamm.* **67,** 606-10.

Templeton, A.R. 1983. Phylogenetic inference from restriction endonuclease cleavage site maps with particular reference to the evolution of humans and the apes. *Evolution* **37,** 221-44.

Tombs, M.P. 1968. Horizontal gel-slab electrophoresis. In *Chromatographic and Electrophoretic Techniques,* 2nd edn, vol. 2, ed. I. Smith, pp. 443-52. Heinemann, London.

Upholt, W.B. 1977. Estimation of DNA sequence divergence from comparison of restriction endonuclease digests. *Nucleic Acids Res.* **4,** 1257-65.

Vawter, L. & Brown, W.M. 1986. Nuclear and mitochondrial DNA comparisons reveal extreme rate variation in the molecular clock. *Science* **234,** 194-6.

Vergani, D.F. 1985. *Comparative study of populations in Antarctica and Patagonia of the Southern Elephant Seal,* Mirounga leonina (*Linne, 1758*) *and its Methodology.* Publ. 15, Instituto Antartico Argentino, Buenos Aires.

Vergani, D.F., Spairani, H.J. & Aguirre, C.A. 1986. *Immobilization of Crabeater Seals,* Lobodon carcinophagus, *with the use of Xilazine Hydrochloride at 25 De Mayo Island (Antarctica) and Identification of*

Polymorphism in Transferrins. Contribution No. 317. Instituto Antartico Argentino, Buenos Aires.

Wang, H.C. & Federov, S. 1972. Banding in human chromosomes treated with trypsin. *Nature New Biol.* **235**, 52-3.

Weir, B.S. 1990. *Genetic Data Analysis.* Sinauer Associates, Sunderland, Massachusetts.

Wilson, A.C., Cann, R.L., Carr, S.M., George, M., Jr., Gyllensten, U.B., Helm-Bychowski, K., Higuchi, R.C., Palumbi, S.R., Prager, E.M., Sage, R.D. & Stoneking, M. 1985. Mitochondrial DNA and two perspectives on evolutionary genetics. *Biol. J. Linn. Soc.* **26**, 375-400.

10

Collection of material for the determination of organochlorine and heavy metal levels

P.D. SHAUGHNESSY

Introduction

When Antarctic seals are killed for the collection of biological material, the opportunity should be taken to collect tissues for determination of the levels of organochlorine pesticide residues (DDT and its metabolites DDE and TDE, dieldrin, hexachlorocyclohexanes, chlordane compounds, toxaphene and others), the polychlorinated biphenyls (PCBs) and heavy metals (particularly mercury and cadmium). Such contaminants may affect the conservation status of marine mammal populations (Reijnders, 1988) and may be useful for discriminating between populations (Aguilar, 1987). The material collected for these determinations and the collecting techniques are briefly outlined in this chapter. Anthropogenic hydrocarbons in the Antarctic environment have been reviewed by Hikada & Tatsukawa (1981).

General considerations

In order to establish baseline data on environmental pollutant levels in seal species, it is advisable to assess and standardize the ages and sexes of animals being taken, because environmental pollutant residues may be concentrated and metabolized differently depending on age, sex and reproductive condition of individual animals. When attempting to compare the relative concentrations of pollutants among species or subpopulations care should be taken to select tissues of animals of the same age and sex class to eliminate some of the variability.

If the animal is killed with a gun, no samples should be taken near the bullet entry or path. All tissues should be sampled from a consistent locality; Holden (1975) has suggested that the distribution of these contaminants may not be homogeneous.

Organochlorine residues

Because these contaminants tend to accumulate in tissues with a high lipid content, the highest levels are found in blubber, and this is a preferred material for collection. Blubber should be sampled from the same location on each seal body that is being examined. Addison *et al.* (1973) and Addison & Smith (1974) collected blubber from the mid-dorsal region of harp and ringed seals, while Cape fur seals and northern fur seals have been sampled from the anterior end of the sternum by Henry (1976) and Anas & Worlund (1975) respectively; this latter site is recommended. If samples are taken from the ventral side of adult females, care should be taken not to include the mammary gland tissue with the blubber. Other body tissues such as liver, kidney, muscle and brain have also been examined in some pinnipeds. The levels of organochlorines have been lower than those in blubber (Heppleston, 1973; Holden, 1975; Gilmartin *et al.*, 1976) but the sample variance also tends to be less than in blubber, lending some argument for analyzing tissues other than blubber. Since some organochlorine residues are metabolized in the liver, this tissue should probably be avoided until there is a better understanding of the pharmacodynamics of organochlorine residues in seals.

High concentrations of chlorinated hydrocarbons appear to be involved in reproductive dysfunctions in pinnipeds. Associative relationships have been reported between high concentrations of organochlorine residues and premature birth in California sea lions (De Long, Gilmartin & Simpson, 1973; Gilmartin *et al.*, 1976), uterine occlusions in ringed seals (Helle, Olsson & Jensen, 1976; also see Bergman & Olsson, 1985), and decreased reproductive rates in common (harbour) seals (Reijnders, 1980, 1982) which is caused by feeding on fish polluted with PCBs (Reijnders, 1986). Thus data on reproductive success (observed or interpreted from examination of the ovaries), for populations sampled, are desirable for correlation with chlorinated hydrocarbon concentrations. Also of possible reproductive consequence is the finding of detectable levels of these toxicants in the lipid-rich milk of many marine mammals.

Samples of about 10 g of desired tissue should be stored in either aluminium foil or glass jars (with aluminium foil lining the screw cap) at −20 °C. Alternatively, but less desirably, samples can be preserved in 10% aqueous formalin. Since rubber and plastic gloves may contain interfering compounds, they should not be worn during collection. Likewise, samples should not be placed in plastic bags. Instead, samples should be thoroughly wrapped in foil (if possible), placed in a clean glass jar and appropriately labelled externally. If samples are to be archived for longer than one year, it is best to store them in glass bottles with aluminium foil lining the lids, as teflon sheeting wrappings on tissues will break down after long-term storage. However, long-term storage may lead to changes in pollutant concentrations and ratios (Aguilar, 1987).

The levels of some organochlorine residues have been shown to be related to age in harp seals (Addison *et al.*, 1973; Frank, Ronald & Braun, 1973), male ringed seals (Addison & Smith, 1974) and Cape fur seals (Henry, 1976). Consequently the sex of each seal sampled and its age class (or better still its age, if this is available) should be recorded. The reproductive state of females should also be recorded as this is another variable which may affect the level of pesticide residues (Addison *et al.*, 1973). Blubber thickness (chapter 8, p. 134) at the sternum should be recorded as it is a measure of the total body fat (chapter 14).

Data from other species indicate that as body lipids are depleted (due to disease or fasting) the total amount of lipophilic toxicants remains the same, resulting in an increase in their concentration and possibly some mobilization within the tissues. Holden (1972) found the highest concentrations of DDT in seals that were either dead, dying or in poor condition and Drescher, Harms & Huschenbeth (1977) reported elevated levels of organochlorine residues in emaciated harbour seals.

Heavy metals

Concentrations of heavy metals have been measured in a variety of tissues of pinnipeds, but primarily in liver, kidney and muscle, and less frequently in brain and blubber. Higher concentrations of mercury have been found in liver than in other tissues in most pinnipeds examined, such as the common (harbour) seal (Roberts, Heppleston & Roberts, 1976), grey and common (harbour) seals (Holden, 1975), northern fur seals (Anas, 1971) and in grey, harbour (common), harp and hooded seals from eastern Canada (Sergeant & Armstrong, 1973). The level of mercury in liver has been shown to be correlated with age in several pinnipeds such as the northern fur seal (Anas, 1971), ringed seals (Smith & Armstrong, 1975), grey seals (Sergeant & Armstrong, 1973) and common (harbour) seals (Roberts *et al.*, 1976). Therefore, as with the sample collection of organochlorine residues, the age and sex of the seals sampled should be recorded. Cadmium concentrations are frequently higher in kidney tissues.

Reports indicate that about 80% of the mercury in seals is of the elemental form rather than the organic methyl mercury form which predominates in fish species (Koeman *et al.*, 1975; Smith & Armstrong, 1975;

Roberts *et al.*, 1976). Determining whether a metal is in its organic or inorganic form can be an aid in interpreting its physiological significance. The liver and kidneys, being the body's principal detoxifiers and the main excretion routes for heavy metals, are therefore, the most important tissues to sample.

Because of the likelihood of contamination by heavy metals, samples should be collected as rapidly as possible with a minimum of handling. Samples should be dissected out with stainless steel scalpels and forceps (preferably plastic-tipped). These should be cleaned before use, rinsed with distilled or demineralized water and dried with a clean white paper tissue (coloured paper contains contaminants). Instruments can be cleaned beforehand and stored in a closed plastic bag. If desired, blood can be washed off the samples with distilled or demineralized water.

About 20 g of each tissue should be collected and rapidly frozen to −20 °C in pre-cleaned plastic containers, plastic bags or glass jars. Metal-capped containers must not be used. Containers should be pre-cleaned by rinsing with reagent grade hydrochloric acid diluted to 70% volume, followed by a thorough rinse with distilled or demineralized water. Preservatives must not be added to the samples as they contain unacceptably large amounts of heavy metals. The deep-freeze must be kept clean to avoid, for example, rust flakes accidentally contaminating the samples.

Acknowledgements

The assistance of J.L. Henry, H.O. Fourie and M.J. Orren is gratefully acknowledged.

References

Addison, R.F., Kerr, S.R., Dale, J. & Sergeant, D. 1973. Variation of organochlorine residue levels with age in Gulf of St Lawrence harp seals (*Pagophilus groenlandicus*). *J. Fish. Res. Bd. Can.* **30**, 595–600.

Addison, R.F. & Smith, T.G. 1974. Organochlorine residue levels in Arctic ringed seals: variation with age and sex. *Oikos* **25**, 335–7.

Aguilar, A. 1987. Using organochlorine pollutants to discriminate marine mammal populations: a review and critique of the methods. *Mar. Mammal Sci.* **3**, 242–62.

Anas, R.F. 1971. Mercury in fur seals. In *Mercury in the Western Environment*, ed. D.R. Buhler, pp. 91–6. Oregon State University Press, Oregon.

Anas, R.E. & Worlund, D.D. 1975. Comparison between two methods of subsampling blubber of northern fur seals for total DDT plus PCBs. *Pest. Monit. J.* **8**(4), 261–2.

Bergman, A. & Olsson, M. 1985. Pathology of Baltic grey seal and ringed seal females. *Fin. Game Res.* **44**, 47-62.

DeLong, R.L., Gilmartin, W.G. & Simpson, J.G. 1973. Premature births in California sea lions: association with high organochlorine pollutant residue levels. *Science* **181**, 1168-70.

Drescher, H.E., Harms, U. & Huschenbeth, E. 1977. Organochlorines and heavy metals in the harbour seal, *Phoca vitulina*, from the German North Sea coast. *Marine Biology* **41**, 99-106.

Frank, R., Ronald, K. & Braun, H.E. 1973. Organochlorine residues in harp seals (*Pagophilus groenlandicus*) caught in eastern Canadian waters. *J. Fish. Res. Bd. Can.* **30** 1053-63.

Gilmartin, W.G., DeLong, R.L., Smith, A.W. & Sweeney, J.C. 1976. Premature parturition in the California sea lion. *J. Wildl. Dis.* **12**, 104-115.

Helle, E., Olsson, M. & Jensen, S. 1976. PCB levels correlated with pathological changes in seal uteri. *Ambio* **5**, 261-3.

Henry, J.L. 1976. *Pesticide residues in the Cape fur seal*. First Interdisciplinary Conference on Marine and Freshwater Research in Southern Africa, Port Elizabeth, South Africa, S122, July 1976.

Heppleston, P.B. 1973. Organochlorines in British grey seals. *Mar. Pollut. Bull.* **4**, 44-5.

Hikada, H. & Taksukawa, R. 1981. Review: environmental pollution by chlorinated hydrocarbons in the Antarctic. *Antarctic Record* **71**, 151-64.

Holden, A.V. 1972. Monitoring organochlorine contamination of the marine environment by analysis of residues in seals. In *Marine Pollution and Sea Life*, ed. M. Ruivo, pp. 266-72. Fishing News (Books), London.

Holden, A.V. 1975. The accumulation of oceanic contaminants in marine mammals. *Rapp. P.-v. Reun. Cons. int. Explor. Mer* **169**, 353-61.

Koeman, J.H., van de Ven, S.M., deGroeij, J.J.M., Tjioe, P.S. and van Haaften, J.L. 1975. Mercury and selenium in marine mammals and birds. *Sci. Tot. Environ.* **3**, 279-387.

Reijnders, P.J.H. 1980. Organochlorine and heavy metal residues in harbour seals from the Wadden Sea and their possible effects on reproduction. *Neth. J. Sea Res.* **14**, 30-65.

Reijnders, P.J.H. 1982. On the ecology of the harbour seal *Phoca vitulina* in the Wadden Sea: population dynamics, residue levels, and management. *Vet. Quart.* **4**, 36-42.

Reijnders, P.J.H. 1986. Reproductive failure in common seals feeding on fish from polluted coastal waters. *Nature* **324**, 456-7.

Reijnders, P.J.H. 1988. Ecotoxicological perspectives in marine mammalogy: research principles and goals for a conservation policy. *Mar. Mammal Sci.* **4**, 91-102.

Roberts, T.M., Heppleston, P.B. & Roberts, R.D. 1976. Distribution of heavy metals in tissues of the common seal. *Mar. Pollut. Bull.* **7**, 194-6.

Sergeant, D.E. & Armstrong, F.A.J. 1973. Mercury in seals from eastern Canada. *J. Fish. Res. Bd. Can.* **30**, 843-6.

Smith, T.G. & Armstrong, F.A.J. 1975. Mercury and selenium in ringed and bearded seal tissues from Arctic Canada. *Arctic* **31**, 75-84.

11
Age determination

T.S. McCANN

Introduction

Age determination is of fundamental importance in many ecological and behavioural studies, particularly for studies of growth rates and population dynamics. The most extensively used method is examination of incremental lines on or in teeth (developed independently by Scheffer, 1950, and Laws, 1952) and this chapter concentrates particularly on this. Other methods of age determination using incremental structures in nails and bones and methods giving relative age, for example body length, eye lens weight, suture closure, baculum development, ovarian structure and pelage characteristics, are dealt with in much less detail.

Tooth structure and methods of tooth preparation are described. This is followed by species accounts covering choice of tooth, preparation and interpretation of teeth from Antarctic seals. Some validation studies on the reliability of age determination using teeth are mentioned, followed by a description of the other methods of age determination referred to above.

More extensive reviews of mammalian age determination techniques can be found in Laws (1962), Klevezal & Kleinenberg (1967), Jonsgard (1969), Morris (1972, 1978), Spinage (1973), Scheffer & Myrick (1980) and Fancy (1980).

Teeth

Tooth structure

The permanent tooth comprises a crown exposed above the gum and a root contained in the tooth alveolus of the jawbone. The main tooth substance is dentine which in the crown is covered with a layer of enamel

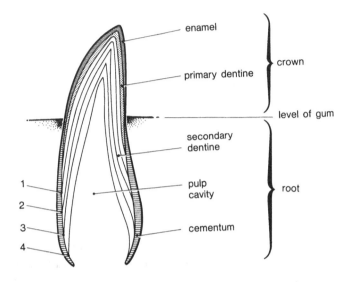

Fig. 11.1 Diagrammatic longitudinal section of a canine tooth from a seal aged four years.

and in the root is covered by one or more layers of cementum. The dentine is present as primary or pre-natal dentine (present at birth) and secondary dentine (deposited after birth). The hollow centre of the tooth, into which the secondary dentine is deposited, is called the pulp cavity (see Fig. 11.1. A canine tooth is illustrated but the description applies to all permanent teeth). In older animals the pulp cavity may become partially or completely filled with secondary dentine, making age determination from the dentine useful only to the age at which the cavity closed or dentine deposition ceased.

Seals grow as they age and the growth patterns in the teeth are directly related to the skeletal growth of the animal. In some species, such as the elephant seal, the pulp cavity remains open throughout life and almost all growth occurs in the dentine. However, in the majority of species, the pulp cavity closes at some stage which terminates tooth growth. Growth of the cementum layer fills the gap between the expanding tooth socket and the tooth.

Dentine

During the life of a seal the rate at which dentine is deposited and its mineralization varies seasonally and the rate of deposition decreases

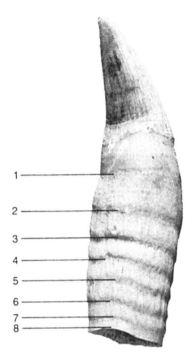

Fig. 11.2 Canine from 8-year-old male Antarctic fur seal, showing external annual ridges.

with age. In Antarctic seals the root of the canine tooth grows in length throughout life or until the pulp cavity closes. The seasonal changes in rate of deposition of dentine results in uneven growth of the root. This causes formation of annual ridges which consist of ring-like swellings on the outer wall of the root (Fig. 11.2). These are directly associated with the annual layers in the dentine.

A section of a tooth will reveal bands or layers in the dentine of differing optical qualities. The differences in optical density are caused by differences in the physico-chemical properties of the dentine deposited at different times of the year. There is still debate on the precise relationship between optical density and the physico-chemical properties of dentine (McLaren, 1958; Carrick & Ingham, 1962; Laws, 1962; Klevezal & Kleinenberg, 1967), but for the purposes of age determination it is probably not necessary to know the exact mineral composition of the incremental layers. Environmental factors influence the rate and type of dentine deposited and it appears that the density of dentine correlates most closely with nutrition (Scheffer, 1950; Laws, 1953a, 1962; McLaren, 1958; Scheffer & Peterson, 1967; Stewart, 1983; Bengtson, pers. comm.).

The layers in the dentine are of two types – translucent and reflective. However, the appearance of a band depends on the conditions under which it is viewed. A translucent substance will appear light or clear in transmitted light but dark in reflected light. Opaque bands appear dark in transmitted light but light or white in reflected light. It is, therefore, important that the conditions under which a tooth is 'read' are clearly stated. The layers of dentine are usually found in alternating broad and narrow layers. Generally broad layers of dentine form during growth and narrow layers in seasons of retarded growth, regardless of their respective light transmitting qualities (Klevezal & Kleinenberg, 1967; Morris, 1978).

In an effort to standardize terminology the International Conference on Determining Age of Odontocete Cetaceans (Perrin & Myrick, 1980) called the layers, lines, bands or zones formed within a year the incremental growth layers (IGLs). The groups of IGLs formed during the course of a year were called growth layer groups (GLGs). These may be recognized by virtue of their cyclic repetition. The decision as to which features are included in the GLGs is made on the basis of an inspection of the overall pattern of dentinal layering, not simply on a count of IGLs (Bowen, Sergeant & Øritsland, 1983).

Although the lack of a standard terminology has caused some confusion, some terms are more versatile for describing particular features than the term IGL. For example, within a year a seal usually deposits two (sometimes more) main zones of dentine. Within these zones there may be narrow bands of contrasting optical density termed accessory layers (Christensen, 1973; Stewart, 1983) or in elephant seals 'resting lines' (Carrick & Ingham, 1962; McCann, 1981). These narrow bands appear most frequently in the earlier, broader rings of dentine and are distinguished from major (seasonal) annulations on the basis of size, position and their irregular occurrence. They are usually disruptive of the overall annual pattern. Describing such structures in terms of IGLs could be confusing. Written descriptions of dentinal structure should refer to pictures or diagrams where possible which would help to reduce confusion and ambiguity.

Cementum

Cementum is deposited on the outer surface of the tooth, so that the most recent layers are represented by the outermost layers of the cementum. This is in contrast to the dentine, where the most recent layers are the innermost. Cementum has a different structure than dentine. When viewed with transmitted light (or in stained thin sections) there is first a broad, translucent band and then a narrow, dark band; the two layers constitute one year's growth, a GLG, but it reflects a gradual change in each IGL throughout the year and then an abrupt change in optical density at the start of the following year. Cementum can be deposited on the root of the tooth throughout life. As with dentine, deposition in early years, when there is rapid growth before the attainment of sexual maturity, tends to be greater than in later years, and enables us to estimate the age of maturity.

Methods

Collection and storage

Teeth are usually collected whole by boiling the jaws in water to loosen the flesh and then extracting the teeth by hand†. Jaws stored fresh or frozen require about one hour's boiling. Storage in formalin fixes the periodontal membrane and jaws stored this way require about five hours' boiling to loosen the teeth. Low & Cowan (1963) suggested that boiling may reduce the stainability of some tooth tissues and this needs to be checked for Antarctic seals, but Turner (1977) found no significant loss in definition between incisors extracted by boiling and those extracted by soaking the mandibles. In elephant seals the pulp cavity normally remains open above the level of the gum line and Laws (1953a, 1960) collected teeth by sawing off the crowns with a hack saw.

Teeth and tooth sections have been stored successfully in a number of ways. Teeth stored in a very dry atmosphere tend to crack which can make sectioning and reading difficult but whole teeth have been stored dry for years quite successfully. Freezing does not affect teeth and can be used as a means of storage. Teeth have also been stored in formalin or a mixture of 70% alcohol and glycerol or equal parts alcohol, glycerol and water. When mixing alcohol and glycerol for tooth storage the precise ratio of parts is not important.

Tooth sections should be stored wet unless they are mounted permanently in DPX, Permount or similar mountant. Sections stored dry can become opaque and usually crack which can make reading of the section difficult or impossible. The alcohol/glycerol mixture is best for storing sections, and crowns of elephant seal teeth which were stored dry for more than 30 years produced good sections after soaking in alcohol/glycerol.

† See note added in proof (p. 227).

External tooth structure – ridges

Ridges on the roots of teeth are clearest in the early years of life when the tooth grows fastest. In later years the growth increment is smaller, producing correspondingly less pronounced ridges, and the deposition of cementum tends to mask the ridges. Whole teeth should be cleaned to reveal the ridges. In older seals the ridges can be brought into greater relief by removing the cementum with an abrasive substance, or chemically by dissolving in acid. This can increase the readability and the upper age to which the tooth can be aged by ridges.

Not all ridges on a root are annual. In Antarctic fur seals only one ridge is formed each year but in elephant seals two ridges are formed. In immature animals these correspond to the annual autumn–winter haul-out to 'rest' and to the annual moult. In mature elephant seals they correspond to the breeding season haul-out and the moult. Knowledge of the seasonal cycle of breeding and moulting or feeding and not-feeding is essential for interpretation of tooth structure in any species.

Internal structure

Sectioning Teeth have been sectioned to reveal the internal structure in a number of ways: by hacksaw (Laws, 1953a; Stirling, 1969) hand-held in portable circular saws, in double-bladed circular saws or with a diamond-edged circular saw on a sliding saw table (Fisher & Mackenzie, 1954; Payne, 1978). Sections can be mounted for cutting on wood blocks using dental wax. Sections or cut surfaces produced by the cruder methods need to be polished through a graded series of carborundum powders or similar abrasives to obtain a smooth surface free of saw cuts, and in the case of thin sections to grind the section to the appropriate thickness. Use of a high speed circular saw with water as a coolant and lubricant produces cut surfaces or thin sections with smooth surfaces which need no polishing or other treatment before reading.

The transmission of light and the definition of IGLs is related to the thickness of the section. To achieve optimal thickness, trial cuts must be made. Frequently quoted thicknesses are 120–$250\,\mu$m or even to $400\,\mu$m (0.4 mm). In general, the thinner the section the better its readability, although extremely thin sections sometimes appear to lose internal structure. Very thin longitudinal sections sometimes curl which makes mounting and reading difficult. Transverse sections cut too thinly can break up if the tooth is brittle. The sectioning properties of teeth are improved if they are stored for some days in an alcohol/glycerol mixture.

Sections can be viewed in water or an alcohol/glycerol mixture or can be permanently mounted. In this case, the section is taken through a series of alcohols up to absolute (100%) then transferred to xylene and mounted under a coverslip using DPX, Permount or similar mountant. Mounted sections are easy to store. A binocular microscope with magnifications of between ×6 and up to about ×50 is most useful. Polarizing filters can enhance the patterns and facilitate reading and polythene petri dishes produce colourful birefringence which may also be useful.

Etching The different IGLs of teeth can be etched at different rates using calcium-dissolving solutions. Etching produces a series of ridges and valleys across the cut surface of the tooth corresponding to IGLs laid down at different times of the year. Teeth to be etched should be sectioned and the cut surfaces polished (if necessary). The etching process described by Stirling (1969) works well on many species. The prepared tooth surface is: (1) etched for 22 h in a 5 part formic acid, 95 parts 10% formalin solution, (2) washed for 4 h in flowing water and (3) neutralized by soaking for 4 h in a solution of three to four drops of 0.88% ammonia in 75 ml of water for 4 h. Etched teeth can sometimes be read without further treatment but the etching is usually followed by staining. In this case after step (3) above, the tooth is stained for 10 min in a 0.25% solution of thionine at 37 °C and left to air-dry.

Mattlin (1978) used a modification of this method for preparing New Zealand fur seal teeth and his modified method was used on Ross seal canine teeth (M.N. Bester, pers. comm.). In this method, after neutralizing in 0.88% ammonia, the teeth were stained for 5 s in a 0.125% solution of thionine at 37 °C and left to air-dry.

When light is projected across the etched surface, light and dark lines, representing reflections and shadows from the ridges and hollows, become evident. The lighting of a specimen is critical for accurate counting.

The preparation of cellulose acetate peels from etched teeth has produced promising results in leopard seal, elephant seal, Weddell seal and Antarctic fur seal (B.R. Mate, pers. comm.) and should be equally useful in Ross seals. The freshly etched and dried tooth is held in position with plasticine or some similar substance so that the etched surface is level. A piece of thin acetate, frosted on one side, is trimmed to the approximate size of the prepared surface. The surface of the tooth is flooded with acetone and then the acetate sheet is carefully placed (frosted side down) on the tooth by starting at one end and laying it on. As the acetone comes in contact with the frosted acetate, the acetate should appear clear. Practice in applying the acetate will produce high quality peels. Excess acetate will melt over the sides of the tooth leaving curled edges which make flat storage of the peels difficult. Some curling of the edges can be rectified by inverting the tooth after most of the acetone has evaporated and placing it on a flat surface.

The acetone softens the acetate which then fills the etched annuli. When the acetone has evaporated the acetate hardens and can be peeled off, removing with it the microtopography of the etched surface of both the dentine and the cementum. The best detail is obtained by allowing the peel to cure overnight before removing it.

The annuli on the peel can be observed under a dissecting microscope and clarity can be enhanced by lightly dusting the peel with fine dark powder. Peels are easily kept as a permanent record by taping them to file cards with other information about the specimen or mounting them between two glass slides taped at the ends.

Decalcifying Decalcifying and staining has been used on crabeater seal (Bengtson & Siniff, 1981) and Weddell seal teeth (W. Testa, pers. comm.). Teeth were decalcified in a buffered 25% formic acid solution. Decalcification progresses slowly and no damage is done if the tooth remains in the acid too long. Dried teeth can be decalcified in 7 to 10 days while teeth stored in formalin may take 15 to 20 days. The decalcification point can be determined by checking for calcium ions in the formic acid solution (Marks & Erickson, 1966; Stirling, Archibald & DeMaster, 1977). The decalcified teeth are washed for 12–16 h in running tap water to neutralize the acid in order to ensure good staining reaction. If they are to be stored before sectioning they should be stored in 70% alcohol.

Other decalcifying agents which have been used are a 5% solution of 67% nitric acid (Marks & Erickson, 1966; Grue & Jensen, 1973); formic acid/sodium citrate solution (Miller, 1974) and 5% hydrochloric acid (Novakowski, 1965).

Prior to sectioning, the teeth should be soaked in water for at least 24 h to hydrate the tissue and then embedded in a mounting compound and cut on a cryostat at $-10\,^{\circ}\text{C}$ to $-20\,^{\circ}\text{C}$ into longitudinal sections 10 μm thick. Cut sections should be placed in water at pH 8–9 for at least 20 min before being affixed to a glass slide with egg albumen. Drying takes 30–45 min at room temperature.

Sections can be stained in a filtered 0.32% aqueous solution of Toluene blue or in Erlich's haematoxylin or with Giemsa stain (Stone, Roscoe & Weber, 1975).

Species accounts

Crabeater seal - Lobodon carcinophagus

Choice of tooth Up to 16 to 20 years the dentine of the canine tooth can be used to age a tooth but closure of the pulp cavity renders this method inaccurate for older animals (Laws, 1958). The canine shows relatively little cementum deposition. In reading cementum, longitudinal sections are generally preferable to cross sections but the canine tooth is curved in two planes and so a sagittal longitudinal section in one plane does not follow the longitudinal axis of the tooth and gives a distorted impression of tooth structure. R.M. Laws (pers. comm.) first investigated the possibility of using post-canine teeth for aging and found that the third post-canine (PC3) gave the best results. If PC3 is not available PC1, PC2 or PC4 can be used (R.M. Laws, pers. comm.; Bengtson & Siniff, 1981; Bengtson & Laws, 1985).

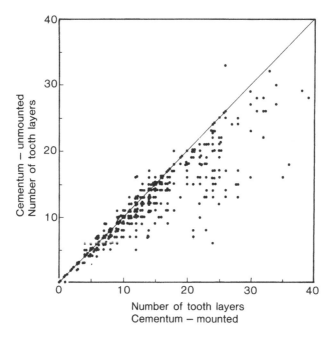

Fig. 11.3 Comparison of the number of recognizable tooth layers in the cementum of crabeater seal teeth when mounted and unmounted. The diagonal line indicates the relationship if layers were equally recognizable in mounted and unmounted sections. The distribution actually recorded indicates that more layers are recognizable in mounted sections. (Unpublished data from R.M. Laws.)

Tooth preparation To obtain thin sections, the post-canine is mounted flat on a block of wood using dental wax and cut root first.

Cementum layers are more clearly seen if the section is mounted in DPX or similar mountant (Fig. 11.3). It is also possible, and may be desirable, not to mount the section but to examine it in the alcohol/glycerol mixture (or water) in a petri dish. Crabeater seal teeth have also been decalcified and stained (Bengtson & Siniff, 1981; see section 'Internal structure', but this method is not preferred (J.L. Bengtson, pers. comm.).

Age determination Both dentine and cementum layers can be read on the same tooth, but cementum probably gives a better estimate of age (Fig. 11.4). As the post-canines are double-rooted there are four cementum thicknesses to choose from to obtain the best count (Fig. 11.5).

In younger animals (Fig. 11.6a) a cementum layer may be very thin, incomplete or difficult to interpret, in which case reference is made

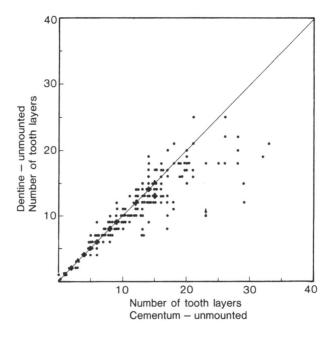

Fig. 11.4 Comparison of the number of recognizable tooth layers in the cementum and dentine of unmounted sections of crabeater seal teeth (conventions as in Fig. 11.3). This suggests that more layers are recognizable in cementum than in dentine. (Unpublished data from R.M. Laws.)

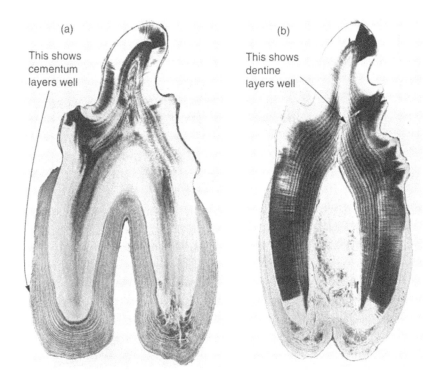

(a)

This shows
cementum
layers well

(b)

This shows
dentine
layers well

Fig. 11.5 Longitudinal sections of post-canine teeth from crabeater seal to show (a) cementum layers and (b) dentine layers. (Photographs courtesy R.M. Laws.)

to the dentine layers to confirm cementum counts. In exceptional teeth the dentine layers might be clear up to 15 years or more (Fig. 11.5b). Cementum is laid down first on the inside of the roots and the layers there are more conspicuous in the first two years. There are usually very clear layers, but often there are only a few places along the length of the root where counts are possible. Sometimes it is necessary to read lines over part of the cementum at one point, to follow a line along and complete the count elsewhere.

The early lines tend to be broader and more diffuse and the distance between them greater than in adult years. A transition zone between early, more irregular broadly spaced layers and the narrower more regular layers of later years may indicate change in growth concomitant with the attainment of sexual maturity. In some cases the change in the cementum deposition is clearly obvious and in others it is not.

(a)

(b)

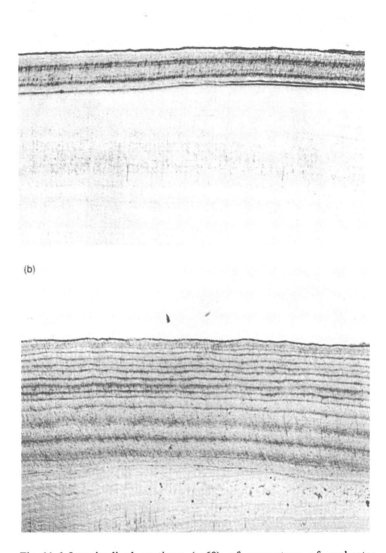

Fig. 11.6 Longitudinal sections (×60) of cementum of crabeater seal teeth, (a) from 2-year-old female, (b) from 15-year-old female. (Photographs courtesy R.M. Laws.)

Weddell seal – Leptonychotes weddellii

Choice of tooth In the Weddell seal the roots of the canine teeth are nearly closed at maturity, but dentine deposition continues to occur inside the pulp cavity (Mansfield, 1958). For cementum annuli the preferred teeth are lower post-canines 2 or 3, although canines are adequate (W. Testa, pers. comm.).

Tooth preparation Stirling (1969) used etching and staining techniques (see section 'Internal structure') to age Weddell seal canines by counts of dentinal annuli.

For counts of cementum layers the teeth may be decalcified. The decalcified teeth are imbedded in paraffin wax and sectioned longitudinally, then stained with Giemsa stain (Stone *et al.*, 1975). However, mounted sections without decalcification will probably also give satisfactory results.

Age determination In the dentine, two bands of differing density are laid down over the course of a year. In etched teeth, one year is represented by a ridge and hollow. This pattern is repeated each year, but the bands become progressively narrower as the animal ages. Although the tooth cavity remains open, after 10 years of age the dentine layers may become too narrow to distinguish in some specimens (Stirling, 1969). Searching is important and the lines of dentine must sometimes be traced from the cone, where they are most distinct, and then back and forth along the section to follow the clearest progression.

Cementum annuli show up very well using the staining method described above and are read in the usual way but the technique renders dentinal annuli practically unreadable (W. Testa, pers. comm.).

Leopard seal – Hydrurga leptonyx

Canine teeth have been used for age determination (Laws, 1957) but recent studies have used post-canines, usually PC3 (J.L. Bengtson, pers. comm.). A thin longitudinal section of the post-canine is cut and mounted as for crabeater seal teeth. Cementum on the roots is examined as in crabeater seals.

Ross seal – Ommatophoca rossii

Canine teeth have been used to age Ross seals, usually an upper canine (M.N. Bester, pers. comm.). Post-canines have not, apparently, been examined but this would probably be fruitful. Cementum is poorly developed in Ross seal canines and age determination is by dentinal layers. Teeth have been prepared by Mattlin's (1978) modification of Stirling's (1969) etching and staining procedure – (M.N. Bester, pers. comm.). The etching produced ridges and 'valleys' in the dentine and the structure and interpretation is similar to the Weddell seal. Etched layers were counted using a binocular microscope and light shining across the etched surface of the tooth.

Elephant seal – Mirounga leonina

Choice of tooth The pulp cavity of the canine tooth remains open throughout life in elephant seals and both upper and lower canines are suitable for aging; lower canines are more easily obtained.

Tooth preparation Canines of elephant seals grow in length throughout life and in many specimens the external appearance can be compared with the internal structure.

The method of using the internal structure of teeth for age determination was first used on elephant seals (Laws, 1952, 1953a). Laws noted that the pulp cavity of the canine extends above the level of the gum so that if the tooth is sawn off at gum level all of the dentinal layers will be included. This method of tooth collection is easy (and important if collection is by commercial sealers, Laws, 1960) and can be employed successfully, but it is recommended that whole teeth are collected.

The simplest method is to cut a transverse section of the tooth ensuring that the cut is above the ridge showing the limit of the first year's growth (Fig. 11.7a). In some older females the pulp cavity near the tip fills with dentine but using whole teeth permits another cut to be made. The tip of the tooth can then be inverted and mounted on a piece of clay or plasticine so that the surface is parallel with the bench top. Light is then shone on the side of the tooth, with a shield preventing the light from reflecting off the cut surface, and the light is transmitted up through the tooth revealing the internal structure. Alternatively, light can be reflected off the cut surface.

Thin transverse sections cut to about 300–400 μm can be viewed in water or alcohol/glycerol with transmitted light.

Carrick & Ingham (1962) used silver nitrate staining to interpret the structure of teeth from animals of known age up to eight years. The sectioned tooth was placed overnight in 5% silver nitrate then rinsed and exposed to daylight until the cut surface appeared rather darker than

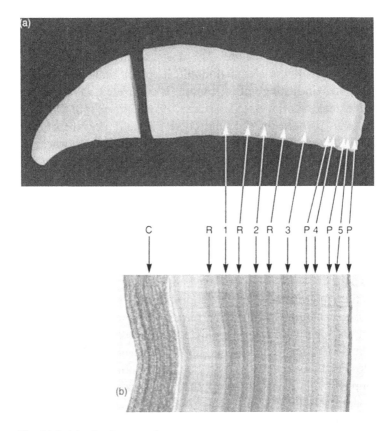

Fig. 11.7 (a) Canine tooth (×2.5) of 6-year-old female elephant seal, showing ridges on the root and location of transverse section. (b) Transverse section from (a) viewed (×40) with transmitted light. Dark lines are dense dentine laid down while the animal is ashore. C = cementum; R = rest line; 1, 2, etc. = annual moult; P = pupping.

desired; this took less than 10 min. The stain was fixed in 10% sodium thiosulphate solution for about 10 min and the tooth was washed in tap water for one hour.

Silver nitrate stains the calcium phosphate of dentine a blackish colour in proportion to its density. They found that bands laid down while animals were ashore breeding, moulting or 'resting' were darker than bands laid down at sea. The same effect of dark and light bands, dark bands representing haul-out periods, is found using unstained sections viewed with transmitted light, either cut surfaces or thin sections; staining is not necessary and is not now used.

Age determination, interpretation The life cycle of the elephant seal is more complex than in other Antarctic seals and this is reflected in the structure of the dentine. The most constant feature in the life cycle of elephant seals is the annual moult. Elephant seals moult on land where they remain fasting until the moult is completed, although males sometimes make short trips to sea (Laws, 1956; Carrick *et al.*, 1962). Immature animals of both sexes do not come ashore during the breeding season (September to mid-November) but moult shortly afterwards, from mid-November to January. Most immatures haul out to 'rest' once or twice at other times of the year and there is a pronounced autumn–winter haul-out of immatures from April to June (Carrick *et al.*, 1962). 'Accessory lines' or 'resting lines' in the dentine represent such haul-outs. On reaching maturity there is a change in haul-out pattern. Females which have become pregnant for the first time moult with the immatures in December but then stay at sea until the following October when they come ashore to pup. The moult is then delayed until January or February. This pattern is then maintained. The change in habits is represented by a change in tooth structure. In immature years the winter haul-out line is roughly mid-way between the preceding and succeeding moult lines. In mature females the line representing pupping is narrow and discrete and is very close to the succeeding moult line. This pattern of two close lines, the pupping line being narrower than the succeeding moult line, is then maintained (Figs 11.7b, 11.8).

In males the pattern is similar. Immature males have an autumn–winter haul-out represented by accessory lines. This is omitted in mature animals and a pattern representing the breeding haul-out in October and the moulting haul-out in March is established. The breeding and succeeding moult haul-out lines are not as close together as they are in females.

From the foregoing it can be seen that not only the age of a specimen but also the age at which it establishes the mature pattern of haul-out, that is, pupping in females and attendance during the breeding season in males, can be determined from a tooth section.

Antarctic fur seal - Arctocephalus gazella

Age determination in the Antarctic fur seal has been described in detail by Payne (1978). Payne had a reference collection of teeth from animals whose age, ranging from 1–11 years, was known from tags applied to them as pups.

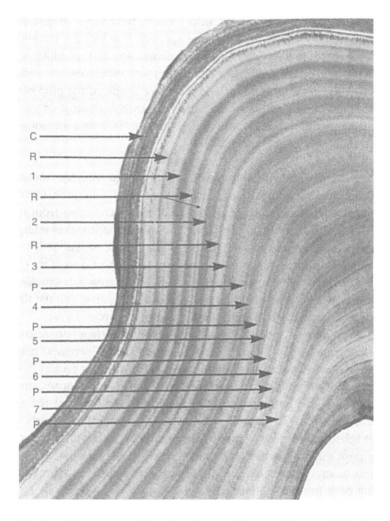

Fig. 11.8 Transverse section of canine tooth from a 14-year-old female elephant seal, viewed (×33) in transmitted light. C – cementum; R = rest line; 1, 2, etc. = annual moult; P = pupping.

Choice of tooth In males the pulp cavity of the canine tooth remains open throughout life, allowing the continual deposition of dentine to take place. The lower canines are flattened laterally and are very curved which makes sectioning and reading difficult. Upper canines should be collected. In females upper or lower canines can be used.

Tooth preparation Both whole teeth and longitudinal thin sections can be used to determine age. For thin sections the entire tooth is mounted on a wooden block with dental wax and the block is clamped in position on the sliding saw table. The tooth is cut root first and a section produced which should include the entire length of the pulp cavity. The sections can be examined in water or alcohol/glycerol or can be permanently mounted. The use of a polarizing filter on a stereomicroscope considerably improves readability.

Age determination, interpretation The simplest means of age determination is to count the external ridges on the root. Over 90% of females less than seven-years-old can be correctly aged from the ridges (Payne, 1978). In males the larger size of the tooth makes interpretation easier and most teeth less than nine-years-old and occasionally teeth from animals up to 12-years-old can be read from the ridges (Fig. 11.2).

The ridge marking the end of the first year's growth is usually less pronounced than subsequent ridges in teeth from males, and the first major ridge is normally that marking the end of the second year.

If age cannot be determined by examination of ridges on the root, a thin section must be cut (Fig. 11.9). Within each annual increment many layers are visible with various discontinuities. Deposition is greater in early life and with the changing pattern of deposition, interpretation can be complex. Frequently the most useful approach is to compare the groups of internal lines with the external annual ridges, which are clearly visible in longitudinal sections. Rotating the polarizing filter often renders annual groups more discernible. In older animals the regular pattern of deposition of later years can often make counting relatively simple and by working back can help to elucidate the structure of younger years.

The primary dentine has a homogeneous appearance and lacks the lines of the secondary dentine so that a neonatal line is usually clearly visible. Frequently the suckling lines, which lie immediately subsequent to the neoneatal line, can be seen (see Fig. 11.9; Scheffer & Peterson, 1967; Payne, 1978). Annual groups of suckling lines can also be seen in adult female teeth (Fig. 11.10), representing the regime of feeding and fasting during the pup-rearing period. The number of feeding trips made in a season directly relates to their duration (Bengtson, 1988) and this is related to food availability (Doidge, McCann & Croxall, 1986), so that examination of the suckling lines can give retrospective information on foraging activity (Bengtson, 1988).

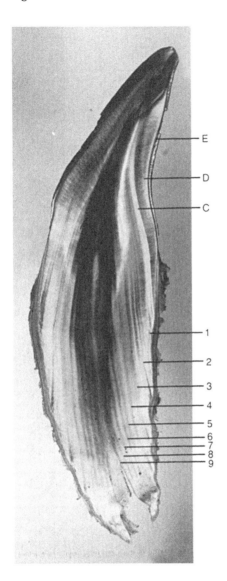

Fig. 11.9 Transverse section of canine tooth from a 9-year-old female Antarctic fur seal. C = cementum; D = dentine; E = enamel; 1–9 = successive annual increments in the dentine. (Photograph courtesy J.L. Bengston.)

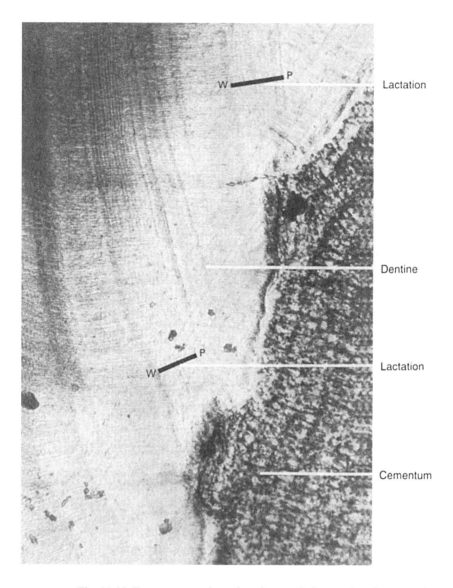

Fig. 11.10 Transverse section of canine tooth from a female Antarctic fur seal showing annual groups of suckling lines. P = pupping; W = weaning. (Photograph courtesy J.L. Bengtson.)

In females the pulp cavity closes at about 15 years of age and no more dentine is deposited, leaving the cementum as the only means of age determination. Interpretation of the cementum is straightforward and with a good section it is simple to count the layers, one wide translucent layer and one narrow opaque layer being added each year. However, readable sections are not obtained from every tooth and there seems to be some variation in the age at which the first cementum layer is deposited. To a certain extent this can be determined for each tooth by checking to see how many years' dentine deposition are covered by the first cementum layer. In some cases a tooth may give an apparently clear-cut cementum count which may be two or three years short of its known age or age assessed from dentinal layers. After the pulp cavity closes and no more dentine is deposited, the cementum is the only means of age determination.

Reliability, validation

The determination of a seal's age by examination of its tooth structure is the most precise method available to us; however, the method is not without error. Where known-age specimens are available it is nearly always possible to relate tooth structure by counts of annuli or ridges to the known age of the specimen, that is, to validate the technique. Known-age material for age determination studies has usually come from seals tagged or branded as pups.

Carrick & Ingham (1962) were able to examine teeth from elephant seals of known age and history and this greatly assisted their interpretation of dentinal structure. The other Antarctic phocids have a simpler dentinal structure and cementum structure is more or less the same for all species and interpretation is straightforward. Age determination of Antarctic fur seals was based on examination of known-age material (Payne, 1978) and tetracycline marking has confirmed the interpretation of microdentinal structure in this species (J.L. Bengtson, pers. comm.; BAS, unpub. data). However, teeth of the same age can vary in their structure and problems inevitably arise in the examination of unknown age teeth. (See also chapter 4, section 'Vital stains'.)

Anas (1970) quantified the error involved in assigning an age to known-age teeth of northern fur seals. He noted that among four readers all made errors and that the error increased with increasing age of the tooth, but that the most experienced reader made the fewest errors. Few published accounts of age determination deal with the error involved in determining the age from teeth. Stewart & Lavigne (1979) noted the lack of critical reference to possible errors associated with estimating age from tooth sections. They presented a standardized technique of repeated blind readings of harp seal teeth and a statistical basis for deleting mistakes in readings ('outliers'). This involved repeated blind readings of the teeth, using three readings when the results were the same for each reading. If the three readings were not the same the mean and 95% confidence intervals were calculated from five readings. Individual readings beyond the 95% confidence intervals ('outliers') were discarded as reading errors. The final age estimate was the median of the remaining values. The technique of multiple blind readings is recommended. Bowen *et al.* (1983) discussed validation of age estimation in the harp seal using known-age samples and described the errors involved. Dapson (1980) presented guidelines for statistical usage in age determination studies and suggested that in addition to a detailed methodology and interpretation of annuli, reports should include estimates of accuracy and should not assume that counts represent a faultless age estimation procedure.

The above cautions notwithstanding, age determination by reference to tooth structure is of wide applicability and utility.

Nail markings

The seasonal changes in the growth and deposition of material in teeth is also reflected in changes in the nails of seals. These have been correlated with age in a number of species (Plehanov, 1933; Laws, 1953a, 1962; McLaren, 1958; Hofman, 1975 in Bengtson & Siniff, 1981). Bands in the nails appear dark and light, a pair of bands denoting one year's growth, and slight ridges usually occur on the outside of the claw. Wear at the tip of the nail gradually eliminates bands so that in older animals claw markings do not reveal the true age. The age at which this occurs varies between species and between individuals of a given species but the method has some use in younger animals. A major advantage is that nails can be drawn from live animals.

Bengtson & Siniff (1981) compared ages in crabeater seals determined from cementum annuli and nail markings (Fig. 11.11). The nails were dried for storage then softened in a 0.025% solution of potassium hydroxide. Softened nails were sectioned longitudinally by hand into sections approximately 0.5 mm thick, then mounted with 'Permount' on a glass slide with a coverslip. Results from the two techniques were 'somewhat similar' in the first two years but after that were often widely divergent. They concluded that nails were a poor indicator of age in crabeater seals. The

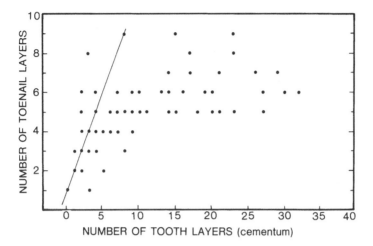

Fig. 11.11 Relationship between number of tooth layers to toenail layers in female crabeater seals. If the two methods were equally accurate in estimating age, their estimates would lie along the straight line. (From Bengtson & Siniff, 1981.)

technique may be more successful in Weddell seals (W. Testa, pers. comm.).

Laminated bones

Laminae in the dentary of seals were reported by Chapskii (1952), and Laws (1953a) recorded laminae in the tympanic bullae of elephant seals which correlated with age determined from the teeth.

Such laminae have since been discovered in the bones (principally the dentary) of a wide variety of mammals and the techniques for preparing and interpreting such structures are extensively reviewed by Klevezal & Kleinenberg (1967).

Bone resorption, which eliminates some layers, is known to occur but affects different bones in different ways. The method might be used as an alternative, or in addition to tooth layers but the validity needs to be established for each species.

Body length

Seals grow in length throughout life so that body size is an indicator of relative age. However, growth occurs at a progressively decelerating rate and the variation within a year class, and the overlap between year classes, especially among older animals, is so great that body length cannot be used reliably to assign a seal to a particular age (Laws, 1953b; Bryden, 1972; Bengtson & Siniff, 1981).

In the field it may be necessary to assign relative ages, for example in behavioural studies, and body length can be useful, especially if combined with other features, such as scars, which accumulate throughout life.

Eye lens weight

As with body length, the weight of the eye lens increases throughout life in an individual. Aging animals by measuring eye lens dry weight does not seem to have been tested in Antarctic seals but has been tried in northern fur seals and harp seals. Bauer, Johnson & Scheffer (1964) examined lenses of northern fur seals aged 72 days to about 21 years and concluded that they were of little value in estimating age beyond the second year. Stewart (1983) found a correlation between age and average lens weight in harp seals, but the predictive equation had wide confidence intervals. For example, a lens weight of 7.2 g indicated an age of 0–15.9 years. Similar variation is likely to be found in Antarctic seals and the method cannot be recommended.

Suture closure

The sutures between the major bones of the skull close progressively as the animal ages and a number of early studies used this as a means of grading animals on the basis of 'suture age'. Laws (1962) gives a review of such work up to that date. Study of known-age animals has shown suture closure to be very variable (Scheffer & Wilke, 1953; King, 1972) and is an indication of age within wide limits only, and this probably applies to most, if not all, other species. King (1972), in a detailed study of phocid skulls, found no signs of fusion of the sutures used to estimate 'suture age' in a known-age series of male elephant seals, and partial fusion only of one suture in females.

Baculum development

The baculum increases in length and weight with age and there is a thickening of the proximal end, becoming especially pronounced at maturity. Hewer's (1964) detailed study of the grey seal showed that baculum development was a rough guide to age, particularly in younger animals, and was a good guide to sexual maturity. Once sexual maturity was attained individual variation was too great to be of any use in age determination. Also the method is only applicable to males.

Ovarian structure

After ovulation the corpus luteum eventually regresses to form a hard fibrous body, the corpus albicans. In whales corpora albicantia persist throughout life so that a count of corpora albicantia, divided by the annual accumulation rate, plus the number of years taken to reach sexual maturity, is equal to the age of the whale in years (Laws, 1961).

In seals the corpora albicantia persist for a variable number of years (depending upon species) as hard scars of a whitish or brownish colour, but eventually become indistinguishable from the other tissues of the ovary. Thus a balance is reached between production and resorption. In addition to this fundamental limitation as a means of ageing seals the method can only be used on mature females. If another reliable method is used to estimate the age (e.g. teeth) or if the age is known, corpora accumulation may be used in some circumstances to estimate the age at which the animal first ovulated (see chapter 12).

Pelage and general appearance

Pelage can be useful in distinguishing pups from older age groups but there do not appear to be any gradual changes in adult pelage which could be used for age determination. The lanugo of Antarcic phocids ranges from the light grey-brown of crabeater and Weddell seals to the black of elephant seals. The moult takes place in the latter half of the lactation period, which ranges from three to six weeks in different species, after which the adult pelage of shorter, stiffer hairs appears. Antarctic fur seals have a black, shiny lanugo and begin to moult this at about seven weeks, moult being complete by about three months.

The adult pelage in this species darkens with age, with males darkening more quickly than females. Pelage colour and body size can be used to age juvenile male fur seals fairly accurately up to about 4 years. However, this can only be done with much experience and a reference set of tagged animals of known age. In elephant seals relative ages could, with experience, be assessed by general appearance, particularly degree of scarring (Laws, 1953b), and primiparae were fairly successfully distinguished from

multiparae by the same means (McCann, 1982), but accurate age determination by these means is not possible.

Acknowledgements

I thank R.M. Laws, I.L. Boyd and J.P. Croxall for constructive comments on the manuscript. For supplying information on particular techniques of age determination in particular species I thank J.L. Bengtson,
M.N. Bester, R.M. Laws, B.R. Mate, R.E.A. Stewart, I. Stirling and
J.W. Testa.

References

Anas, R.E. 1970. Accuracy in assigning ages to fur seals. *J. Wildl. Mgmt.* **34**,
844-52.
Bauer, R.D., Johnson, A.M. & Scheffer, V.B. 1964. Eye lens weight and age
in the fur seal. *J. Wildl. Mgmt.* **28**, 374-6.
Bengtson, J.L. 1988. Long-term trends in the foraging patterns of female
Antarctic fur seals at South Georgia. In *Antarctic Ocean Resources and
Variability*, ed. D. Sahrhage, pp. 286-91. Springer-Verlag, Berlin.
Bengtson, J.L. & Laws, R.M. 1985. Trends in crabeater seal age at maturity:
an insight into Antarctic marine interactions. In *Antarctic Nutrient Cycles
and Food Webs*, ed. W.R. Siegfried, P.R. Condy & R.M. Laws
pp. 669-75. Springer-Verlag, Berlin.
Bengtson, J.L. & Siniff, D.B. 1981. Reproductive aspects of female crabeater
seals (*Lobodon carcinophagus*) along the Antarctic Peninsula. *Can. J.
Zool.* **59**, 92-102.
Bowen, W.D., Sergeant, D.E. & Øritsland, T. 1983. Validation of age
estimation in the Harp Seal, *Phoca groenlandica*, using dentinal annuli.
Can. J. Fish. Aquat. Sci. **40**, 1430-41.
Bryden, M.M. 1972. Body size and composition of elephant seals (*Mirounga
leonina*): absolute measurements and estimates from bone dimensions. *J.
Zool., Lond.* **167**, 265-76.
Carrick, R. & Ingham, S.E. 1962. Studies on the southern elephant seal,
Mirounga leonina (L.). II. Canine tooth structure in relation to function
and age determination. *C.S.I.R.O. Wildl. Res.* **7**, 102-18.
Carrick, R., Csordas, S.E., Ingham, S.E. & Keith, K. 1962. Studies on the
southern elephant seal, *Mirounga leonina* (L.). III. The annual cycle in
relation to age and sex. *C.S.I.R.O. Wildl. Res.* **7**, 119-60.
Chapskii, K.K. 1952. [Determination of the age of some mammals from the
microstructure of the bones.] *Izv. Estestv.-Nauchn. in-ta im. P.F.
Lesgafta.* **25**, 47-66.
Christensen, I. 1973. Age determination, age distribution and growth of
bottlenose whales *Hyperoodon ampullatus* (Forster), in the Labrador Sea.
Norwegian J. Zool. **21**, 331-40.
Dapson, R.W. 1980. Guidelines for statistical usage in age-estimation
techniques. *J. Wildl. Mgmt.* **44**, 541-8.

Doidge, D.W., McCann, T.S. & Croxall, J.P. 1986. Attendance behaviour of Antarctic fur seals. In *Fur Seals: Maternal Strategies on Land and at Sea*, ed. R.L. Gentry & G.L. Kooyman, pp. 102-14. Princeton University Press, Princeton.

Fancy, S.G. 1980. Preparation of mammalian teeth for age determination by cementum layers: a review. *Wildl. Soc. Bull.* **8**, 242-8.

Fisher, H.D. & Mackenzie, B.A. 1954. Rapid preparation of tooth sections for age determinations. *J. Wildl. Mgmt.* **18**, 535-7.

Grue, H. & Jensen, B. 1973. Annular structure in canine tooth cementum in red foxes (*Vulpes vulpes* L.) of known age. *Dan. Rev. Game Biol.* **8**, 1-12.

Hewer, H.R. 1964. The determination of age, sexual maturity, longevity and a life-table in the grey seal (*Halichoerus grypus*). *Proc. Zool. Soc. Lond.* **142**, 593-624.

Hofman, R.J. 1975. *Distribution Patterns and Population Structure of Antarctic Seals*. Ph.D. Thesis, University of Minnesota, Minneapolis.

Jonsgard, A. 1969. Age determination of marine mammals. In *The Biology of Marine Mammals*, ed. H.T. Andersen, pp. 1-30. Academic Press, New York.

King, J.E. 1972. Observations on phocid skulls. In *Functional Anatomy of Marine Mammals*, ed. R.J. Harrison, pp. 81-115, Academic Press, London.

Klevezal, G.A. & Kleinenberg, S.E. 1967. [Age determination of mammals by layered stucture in teeth and bones.] *Fish. Res. Bd. Can.*, Translation Series No. 1024, 1969.

Laws, R.M. 1952. A new method of age determination for mammals. *Nature* **169**, 972-4.

Laws, R.M. 1953a. A new method of age determination in mammals with special reference to the elephant seal (*Mirounga leonina*, Linn.). *F.I.D.S. Sci. Rep.* No. 2. pp. 1-12.

Laws, R.M. 1953b. The elephant seal (*Mirounga leonina* Linn.). I. Growth and age. *F.I.D.S. Sci. Rep.* No. 8. pp. 1-62.

Laws, R.M. 1956. The elephant seal (*Mirounga leonina* Linn.). II. General, social and reproductive behaviour. *F.I.D.S. Sci. Rep.* No. 13, pp. 1-88.

Laws, R.M. 1957. On the growth rates of the leopard seal, *Hydrurga leptonyx* (De Blainville, 1820). *Saugetierk. Mitt.* **5**, 49-55.

Laws, R.M. 1958. Growth rates and ages of crabeater seals, *Lobodon carcinophagus* (Jacquinot & Pucheran). *Proc. Zool. Soc., Lond.* **130**, 275-88.

Laws, R.M. 1960. The Southern Elephant seal (*Mirounga leonina* Linn.) at South Georgia. *Norsk. Hvalf-Tid.* **49**, 466-76, 520-42.

Laws, R.M. 1961. Reproduction, growth and age of southern fin whales. *Discovery Rep.* **31**, 327-486.

Laws, R.M. 1962. Age determination of pinnipeds with special reference to growth layers in the teeth. *Z. Saugetierk.* **27**, 129-46.

Low, W. & Cowan, I. McT. 1963. Age determination of deer by annular structure of dental cementum. *J. Wildl. Mgmt.* **27**, 466-71.

Mansfield, A.W. 1958. The breeding behaviour and reproductive cycle of the Weddell seal (*Leptonychotes weddelli* Lesson). *F.I.D.S. Sci. Rep.* No. 18, pp. 1-41.

Marks, S.A. & Erickson, A.W. 1966. Age determination in the black bear. *J. Wildl. Mgmt.* **30**, 389-410.

Mattlin, R.H. 1978. *Population Biology, Thermoregulation and Site Preference of the New Zealand Fur Seal*, Arctocephalus forsteri *(Lesson, 1828), on the Open Bay Islands, New Zealand.* Ph.D. Thesis, University of Canterbury, Christchurch.

McCann, T.S. 1981. *The Social Organization and Behaviour of the Southern Elephant Seal*, Mirounga leonina *(L.).* Ph.D. Thesis, University of London, London.

McCann, T.S. 1982. Aggressive and maternal activities of female Elephant seals (*Mirounga leonina*). *Anim. Behav.* **30**, 268–76.

McLaren, I.A. 1958. The biology of the ringed seal (*Phoca hispida* Schreber) in the eastern Canadian Arctic. *Fish. Res. Bd. Can. Bull.* No. 118, pp. 1–97.

Miller, F.L. 1974. Age determination of caribou by annulations in dental cementum. *J. Wildl. Mgmt.* **38**, 47–53.

Morris, P. 1972. A review of mammalian age determination methods. *Mammal Rev.* **2**, 19–104.

Morris, P. 1978. The use of teeth for estimating the age of wild mammals. In *Development, Function and Evolution of Teeth*, ed. P.M. Butler & K.A. Joysey, pp. 483–94. Academic Press, New York.

Novakowski, N.S. 1965. Cemental deposition as an age criterion in bison, and the relation of incisor wear, eye-lens weight, and dressed carcass weight to age. *Can. J. Zool.* **43**, 173–8.

Payne, M.R. 1978. Population size and age determination in the Antarctic fur seal *Arctocephalus gazella*. *Mammal Rev.* **8**, 67–73.

Perrin, W.F. & Myrick, A.C. (Eds) 1980. *Age Determination of Toothed Whales and Sirenians*. Rep. Int. Whal. Comm. Spec. Issue No. 3.

Plehanov, P. 1933. [The determination of age in seals]. *Sovetskii Sever* **4**, 111–14.

Scheffer, V.B. 1950. Growth layers on the teeth of Pinnipedia as an indication of age. *Science* **112**, 309–11.

Scheffer, V.B. & Myrick, A.C. 1980. A review of studies to 1970 of growth layers in the teeth of marine mammals. In *Age Determination of Toothed Whales and Sirenians*, ed. W.F. Perrin & A.C. Myrick, pp. 51–63. Rep. Int. Whal. Comm. Spec. Issue No. 3.

Scheffer, V.B. & Peterson, R.S. 1967. Growth layers in teeth of suckling fur seals. *Growth* **31**, 35–8.

Scheffer, V.B. & Wilke, F. 1953. Relative growth in the northern fur seal. *Growth* **17**, 129–45.

Spinage, C.A. 1973. A review of the age determination of mammals by means of teeth, with special reference to Africa. *East Afr. Wildl. J.* **11**, 165–87.

Stewart, R.E.A. 1983. *Behavioural and Energetic Aspects of Reproductive Effort in Female Harp Seals*, Phoca groenlandica. Ph.D. Thesis, University of Guelph, Guelph.

Stewart, R.E.A. & Lavigne, D.M. 1979. *Age Determination in Harp Seals. International Workshop on Biology and Management of Harp Seals.* Working paper HS/WP10 736. University of Guelph, Guelph, Ontario, Canada.

Stirling, I. 1969. Tooth wear as a mortality factor in the Weddell seal, *Leptonychotes weddelli*. *J. Mammal.* **50**, 559–65.

Stirling, I., Archibald, W.B. & DeMaster, D. 1977. Distribution and abundance of seals in the Eastern Beaufort Sea. *J. Fish. Res. Bd. Can.* **34**, 976–88.

Stone, W.B., Roscoe, D.E. & Weber, B.L. 1975. Use of Romanowski stains to prepare tooth sections for aging mammals. *N.Y. Fish Game J.* **22**, 156-8.

Turner, J.C. 1977. Cemental annulations as an age criterion in North American sheep. *J. Wildl. Mgmt.* **41**, 211-17.

Note added in proof:

In a recent study, Arnbom *et al.* (1992) showed that teeth could be extracted from live Antarctic fur seals and southern elephant seals under anaesthesia (Baker *et al.*, 1990; Boyd *et al.*, 1990) and that growth layers in these teeth could provide an estimate of age. In Antarctic fur seals, post-canine teeth can be extracted from females and incisors from males whereas, in southern elephant seals, only incisors can be extracted as other teeth are too deeply rooted. Ten mg of the local anaesthetic lignocaine should be administered by injection close to the root of the tooth. Dental elevators are then used to loosen the tooth from the alveolus and periodontal ligament. Removal of teeth by this method takes 1-2 min and the seal should then be given a single dose of a long-acting formulation of an antibiotic such as oxytetracycline. Seals which have been recaptured at intervals from a few days to 4 years after removal of a tooth have shown no signs of infection or distress related to the removal of the tooth.

Arnbom, T.A., Lunn, N.J., Boyd, I.L. & Barton, T. 1992. Aging live Antarctic fur seals and southern elephant seals. *Mar. Mamm. Sci.* **8**, 37-43.

Baker, J.R., Fedak, M.A., Anderson, S.S., Arnbom, T. & Baker, R. 1990. Use of tiletaminezolazepam mixture to immobilise wild grey seals and southern elephant seals. *Vet. Rec.* **126**, 75-7.

Boyd, I.L., Lunn, N.J., Duck, C.D. & Barton, T. 1990. Response of Antarctic fur seals to immobilization with ketamine, a ketamine-diazepam or ketamine xylazine mixture and Zoletil. *Mar. Mamm. Sci.* **6**, 135-45.

12
Reproduction

R.M. LAWS AND A.A. SINHA

Introduction

An understanding of the salient features of the reproductive cycle in seals is of interest not only to the specialist, but also to workers in other fields, and for comparative purposes. The complex social behaviour shown by many seals is interpretable only in the light of the reproductive status of the animals concerned; an understanding of seal population dynamics can be reached only if the reproductive success of the members of the population is known; population changes can be monitored by observing changes in the mean age of puberty. These examples could be further enlarged, but the undoubted importance of the subject is attested by the large volume of work upon it. This chapter first provides an introduction to the reproductive cycle and the organs of reproduction which should enable newcomers to the field to find their way about and then describes some methods for studying foetal growth, estimating pregnancy rate and observations of birth and lactation. A method of establishing paternity particularly in relation to pups, by DNA fingerprinting is described. Puberty and sexual maturity are defined and a method described for calculating the mean age at sexual maturity. Finally, some additional field techniques and basic methods of analysis are described.

Although reproductive parameters have been established for some populations of certain species for a number of years it should be emphasized that there is considerable variation between populations or with latitude, and within unit populations there may be changes with time. Further studies are necessary.

The annual cycle

Antarctic seals show a very regular annual cycle imposed by the extreme seasonality of the polar environment. The principal environmental factors affecting them are the fast ice, the pack ice, its seasonal advance and retreat and the effects of these on primary and secondary production – and so on availability of food for the seals, which in turn affects their annual reproductive cycle.

The typical female Antarctic phocid seal hauls out in the spring on ice or land to give birth to a single pup. Birth is usually five to seven days after haul-out. Pupping is synchronous, most births occurring within a spread of about a month during September to November (Fig. 12.1). Fur seals haul-out a little later but pupping is equally synchronous.

The Antarctic phocids have a short lactation period of three to six weeks. At this time the lactating female fasts completely (elephant, crabeater, leopard seals) or partially (Weddell seal). In the fur seal, lactation extends over several months and the female makes a series of feeding trips to sea, each lasting about a week, with two to three days' suckling in between. In both phocids and otariids, immediately following parturition there is a rapid regression of the corpus luteum of pregnancy in one ovary, accompanied by rapid growth of follicles in the opposite ovary, one of which ovulates to initiate the next cycle.

Ovulation is at about 19 days post-partum in the elephant seal, about 48 days post-partum in the Weddell and leopard seals, and eight days post-partum in the fur seal; the timing of this event in the other three species is not yet known. The ovulated follicle rapidly develops into a corpus luteum and the ovum passes down the fallopian tube to the uterus. If fertilization has occurred the ovum then develops into a blastocyst, but initially no further. There is a delay of 2–4.5 months, according to the species, during which the blastocyst enlarges slowly. The corpus luteum, having attained the size of a walnut, shows no further macroscopic change until the embryo implants into the tissue lining the uterus, but then it enlarges and persists throughout the remainder of pregnancy; it may continue to grow in size. If fertilization has not occurred, the corpus luteum probably persists for a short time and then rapidly regresses to form a scar, the corpus albicans. In some, possibly all, species implantation of the embryo is correlated with the end of the moult period. In a maximally reproducing population the whole cycle from one parturition to the next occupies approximately 12 months, but the period of active foetal growth occupies only seven to eight months. The left and right ovaries and uterine horns alternate in function. (Fuller details are presented in Amoroso *et al.*, 1965; Laws, 1956a, b, 1984.)

The diagram (Fig. 12.2) illustrates the annual cycle of the southern elephant seal. The major variations shown by the other species are in the

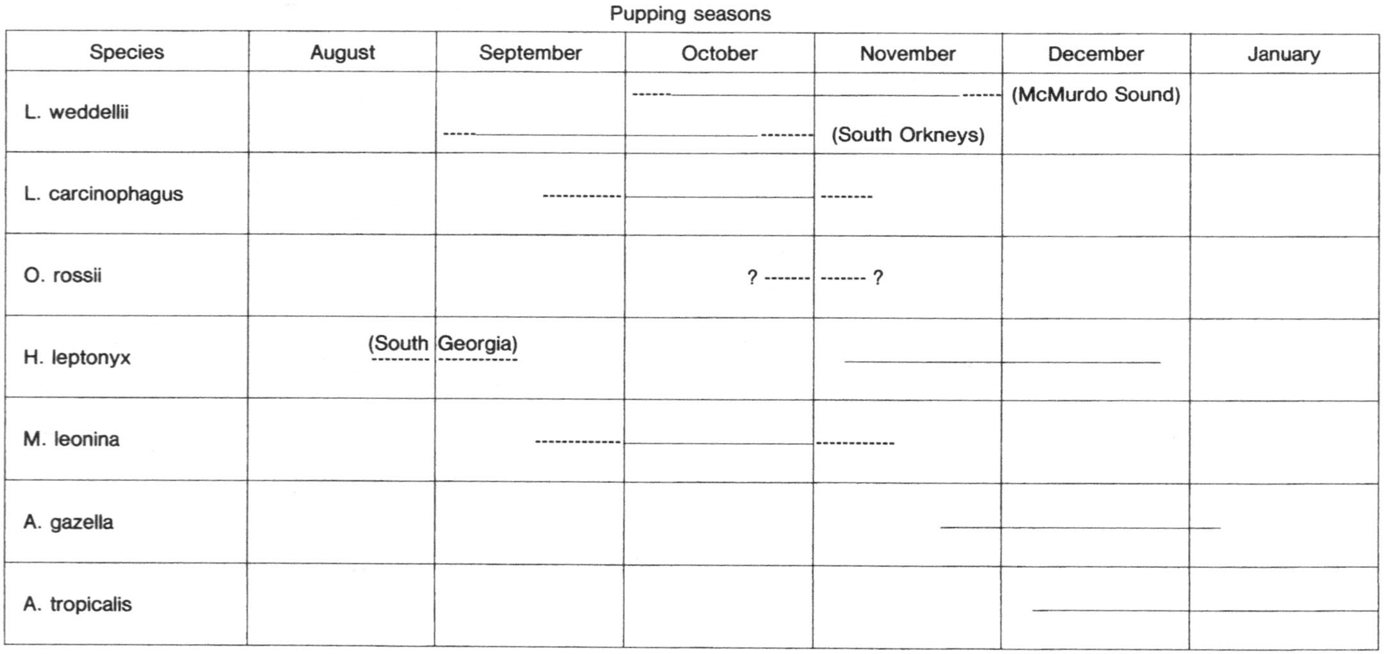

Pupping seasons

Species	August	September	October	November	December	January
L. weddellii			(South Orkneys)		(McMurdo Sound)	
L. carcinophagus						
O. rossii			?	?		
H. leptonyx	(South Georgia)					
M. leonina						
A. gazella						
A. tropicalis						

Fig. 12.1 Pupping seasons for seven species of Antarctic seals. (From Laws, 1984.)

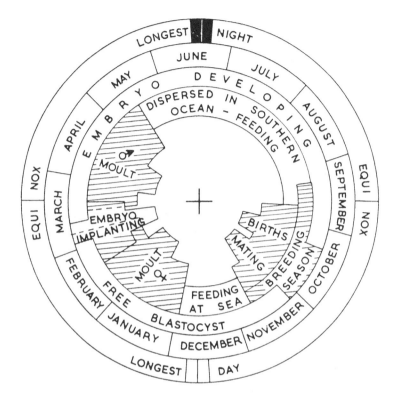

Fig. 12.2 Diagram to illustrate the annual cycle of adult southern elephant seals at South Georgia. (From Laws, 1960.)

length of the periods from birth to mating, and from conception to implantation (Laws, 1984). Other species do not have such obvious or extensive moults as the elephant seal. The separate age groups within the female population may have slightly different annual chronologies. Very little is known of the cycle in females which for one reason or another have not produced pups during the breeding season.

The reproductive cycle of the male is closely correlated with that of the female, but males are reproductively active (as judged by the presence of active sperm in their reproductive tract) over a much longer period in the breeding season than the duration of oestrous in individual females. In the highly polygynous elephant seals and fur seals, adult males haul-out at the beginning of the breeding season before the first females come ashore. They are present on the beaches until after the last adult female has left, although there seems to be a tendency for younger males to replace older ones at the end of the season.

Agonistic behaviour during the breeding season is characteristic of most seals, but is most highly developed in those that breed in large aggregations. The correlation between sexual dimorphism, male intra-specific aggression and polygyny has been described by Bartholomew (1970).

The male reproductive system

Gross anatomy

The reproductive system of pinnipeds is the basic carnivore pattern with a few accessory glands. Accounts of the general anatomy have been given by Harrison, Matthews & Roberts (1952), Laws (1956b), Amoroso *et al.* (1965), Harrison (1969).

Testes

In phocid seals the testes are situated in the groin, external to the abdominal muscles, but covered by the posterior part of a superficial muscle (the paniculus carnosus), at about the level of the knee joint, or slightly posterior to it. In young seals, a prominent lymph node lying immediately below and medial to the testis can easily be mistaken for it. The testes of otariid seals are generally said to be scrotal, but in the Antarctic fur seal at least, this is true only for part of the time. Usually the testes are withdrawn from the scrotum into the inguinal position, descending only when the seal is suffering from heat stress as a result of sustained activity such as territorial fighting, or on still, bright, sunny days. In the inguinal position the testes are situated as in the phocids. When fur seal testes are withdrawn, the scrotum is visible externally as two areas of hairless skin, lateral to and slightly in front of the anus. Except when occupied by the testes, the scrotum is not pendulous.

The testes are ovoid, circular in cross section and twice as long as broad. They are enclosed in a glistening white tunica albuginea, the posterior part of which adheres to the tunica vaginalis. The duct of the testis, the epididymis, lies along the side of the testis. The ductus deferens arises from the epididymis on the inner border of the testis and passes, together with the testicular ligament and the blood vessels, through an opening – the processus vaginalis – into the abdominal cavity. It then runs forwards and medially until it loops around the obliterated hypogastric artery and passes dorsally to the bladder to enter the prostate gland on its dorsal surface. The testis weight increases with age (Fig. 12.3).

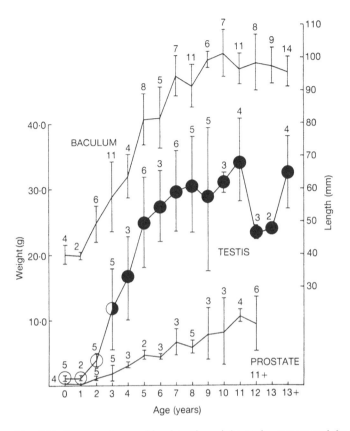

Fig. 12.3 Increase in combined testis weight and prostate weight with age in male sub-Antarctic fur seals collected during the breeding period, and increase in baculum length with age. (Open circles: spermatozoa absent; closed circles: spermatozoa present; divided circles: some males without spermatozoa.) (From Bester, 1990, courtesy Zoological Society of London.)

Prostate

The prostate surrounds the neck of the bladder, just inside or slightly in front of the pubic arch. It is quite a bulky gland, weighing up to 760 g in the elephant seal and 27 g in the Antarctic fur seal. The prostate weight increases with age (Fig. 12.3). Its anatomy is basically the same in all seals.

Penis and baculum

The penial opening lies about two-thirds of the distance from the anus to the umbilicus (Fig. 12.4). The crura penis ligaments are attached broadly over the posterior border of the pubis and immediately turn

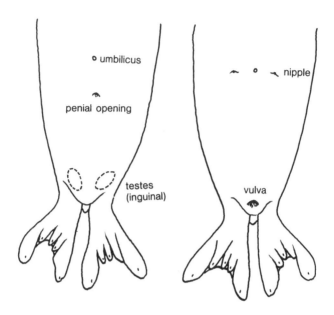

Fig. 12.4 Appearance of male and female genital openings.

downwards and forwards to fuse and form a forwardly-directed corpus cavernosum penis. The structure of the protrusible part of the penis is distinctly different in phocids and otariids.

In phocids, the baculum or os penis, the ossified distal part of the corpus cavernosum penis, does not reach to the end of the penis but is surmounted by a fleshy claviform glans. The baculum itself is irregular in outline and formed of cancellar bone with little obvious sculpturing. The dorsal surface is often keeled and the ventral surface flattened or grooved to receive the urethra. Hamilton (1939) has described the changes in shape of this bone in the leopard seal, and Laws (1956a) and Mansfield (1958) have figured the baculum in the elephant seal and Weddell seal, respectively.

In otariids the baculum has a more complex shape than in phocids. In the Antarctic fur seal it consists of a slightly curved shaft of compact bone with an expanded proximal end, roughened where the fibrous elements of the corpus cavernosum penis are attached, and a bifurcated distal end which lies immediately beneath the distal end of the glans and is covered by a thin layer of epithelium. The urethral surface of the concave side is grooved or flattened. The structure of the glans is highly characteristic – it forms a hollow collar of erectile tissue, in the centre of which the baculum forms a conical prominence.

The baculum of seals grows rapidly during life, often with a sudden increase of length at the attainment of sexual maturity and the size of the bone has been used as an index of age or maturity (Hamilton, 1939; Laws, 1956a; Bester 1990; see also Fig. 12.3). Mansfield (1958) was unable to find any correlation between baculum length and body length in the Weddell seal.

Fine anatomy

Testis and epididymis

Harrison *et al.* (1952), Laws (1956b), Mansfield (1958) and Harrison (1969) have described the histology of the testis and epididymis of Antarctic phocids. The most detailed general account is by Laws (1956b) on the southern elephant seal; he described the development of the tubules and interstitial tissue up to the attainment of sexual maturity, and the yearly cycle in the mature male in which seven phases of tubule activity were described (Fig. 12.5). Tubules attain a maximum diameter of 220 μm during the breeding season and shrink to 140 μm in winter. Sinha, Erickson & Seal (1977a,b) have described the ultrastructure of seminiferous tubules, Leydig cells and epididymis of crabeater, leopard and Ross seals. Interstitial cells of the testis enlarge during the breeding season, but regress and often degenerate after breeding is over (Sinha *et al.*, 1977a). Harrison *et al.* (1952) and Mansfield (1958) showed peak activity (as indicated by testis and seminiferous tubule diameter) in the testes of Weddell and crabeater seals, during October. Sinha *et al.* (1977a) noted associated changes in the vascularization of the tunica albuginea at this time. There is also good evidence of seasonal activity in the sub-Antarctic fur seal (Bester, 1990; Fig. 12.6).

Laws (1956b) described the cycle of the epididymis in the southern elephant seal. Changes in the epididymis tubules are correlated with those in the testis and were attributed to the same hormonal stimuli as those affecting the testis and spermatogenesis. Mansfield (1958) could find no variation in the structure of the epididymis, other than the height of the epithelium, in a series of Weddell seals taken between August and November. Other references to the epididymis are usually confined to the presence or absence of sperm in the tubules.

Prostate

Published accounts of the histology of the prostate of Antarctic seals appear to be lacking. In the Antarctic fur seal there is a marked cycle in

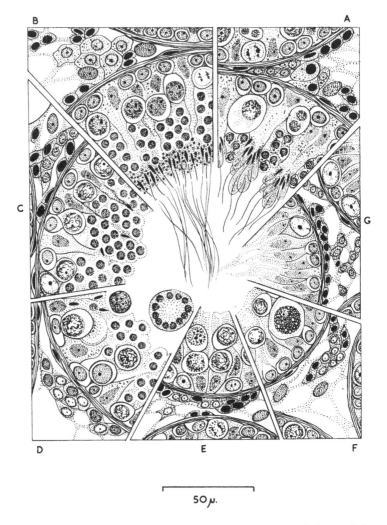

50 µ.

Fig. 12.5 Annual cycle in the histological appearance of the seminiferous tubules in the southern elephant seal. (From Laws, 1956b.)

weight of the prostate which matches that of testis activity. During the
height of the breeding season, in December, the prostate epithelium is
about 330 µm in height and has a lobulated free border from which the
distal parts of the cells are detached in apocrine secretion. During the
moult, in March, the epithelium is cuboidal or only slightly columnar and
about 16 µm high (W.N. Bonner, pers. comm.).

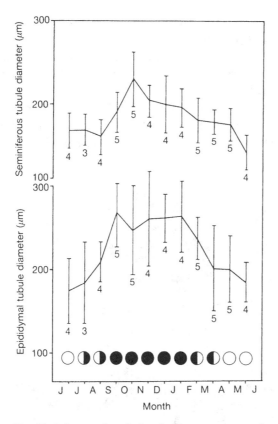

Fig. 12.6 Seasonal variation in the occurrence of spermatozoa in, and outside diameters of, seminiferous tubules in adult male sub-Antarctic fur seals. Sample sizes are given below the means and vertical lines represent one standard deviation. (Open circles: spermatozoa absent; closed circles: spermatozoa present; divided circles: some males without spermatozoa.) (From Bester, 1990, courtesy Zoological Society of London.)

The prostate gland is a highly seasonal accessory organ, since it is largely dependent on testicular androgens for its maintenance. Weighing the prostate may therefore provide a fast and dependable assessment of reproductive state, which could be more sensitive than testis weight.

Female reproductive system

Gross anatomy

As in the male, the reproductive organs of female pinnipeds are typically carnivore in type. They comprise a pair of ovaries and the ducts of the reproductive system. The gross anatomy has been described by Harrison *et al.* (1952) and Laws (1956b). Observations by Sinha & Erickson (1974) have confirmed the generalized features of the reproductive organs for crabeater and leopard seals. Figure 12.7 shows a generalized diagram of the female reproductive tract.

The ovaries lie close to the dorso-lateral wall of the abdominal cavity near the kidneys and are completely enclosed in large peritoneal bursae, which have a small opening near the tip of each uterine horn. The

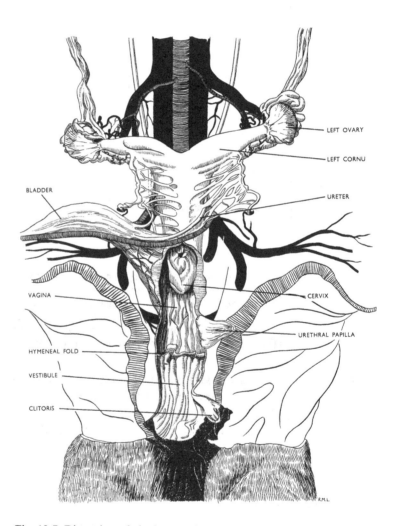

Fig. 12.7 Dissection of elephant seal pup to show the female reproductive organs. (From Laws, 1956b.)

peritoneal opening of the oviduct or fallopian tube is surrounded by complexly folded fimbria, some of which project through a narrow constricted tubo-uterine junction.

The uterus is bicornuate, that is with two horns, but posteriorly they join to form a short common uterus (corpus uterus) and its distal portion, the cervix, projects into the proximal portion of the vagina and opens into the latter through the external os uterus. In the immature animal, the uterus is slender and shows little vascularization. The great distension caused by gestation and birth leaves the parous reproductive tract permanently swollen and with conspicuous blood vessels, particularly on the hind border of the uterine horns. The site of a recent pregnancy is often marked by a placental scar on the lining of the cornu. The vagina is a very tough and fleshy tube, extending from the cervix to the vestibule or urinogenital sinus; the vaginal mucosa is often thrown into deep longitudinal folds. The reproductive organs are attached to the wall of the abdominal cavity by a series of ligaments. The urinary bladder lies ventral to the vagina and can be used as a landmark to identify the right and left sides of the reproductive tract when removed from the carcase. The clitoris, which often contains an os clitoridis, a homologue of the baculum (section 'The male reproductive system'), is also located ventrally.

The mammary glands of seals are abdominal and even when fully active do not appear to distend the general body contour; there is great seasonal variation in the size and extent of the glands depending on whether the female is lactating or not. The nipples lie flush with the surface, but are strongly erected when stimulated by a sucking pup. In Antarctic phocids there is a single pair of nipples (though supernumerary nipples, either singly or as a pair, are not uncommon); fur seals have two pairs of nipples.

The female reproductive organs undergo cyclical changes during the breeding season, but remain regressed in sexually immature and non-pregnant adult animals. The reproductive organs of pinnipeds, as in other mammals, show great variation in size and shape depending on the age, reproductive state and, to a lesser extent, the nutritional state of the animals. It is recommended that the gross morphology of the reproductive organs be carefully recorded before proceeding with the detailed examination of the particular organs under investigation.

The ovary

Detailed descriptions of the ovarian cycle, including histological changes have been provided for Antarctic seals by Harrison *et al.* (1952), Laws

(1956b), Amoroso *et al.* (1965) and Harrison (1969), who have also reviewed the literature. General studies on mammalian ovaries include those of Brambell (1956), Perry (1971) and Harrison & Weir (1977).

General considerations

The size and shape of the ovary depends largely on the age and reproductive status of the female. Because growth of the ovary is generally controlled by pituitary hormones a great deal can be learnt about the hormonal state and functioning of the ovaries by examining the differentiation of the follicles and glandular tissues throughout the annual reproductive cycle. For example the pro-oestrus and oestrus periods are limited to three to six weeks in the year, when follicles in the ovaries develop rapidly and become large Graafian follicles. Usually one, or rarely two, of the Graafian follicles reach the pre-ovulatory stage and ovulate their ova on the ovarian surface. Ova then enter the oviducts with the assistance of the fimbria. As a general rule the ovaries and uterine horns alternate in function.

This sequence usually occurs regularly each breeding season in sexually

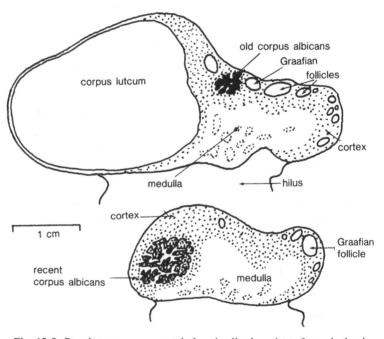

Fig. 12.8 Ovarian structures seen in longitudinal section of a typical pair of ovaries from a pregnant female elephant seal.

mature females. Following ovulation an ovum is fertilized by a sperma-
tozoon to form a zygote; in some cases it may remain as an unfertilized
ovum which eventually degenerates in the uterus. If the female does
not become pregnant the larger follicles and glandular tissues in the
ovary, including the newly-formed corpus luteum, and the uterus,
regress rapidly, usually within one to two weeks, and the female
enters into the prolonged non-breeding season. Much of the ovarian,
uterine and hormonal activities of the animal remain at the lowest
level in non-pregnant sexually mature females during the non-breeding
season.

The ovaries play a central role in the female reproductive phenomena
and are subject to both external and internal regulation. Figure 12.8
illustrates schematically the ovarian structures and functions.

Follicles

Embedded in the connective tissue of the ovary are the follicles, in which
the eggs develop. In pinnipeds, as in most other mammals, there is some
differentiation of small follicles into medium-sized follicles throughout
the year. This activity pattern changes at the onset of the breeding season
in sexually mature animals and in those approaching sexual maturity. At
this time there is an increased growth and differentiation of ovarian
follicles, some of which develop into large Graafian follicles. In the
Graafian follicle the ovum, surrounded by a group of corona cells, lies
bathed in follicular fluid in the cavity of the antrum of the follicle. The
ovum and corona are attached by basal granulosa cells which line
the antrum. Surrounding the follicle and separating it from the stroma
of the ovary is a layer of cells, the theca. Eventually one Graafian follicle,
or rarely more, reaches the pre-ovulatory stage (about 12–15 mm
diameter) and ovulates, probably shortly after oestrus. Immediately after
ovulation the surface of the ovary is marked by a blood-spot at the site
of the ovulated follicle, which later develops into a scar.

Most follicles in the ovary, including some well-developed pre-ovulatory
follicles, do not ovulate and therefore undergo a process known as atresia.
Atretic follicles contain remnants of the degenerating ova and surrounding zona pellucida. Atretic follicles may form small white scars in the
cortex of the ovary, which are characteristically long and thin – and quite
different from corpora albicantia. The cells infiltrating the atretic follicles
sometimes differentiate into interstitial cells which secrete oestrogen and
other steroids. Detailed studies of the interstitial cells have not yet been
made in Antarctic seals.

The corpus luteum

A follicle that has ovulated develops into a secretory body known, on account of its yellow colour, as a corpus luteum. Although details of the process are not fully established for Antarctic seals, it probably takes much the same course as in other mammals. The granulosa cells lining the large pre-ovulatory Graafian follicle proliferate to fill the antrum and in so doing undergo cellular changes to become secretory lutein or luteal cells. Some of the theca interna cells from the periphery of the follicle may also be incorporated during this process. A connective tissue capsule develops and the whole body is now a recognizable corpus luteum.

Corpora lutea (CL) can be classified according to the stage of the reproductive cycle:

1. CL of ovulation. This is usually slightly smaller than the pre-ovulatory follicle from which it is derived, but it enlarges slightly after the invasion and luteinization of the granulosa cells. There is usually a small central cavity containing some material which is usually gelatinous after fixation. The luteal cells are usually small, but have large nuclei with scanty cytoplasm. These cells are roughly 25 to 50% larger than the follicular granulosa cells.

2. CL of delay. As already mentioned, following fertilization and formation of the blastocyst, there is a delay in the attachment of the blastocyst to the uterine epithelium, and in further development. The blastocyst in Antarctic seals remains free for about 1.5 to 4.5 months following fertilization (Laws, 1984). Because of the very close relationship between the status of the blastocyst and the CL, the latter also shows some quiescence. In the elephant seal the luteal cells are oval or polygonal and most are vacuolated. The developing capillary network is confined to the periphery of the gland (Laws, 1956b). The corpus luteum of the period of delay in Antarctic ice-breeding seals still remains to be studied in detail.

3. CL of implantation and development. The luteal cells do not appear to undergo much growth at this stage, but vacuolation is less and the amount of secretory material is greater. Vascularization is more extensive, but less than in some other carnivores (Laws, 1956b). The ultrastructure of luteal cells has been described by Sinha & Erickson (1972, 1974). They possess the features of steroid-secreting cells and are involved in the synthesis of progesterone. However, the CL probably has little or no func-

tional role – at least as a source of progesterone – after the establishment of the placenta. The growth of the CL is paralleled by placental growth and probably therefore by the level of chorionic gonadotrophin (CG). In fact the growth of the CL may be stimulated by CG and if so the presence of a foetus has important consequences for the life history of CL and corpora albicantia (CA). The placenta therefore is not just important as an interface between mother and foetus. The CG from seals appears to be almost identical to human CG, so it should be possible to use a human colorimetric pregnancy test on seals if a urine, serum or blood sample can be obtained. This test could be carried out easily in the field and may even be quantitative. It is not known whether CG is present in seals during the period of delayed implantation. CG has been found in grey seal and sea lion placentae.

4. CL of parturition and lactation. This is the final stage of the corpus luteum. Cells become progressively more shrunken and phagocytosis begins. Invasion of connective tissue begins so that the gland shows a central stellate core of white fibrous tissue.

The corpus albicans

Usually within two to three weeks of birth a follicle will have matured and ovulated in the opposite ovary from the one containing the CL of the previous pregnancy. While the new CL develops, the old one quickly degenerates into a CA.

The regression of the CA has been followed in specimens of the southern elephant seal collected throughout the year (Laws, 1956b), but has not been studied with known history or dated individuals in other Antarctic species. The duration of a recognizable CA may vary greatly between species and individuals, but remnants of the CA may persist for up to three years or more in some species. Initially there is normally a change in the colour of the fresh gland from the characteristic yellow of the CL to a brownish colour (which may have a reddish tinge in fixed material), perhaps associated with shrinkage of the luteal cells. This is accompanied by a rather sudden decrease in diameter, continuing the shrinkage of the CL noticed in the last month or so of pregnancy. The CA continues to shrink as the luteal cells degenerate further and is invaded by macrophages, neutrophils, fibroblasts and connective tissue. In the elephant seal it regresses from an average diameter of about 18 mm at parturition to about 3.5 mm 12 months later (Laws, 1956b). The decline in the size of the CA may be modified by the presence of luteotrophic placental hormones from the following pregnancy as suggested for grey seals. Because of invasion by fibrous tissue, the young CA feels much harder than a CL, which is soft and cheese-like in consistency.

By the second year the CA is recognizable macroscopically as a diffuse fibrous body in the ovary, frequently retaining the stellate core of fibrous tissue developed at parturition. If palpated with finger tips, or finger nails, or stroked with a needle, a characteristic toughness or lumpiness can be felt against the more yielding ovarian stroma. (This is a useful guide when making CA counts.) In section a characteristic grouping of small knots of thickened blood vessels, surrounded by small degenerated luteal cells and fibroblasts can be seen (Fig. 12.8). Note that atretic follicles may form small white scars, but they are essentially different from CAs.

CAs can be formed from the degeneration of any CL, whether of ovulation, delay or further development. It is not yet possible to distinguish between CAs formed from these different types of CLs. Nor is it yet possible to distinguish a CA arising from the degeneration of a CL of a successful pregnancy from that which results from the early degeneration of a CL, whether because it is not associated with conception, or following an abortion.

It is clear that observations on the presence of follicles, CLs and CAs in the ovaries of a seal can give valuable information about its reproductive history, and this technique has been used both in seals (Øritsland, 1970; Bengtson & Laws, 1985) and whales (Laws, 1961). However, the situation is complicated by uncertainty about the duration of the corpora and the inability to distinguish those arising after a successful pregnancy from those arising after the failure of fertilization or implantation, or the resorption or abortion of an embryo or foetus. In addition it is suspected that in Antarctic seals there may be a few CL of pseudopregnancy, as has been reported for several carnivores (Brambell, 1956) while Laws (1956b) has reported the presence of 'accessory' CL in elephant seals.

The reproductive tract

The uterus

The growth, differentiation and function of the uterus is almost entirely dependent on hormones secreted by the ovary, particularly by the interstitial cells and CL. Other parts of the tract, notably the vagina, also show cyclical changes that follow those of the ovary and uterus. Laws (1956b) has provided a complete account of these changes in the uterus and vagina of elephant seals and Mansfield (1958) has described them in the Weddell seal. Harrison *et al.* (1952) have provided descriptions of the uterine mucosa in several Antarctic seal species.

The following is based on Laws' (1956b) description of the elephant seal. Changes in the uterine mucosa are complicated by the differences that exist in mature animals between the two cornua. In the juvenile and non-pregnant seal that has not recently given birth, the uterus is lined by a low cuboidal epithelium, glands are few and there is little secretion. In a mature animal on the approach of ovulation, the epithelial cells lining the cornu into which the egg will be released become taller and columnar, and may develop cilia; uterine glands increase in number and size and become very coiled. The epithelium becomes less active during the period of delay, though the glands continue to produce secretion which may play a role in nourishing the blastocyst. At the time of implantation, the glands, although fewer in number than at ovulation, become very tortuous. As pregnancy develops the glands become fewer but contain much secretion. The attachment of the placenta (see below) affects a large part of the lining of the uterus. When the placenta is expelled at birth this is left in a 'raw' condition, but the great contraction of the muscular part of the uterus reduces its area. The remainder of the uterine epithelium is very oedematous but rapidly regresses to an inactive state.

The vagina

The vaginal epithelium is normally thrown into deep folds and there is often considerable variation in its condition at the crests and at the bases of the folds. At parturition the superficial cells are mucified, but near ovulation cornification occurs. (This is the basis for a vaginal smear technique which has been used to determine oestrus in living seals in the field; see 'Detection of oestrus' section.) The vaginal epithelium regresses during the period of delayed implantation, but mucifies again once implantation has occurred.

Pregnancy

Development of the embryo

Implantation of the blastocyst

In most mammals the ovum is fertilized in the fallopian tube and it takes four to six days for the embryo to travel down the tube and enter the uterus. In seals, attachment and further development of the embryo does not take place immediately. The egg divides to form a hollow ball of cells, the blastocyst, which remains free in the lumen of the uterus for up to 4.5 months. The blastocyst is not entirely inactive; Gibbney (1953) found a 0.3 mm diameter blastocyst in an elephant seal in December, while free blastocysts ranging from 1.6 mm (W.N. Bonner, unpub. data) to 8 mm (Laws, 1956b) have been taken from elephant seals in the latter half of February. Attachment takes place at this time, or in the first half of March. On attachment, a chorio-allantoic placenta develops. The placenta is zonary in shape and forms a complete band around the membranous sac in which the foetus is contained. Along the edges of the placental band, and in its middle where the blood vessels connecting it to the foetus are attached, are areas of bright purplish extravasated blood, the marginal and central haematomata. In young crabeater and leopard seal foetuses these are discontinuous islands (Sinha & Erickson, 1974), but in more developed foetuses the marginal haematomata, and to a lesser extent the central haematomata, are continuous. They probably serve as a source of iron for the foetus, as placental cells can be seen actively phagocytosing the extravasated blood. The main part of the placenta is zonary, labyrinthine and of the endotheliochorial type (Sinha & Erickson, 1974). However, the seal placenta with its haematomata is also an example of partial erosion of the maternal membranes similar to those in the haemochorial type placenta.

Foetal growth

Huggett & Widdas (1951) drew attention to the relationship between foetus weight (W) and its age in days from conception (t), such that:

$$W^{\frac{1}{3}} = a\left(t - t_0\right)$$

where t_0 is the intercept of the linear regression with time axis (approximately 20% of the gestation time in large mammals which do not have delayed implantation) and a is a constant, which they termed the specific foetal growth velocity. If gestation time (t_g) and mean weight at birth (W_g) are known then:

$$a = W_g^{\frac{1}{3}} / \left(t_g - t_0\right)$$

In the seals t_0 is lengthened because it includes the period of delay, and conception dates can only be established by observing mating, or by suitable analysis of dated ovarian collected material. In Antarctic seals where birth dates are more easily established than conception dates, an approximate estimate of implantation date (and the start of foetal growth) is given by calculating:

$$t_g - t_0 = Wg^{1/3}/a$$

adding 15% and subtracting from mean birth date. In practice, whether or not birth weight is known, foetal weights from collected animals can be plotted as $W^{1/3}$, a regression fitted, and estimates of W_g and $t_g - t_0$ obtained if the pupping season is known. Even if the pupping season is unknown, the average date of implantation can be estimated.

Neonatal gonad hypertrophy

Harrison and his co-workers drew attention to the hypertrophy of the gonads of foetal and new-born northern phocid seals (e.g. Harrison *et al.*, 1952; Amoroso *et al.*, 1965). This is probably related to the gonadotrophic activity of the placenta. Similar changes have not been confirmed in Antarctic seals. Bonner (1955) was unable to show a definite decline in testis weight for the southern elephant seal from full-term to two to three weeks old, but in the same period there was a 50% reduction in the weight of the prostate and adrenals.

Pregnancy rate

The pregnancy rate is a basic biological attribute and an essential parameter for population models, which changes in time according to the status of a population. It is, therefore, important in studies of population dynamics and as a management criterion.

The pregnancy rate of a population is the proportion of the total number of adult (i.e. sexually mature, section 'Puberty and sexual maturity') females that are pregnant, expressed as a percentage. If possible, age-specific pregnancy rates should be calculated, that is the percentage pregnant in each age group. Age-specific pregnancy rates initially increase with increasing age and usually decrease in old age groups. Whether or not age-specific rates can be obtained depends on the size of sample available.

Ideally, samples for calculating pregnancy rates should be collected during the months when the embryo is actively growing after implantation of the blastocyst. There are serious problems of sampling bias due to possible segregation of pregnant and non-pregnant females, especially when they have hauled out on land. The proportion of females with an active CL in the ovaries is probably a reasonable approximation to the pregnancy rate, provided they are collected at least a month after the end of the mating period, so as to avoid including CL of ovulation, not associated with pregnancy. The possibility of pseudopregnancy, even if the incidence is low, could affect an estimate derived in this way. Foetal deaths occur during pregnancy so that the pregnancy rate is likely to be higher in early pregnancy than near term. In practice this difference is unlikely to be detectable with the sample sizes that can usually be obtained.

Birth and lactation

Birth is a rapid process in all seals. Laws (1956a) observed that parturition took eight minutes in the elephant seal and that the placenta was expelled within an hour. Occasionally the placenta is expelled at the same time as the foetus, so that it remains attached to the pup until the umbilical cord parts from the umbilicus, usually within two to three days of birth (in elephant, Antarctic fur and grey seals). Antarctic seals have never been observed to eat the placenta, as often happens in other carnivores.

The sex ratio of newborn pups (the secondary sex ratio) should be recorded. In some species it progressively changes during the pupping season, so a spot count may be misleading (Coulson & Hickling, 1961; Stirling, 1971).

Lactation is short in phocids, usually two to six weeks; long in fur seals, around three to four months (Bonner, 1968, 1984; Laws, 1984). The milk of some phocids is very rich in fat and protein and the growth of the pup is correspondingly rapid (Bowen, Oftedal & Boness, 1985). Milk samples, especially if dated relative to parturition would provide valuable information, particularly for leopard and Ross seals.

Routine and accurate paternity analysis

Difficulty of establishing paternity in the field

Maternity in seals is usually relatively easy to determine, because the pup can be observed being suckled by its mother. Paternity, however, is much more difficult to ascertain. The only cues will come from actual observed matings, and even when a mating is seen, there is always a possibility that the putative father is not the only male to have copulated with the female in question. With no evidence as to which animal is the actual father in multiple matings (even if all are sighted), the question of paternity, and hence of reproductive success, has, until now, remained intractable.

DNA fingerprinting

Principle of the method

Recent developments in DNA technology have revealed very powerful methods for studying relatedness, with a resolution capable of accurately assigning a pup to one or both parents (Amos & Dover, 1990; Kirby, 1990). Highly variable regions of DNA are isolated, using specific DNA-cutting enzymes, and size-fractionated on a gel by simple electrophoresis. (See also Chapter 9.) A radio-actively labelled probe is then used to focus on the DNA sequences of interest. The resulting pattern of some 40 bands is effectively unique to each individual and has been termed a DNA 'fingerprint' (Jeffreys, Wilson & Thein, 1985). Paternity can be established because a pup's fingerprint is a random 50:50 mix of bands from both parents. Subtraction of all those attributable to the mother will leave an array of bands, all of which should match bands found in the father. (This is not strictly true, because mutation rates are very high among these sequences and it has been calculated that a band will mutate once every 250 generations. Nevertheless, within the space of one generation, even this phenomenally high rate of mutation may be ignored.)

Collection of material in the field

Application of these techniques requires only small quantities of DNA, and most animal tissues can be used. However, specialized storage tissues, for instance blubber, are probably poor sources. In pilot experiments both skin and blood have proved adequate, blood being the easier to process. Either of these tissues is recommended if material is to be taken from live animals, but blood (even if clotted) or muscle from freshly dead specimens. Useful samples may even be obtained from animals that have been dead for 24 hours or more, especially if prevailing temperatures are below 10°C, as is often the case in the Antarctic. However, the quality of results will undoubtedly decline with increasing delay. In such situations, skin is the preferable tissue because it takes much longer to die and contains many fewer autolysing enzymes.

The minimum quantity of tissue required is about 0.3 ml of blood or a 3 mm cube of tissue. In general, however, rather more than the minimum should be taken, if possible, so as to facilitate work in the laboratory and provide some back-up material. Note that sand and other possible non-biological contaminants, whilst best kept to a minimum, will not noticeably interfere with the results. Conversely, tools being used to sample one animal should be cleaned before being used on the next.

Preservation and storage of material

Samples should be preserved in separate containers in distilled water containing 0.5 M EDIA (ethylenediaminetetracetic acid), 50 mm Tris (trishydroxymethylaminomethane), pH 8.0, and saturated with SDS (sodium dodecyl sulphate). The SDS should be added last and the solution warmed to well above room temperature; about 50°C is fine. On cooling the SDS will often form a dense white precipitate, especially if it is cold. This does not matter so long as the solution is shaken before it is added to the samples. The quantities required for preservation are not very critical, but as a rough guide, 1 ml should be added to each 2 ml of blood, and tissue samples should be placed in a volume twice their own. Under these conditions samples are stable for considerable periods at room temperature; (in a trial, a sample of blood yielded perfectly usable DNA after three months at 25°C). Despite this, it is always best to keep them cool whenever possible and larger pieces of tissue should be cut up to aid penetration of the preservative.

Puberty and sexual maturity

Definitions

The age at puberty is the age at which reproduction first becomes possible and sexual maturity is the age when the animal reaches its full reproductive capacity (Joubert, 1963; Asdell, 1965). Both are more easily determined in the female.

In female seals, puberty is the age at first oestrus, taken to be the first ovulation (determined by examination of the ovaries for CL or CA, or mature follicles – more than 12 mm in diameter). Sexual maturity may be taken as the age at first conception and in seals is usually, but not invariably, the same as the age at puberty.

Puberty in the male is defined as the first production of viable sperm; sexual maturity is less easy to define, because it is complicated by social factors leading to deferred breeding in some species that need to become 'socially mature' in order to meet the definition given above (e.g. elephant seal and fur seals). It is taken here to be the age at which viable sperm are first produced in quantity (as determined by histological examination of the testis or epididymis sections, or from sperm smears). It is usually associated with the attainment of a certain minimum testis, or testis and epididymis, or prostate weight (Fig. 12.3). Once these weights have been correlated with spermatogenesis in an initial sample the organ weights can be used to determine age at sexual maturity for larger samples.

It is now thought that puberty occurs when mammals reach a certain threshold size of body weight, rather than a certain age. Thus, slower-growing animals tend to reach puberty later than faster-growing animals of the same species (e.g. Laws, Parker & Johnstone, 1975). Puberty or sexual maturity is clearly an important aspect of population ecology; ideally it should be established for separate populations of each species, and changes with time should be monitored where possible.

Calculation of the average age at sexual maturity

As mentioned above the age of sexual maturity is an important criterion for evaluating the status of a population (Eberhardt & Siniff, 1977). However, the fact that variation exists among individuals concerning the exact age at which a female first ovulates, or a male produces sperm, makes this criterion difficult to apply. Valid comparisons, in space or time, between two populations cannot be made if different methods have been used, so it is essential that the method of calculating the average age of sexual maturity should be fully described or referenced. The procedure described here (DeMaster, 1978) is general enough to apply to data from both sexes and provides a statistical means of evaluating the significance of the differences between the populations being compared.

Taking the female as an example, the calculation is as follows:

$$f(x) = t(x)/n(x) \tag{1}$$

where

x — age of female
$f(x)$ = estimated probability of ovulating at or before age x
$t(x)$ = number of females in sample of age x who have ovulated
$n(x)$ = number of females of age x in sample

With this definition of $f(x)$, the probability of first ovulating at age x is:

$$P(x) = f(x) - (f(x-1)) \tag{2}$$

where

$P(x)$ = probability of first ovulating at age x, and comes from the relationship:

$$f(x - 1) + P(x) = f(x) \tag{3}$$

The average age of sexual maturity is then calculated as follows:

$$w = \sum_{x=0}^{w} (x)\, P(x) \tag{4}$$

where

x = average age of sexual maturity
w = maximum age in sample
a = first age when f(x) = 1

The estimated variance for x is:

$$v(x) = \sum_{x=0}^{w-1} \frac{f(x)(1 - f(x))}{nx - 1} + \frac{w^2 f(w)(1 - f(w))}{n(w) - 1} \tag{5}$$

and has a 95% confidence interval (CL) described by:

$$95\% \; CL \; = x \; 1.96 \; (v(x))^{\frac{1}{2}} \tag{6}$$

A modified t-test (Snedecor & Cochran, 1967, p. 114) using

$$\sum_{x=0}^{w} n_x$$

as the sample size for a particular sample, can be used to determine the significance of the difference. Age-class samples of approximately 25 would probably be adequate to discern half-year differences in the average age of sexual maturity between two populations.

The appendix (A12.1) to this chapter gives a worked example (DeMaster, 1978).

Back-calculation of age at sexual maturity

Laws (1953) showed that the pattern of the tooth structure of elephant seals could be used to estimate the age at sexual maturity, and McCann (1980) confirmed this by demonstrating that the suckling period was represented in the dentine pattern of females (Chapter 11, section on the elephant seal).

Laws (1977) found that the layers in the cementum of crabeater seal teeth could be used to estimate the ages at which individual mature animals had attained sexual maturity (Chapter 11, section on the crabeater seal). The cementum layers deposited during the immature years, except in the first year of life, are thicker than in later years and measurements confirmed that a transition zone occurred; this is appreciable to the eye under the microscope without measurement. While the correlation between the transition zone and age at sexual maturity holds for individuals, it now appears not to be possible to use this method to demonstrate possible changes in the mean age of puberty by back-calculation as Laws (1977) did. See Bengtson & Laws (1985) for discussion.

Field collecting techniques and basic analysis of material collected

The objective of the studies is assumed to be the determination of the reproductive status of the animals and to use this to estimate population parameters. For both sexes it is important to establish whether the animal is sexually immature or mature; for females, whether they are pregnant (if so, whether in first pregnancy or a later pregnancy), anoestrous, in oestrus, and/or lactating; for males, whether in full spermatogenesis or inactive. Even gross examination can yield valuable information, but this will often be enhanced by the use of microscopical, biochemical or other techniques.

The data from single animals are not particularly informative, but when pooled with others provide the raw data for estimating population characteristics in quantitative terms. The more important of these for management or conservation are: mean age at sexual maturity; pregnancy rate and the timing of the annual reproductive cycle.

A secondary objective is the fuller qualitative understanding of the biology of the species.

Finally, it should be emphasized that reproduction and reproductive success (natality rate, lifetime reproductive success) is fundamentally related to nutrition (Chapter 13) and growth (Chapter 8) and almost certainly to population density (Chapter 2).

General

1. Make notes on the behaviour of the animal before collection and its social status (e.g. solitary, male/female pair, family group (crabeater seal), female suckling pup, harem bull, etc.).
2. See Chapter 8 for general collecting instructions and for instructions for collecting basic reproductive material. Small fixed blocks of tissue are better for subsequent histological examina-

tion than whole organs. Therefore make several slices through the organs before immersing completely in the fixative.

3. If blood is collected (Chapter 9), remember that the serum or plasma can be analysed for hormone levels which may help to elucidate the reproductive data if these are obscure and difficult to interpret. (It should be frozen and stored at $-20°C$ or less). Colorimetric techniques are becoming available which should make it possible to measure levels of some hormones in blood, urine or glandular tissue in the field. This is described in section on enzyme immunoassays.

Anatomical and histological methods – the male

The following basic records and observations can be made in the field, or subsequently in the laboratory. When plotted against age or month they will provide information on sexual immaturity/maturity and on the timing of the annual cycle.

Testis

1. Weigh (either fresh or preserved), fix and store.
2. In material fixed, sectioned and stained for microscope examination: average diameter of 25 seminiferous tubules, which must be circular in cross section, not oval, selected randomly under the microscope. Calculate mean and standard deviation.
3. On the same material: main stage of spermatogenesis (Fig. 12.5, stages A–G).

Epididymis

1. Fix and store.
2. In material fixed, sectioned and stained for microscope examination: average diameter of 25 tubule sections, as above. Calculate mean and standard deviation.
3. Presence or absence of sperm. This can be determined in the field from a smear taken from the tail portion of the epididymis, or in such a smear prepared in the field or in histological sections prepared in the laboratory.

The smear is made by placing a drop of epididymal fluid on a slide, and drawing it out into a thin film using another slide held at an angle, as for a blood smear. Fix in 90% alcohol. It can be left unstained, but preferably should be stained with a blood stain or nigrosin/eosin (Campbell, Dott & Glover, 1956). The nigrosin/eosin method differentiates live and dead sperm. It would also be useful to know whether sperm are motile; this can be asessed by adding 0.9% physiological saline and examining under a microscope in the field. For this purpose a pocket microscope, such as the McArthur, can be used.

Prostate

1. Weigh (either fresh or preserved, fix and store).
2. In material fixed, sectioned and stained for microscope examination: measure epithelium height in 25 tubule sections; calculate mean and standard deviation.

Notes:
(a) Volumes by displacement can be used as an alternative to weights if a suitable balance is not available.
(b) For consistency and comparability all weights should be taken of fresh or preserved material, not a mixture of the two.

Anatomical and histological methods – the female

Detection of oestrus

Vaginal smears have been used in a variety of living animals (on seals by Bigg, 1973), to determine whether or not an animal is in oestrus, on the assumption that the appearance of the vaginal epithelium is an adequate indication of the state of the ovary and hormonal status of the animal. Little research has been done on vaginal smears from Antarctic seals and further correlation with ovarian condition is needed.

At least two slides should be prepared from each specimen. A sterile 15 cm swab is moistened with 0.9% physiological saline, inserted into the vagina to about two thirds of its depth and rotated against the walls. The swab is then carefully removed to avoid contamination and rolled along the length of a glass microscope slide, with even pressure. The slide is then immediately fixed by flushing with 100% alcohol, or spraying with an aerosol fixative. The whole process should take less than 2 min. The slides should be stained as soon as possible with haematoxylin and eosin, or with Wright's stain.

A few additional points should be mentioned. To avoid either cervical contamination, or sampling too close to the urinogenital sinus, the swab should be inserted until it reaches the cervix and then pulled back so that the cotton tip is in the appropriate place. If the swab is held at the same point for all subsequent smears, a reasonably constant depth will be sampled for all females. It may be necessary to load the swab in a plastic pipette to ensure proper depth of penetration. In extremely cold situations, it will be necessary to fix the smear as rapidly as possible to avoid liping due to freezing. Some researchers have found that hairspray serves as an effective and inexpensive alternative to aerosol fixatives, but the aerosol is recommended in extremely cold weather.

In Antarctic seals the vaginal smear detection method may be inconvenient and possibly unsatisfactory as an indicator of oestrus, without further research. The presence of vulval swelling may be at least as good – and much easier in the field – but needs to be investigated.

Ovaries, blastocyst, embryo

1. Weigh both ovaries (fresh or preserved), fix and store.
2. If a CL is present look for a distension of the uterus caused by a foetus; if that is not observed look for a swelling in the uterus, which may only be slight. Record the side of the ovary containing the CL and that of the uterine horn containing the conceptus (they are not always the same).
3. If no swelling is seen, consider flushing the oviduct and/or uterine horn, on the same side of the body as the CL. The oviduct should be flushed only if mating is thought to have occurred recently. For the oviduct, dissect out the tube from the ligament and cut at the uterotubal junction. Take a large syringe (10 ml or more) of 0.9% NaCl and insert at the ovary end of the tube; tie ligature round needle and flush contents into a petri dish. Follow the same procedure with the uterine horn, cutting at the caudal end and inserting syringe into the tubal end. Alternatively look for blastocyst or embryo as below (4); it may be recovered if not too small.
4. If a swelling is present look for the blastocyst or small embryo after making an incision on the lateral aspect of the uterine horn. A hand lens may help identification. Fix with 10% buffered neutral formal-saline. (The commercial formalin or formaldehyde is 37 to 40%.) To make 10 ml of fixative take 10 ml of formalin, mix with 90 ml of distilled water (or tap water if distilled water not available) and add 0.4 g of sodium phosphate monobasic ($NaH_2PO_4.H_2O$) and 0.65 g of sodium phosphate

dibasic (Na_2HPO_4) anhydrous. Using this formula make 1000 ml of fixative. Fix for 14 hours. Wash in tap water for two to four hours and store in 70% ethanol for embedding and histology.

5. If a larger embryo or foetus is present, cut the cornu on the lateral side to expose the conceptus. Look for a zonary band which may have one or more small areas of small sacs containing dark red blood, which are called haematomes. Cut out the foetus and fix the entire conceptus or the cornu if the embryo is small. Measure the foetus (see Fig. 12.9) and collect if less than 15 cm in length. If greater than 15 cm, a standard length measurement should be possible; if not take a crown–rump measurement. Weigh the embryo or foetus as accurately as possible after cutting the umbilical cord.

6. If no embryo is found, look for and measure the width of any placental scar found in the uterus. This will appear as a small

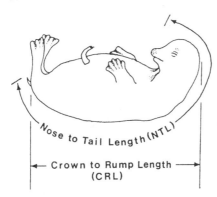

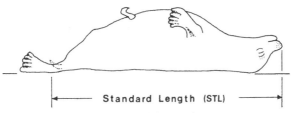

Fig. 12.9 Linear measurements usually taken from foetuses. Crown–rump (CRL) and nose–tail (NTL) are taken along the curvatures, but standard length (STL) is taken in a straight line along the body. (From Smith, 1987.)

haemorrhagic area indicating where the foetus was previously attached.

7. In the laboratory, slice the ovary longitudinally, using a sharp knife to cut slices about 2 mm thick; the cuts should not be complete, so that the ovarian structures can be read like the leaves of a book. (Alternatively a small kitchen meat slicer can be used and the slices threaded on nylon thread.)

   Note the presence of small (< 3 mm diameter), medium (3–12 mm) and large (> 12 mm) follicles; measure the mean diameter of the largest follicle. Count the number of CL and measure the diameter (along two axes at right angles); there is normally only one CL. (Note that adrenocortical tissue may occur quite frequently in the medulla of the ovary (Boyd, 1984)). It has been found in grey, harp and ringed seal ovaries. In cross section it is similar in colour to the CL, but deep in the ovary near or in the hilus and never as large as a CL. It should not be confused with CL tissue but may have intrinsic interest in relation to the reproductive cycle.)

   Count the number of CA and measure the diameter of the largest. See section 'Female reproductive system' for hints on what to look for.

8. In spring the presence of a large follicle probably means that the animal is approaching oestrus; the presence of at least one CL or CA signifies that the animal is sexually mature; the presence of more than one corpus (CL + CA) means that the seal is probably multiparous. One CL and no CA means that it is probably in its first pregnancy if an embryo is present; if no embryo it has probably recently ovulated for the first time.

   With care it may be possible to back-calculate or estimate the age at which sexual maturity was achieved, from the CL + CA count and the present age of the animal, assuming one corpus is formed annually. This is probably a reasonable procedure until two to five corpora have accumulated.

Other tissues

Weighing the placenta after draining fluids off would be useful, because few data exist. Histological sections of uterus and vagina may be prepared and examined because relatively little attention has been paid to seasonal changes in the Antarctic seal species, other than the elephant seal.

Where possible it may be worth collecting the pituitary and pineal glands, since they have received little attention in seals. Frozen endocrine tissue can be subjected to certain analyses which would not be possible using fixed tissue. However, this is a complex and controversial area probably best left to the specialist.

Tissue fixation for immunocytochemistry and electron microscopy

For both sexes immunocytochemistry and electron microscopy may help to elucidate more specialized problems. It is *not* required for studies of population ecology (the primary purpose of this book), but the field worker may be asked to collect material suitable for immunocytochemical studies or electron microscopy and so instructions are given here. Most tissues can be fixed by this method for both immunocytochemistry and electron microscopy, although it is most likely to be used for reproductive studies. (See Hayat, 1973; Glauert, 1977; Sternberger, 1979; Elias, 1982; Polak & Varndell, 1984; Hopwood, 1985).

Instructions for making solutions

1. Prepare the buffers (stock solutions A and B) first; they will keep in cold temperatures for a couple of months.
   Solution A. Sodium phosphate monobasic ($NaH_2PO_4.H_2O$). Take 53.6 g and dissolve in 1000 ml distilled water (or drinking water) and add 3 g of NaCl.
   Solution B. Sodium phosphate dibasic ($Na_2HPO_4.H_2O$). Take 27.6 g and dissolve in 1000 ml of distilled water (or drinking water) and add 3 g of NaCl.
   The above unmixed solutions are stock solutions; they should be kept refrigerated and brought to room temperature before mixing.
2. Take 115 ml of stock solution A and 385 ml of solution B. Mix well by shaking vigorously. This gives 0.2 M sodium phosphate buffer. By doubling the amounts of A and B, 1000 ml of 0.2 M $NaPO_4$ buffer is obtained.
3. Now take 40 g paraformaldehyde powder (trioxymethylene) and dissolve in 500 ml of distilled water. Heat the water to 60°C while stirring. Use a thermometer. If the solution appears cloudy (whitish), add 1 M NaOH with a dropper until it becomes clear.

Allow to cool at room temperature. Now add 500 ml of the 0.2 M buffer (step 2). This gives 1000 ml of paraformaldehyde fixative in 0.1 M buffer, usually at pH 7.3. Add 25 g of sucrose to the fixative, shake to dissolve. *This is your fixative* for both immunocytochemistry and electron microscopy.

4. If your study is more directed towards electron microscopy than towards immunocytochemistry, add 4 ml of 25% glutaraldehyde to 96 ml of paraformaldehyde fixative (prepared in step 3). The glutaraldehyde-fixed specimens are usually not suitable for immunocytochemistry.

5. Place about 5 ml fixative from the step 3 or 4 in a plastic vial; fill as many as needed and keep in a cool place. These vials can be carried into the field wherever a seal is to be killed; if away from base for more than six hours carry 0.1 M phosphate buffer in a large bottle. This can be obtained by diluting 0.2 M phosphate buffer (step 2) with an equal amount of distilled or drinking water.

6. Remove ovary, uterus, CL, testis, epididymis or any other organ desired, *immediately* after the animal is killed. The sooner the tissue is removed the better the fixation. Take out the organ and dissect the relevant parts of it. Now put the tissue on dental wax or a hard plastic cutting board, take a clean razor blade and slice off *thin strips* about 1–2 mm thick and 1–2 cm long for electron microscopy. For immunocytochemistry, the slices should be about 3–4 mm thick and 1–2 cm long. Pick up two or three slices with forceps and place in the vials containing the fixative from step 5. The specimens must be fully immersed in the fixative. Now close the vials, shake a few times and allow fixation to continue for two to six hours. *Label each vial indicating tissue type and animal's specimen number.*

7. At any time after two hours, pour out the fixative and add about 5–10 ml of 0.1 M phosphate buffer (from step 5). If you run out of the 0.1 M phosphate buffer, use phosphate buffered saline (PBS) which can be prepared using the formula ($NaH_2PO_4.H_2O = 0.262$ g; $Na_2HPO_4.7H_2O = 2.173$ g; $NaCl = 8.5$ g, and distilled water 1000 ml). Shake to dissolve before use. This buffer is also available as a mixed powder from commercial suppliers.

8. The specimen can now be stored for shipment in a cool place if possible. *Do not freeze the fixed tissues.*

The samples fixed as above can be processed for immunocytochemistry and electron microscopy. Tissues collected for electron microscopy can also be processed for light microscopy.

Enzyme immunoassays

Hormones can be measured in small blood samples collected from living or dead animals. Although relatively new, the so-called enzyme immunoassays (EIA), often cited as enzyme-linked immunosorbent assays (ELISA) are now used for a wide range of diagnostically important analyses and have, in many cases, replaced radioimmunoassays based on radioisotopes (RIA). The ELISA system is a non-isotopic technique and has many practical advantages over systems such as RIA, and indeed non-isotopic techniques which include fluorescence and luminescence.

The main advantages are that the reagents are stable and easy to use, and the only item of equipment required is a simple colorimeter, of which many cheap ones exist on the market. With the significant advances being made in the field of biotechnology, driven mainly by pressure to provide kits to measure clinically important hormones, ELISA systems are now extremely sensitive and are a true quantitative assay able to measure vanishingly small concentrations of hormone. They can be used in the field with ease and impunity (Ishiwaka, Kawai & Miyai 1981; Stanley *et al.*, 1985). Many suppliers of diagnostic kits market all the appropriate hormone and metabolite assays and are often prepared to supply the component parts individually, if any specific modifications are necessary. Perhaps even more important the company responsible for the research and development will often directly assist in making the assay absolutely specific for the field worker's purposes. Boyd (1991) describes the assay of progesterone and prolactin in Antarctic fur seal females, using an enzyme immunoassay developed for the measurement of these hormones in human plasma ('Scrozyme', Scrono Diagnostics Ltd., Woking, Surrey GU21 5JY UK). His method is reproduced in Appendix 12.2. A disadvantage of the method is the variation in individual levels of circulating hormones referred to below, which requires initial research to establish the extent of such variations, as related to different reproductive states, by species.

RIA, although extremely well tested over a wide range of vertebrates, has the disadvantage of needing bulky and expensive equipment, together with the mounting problems of transporting and disposing of radioactive isotopes. However, there are many well established RIA laboratories which can be used if the blood samples are frozen immediately after collection. Both EIA and RIA require only small volumes of blood – only 1 ml of serum or plasma is necessary to assay a range of hormones.

Any frozen or fresh endocrine glandular tissue can be assayed for hormones (using RIA or ELISA techniques), to obtain a broad indication of the total endocrine activity of the gland. The tissue can also be subjected to cytogenetic or immunocytological techniques which give a much more detailed account of the activity and nature of specified secreting cells. Determining the hormone concentration per unit weight of glandular tissue is often more indicative of the status of the animal, because there are large individual variations in circulating hormone levels within a species, which may make interpretation of levels in blood or body fluids difficult.

References

Amos, B. & Dover, G. 1990. DNA fingerprinting and the uniqueness of whales. *Mammal Rev.*, **20**, 23–30.
Amoroso, E.C., Bourne, G.H., Harrison, R.J., Matthews, L.H., Rowlands, I.W. & Sloper, J.C. 1965. Reproductive and endocrine organs of foetal, new born and adult seals. *J. Zool., London* **147**, 430–86.
Asdell, S.A. 1965. *Patterns of Mammalian Reproduction.* Constable, London.
Bartholomew, G.A. 1970. A model for the evolution of pinniped polygyny. *Evolution* **24**, 546–59.
Bengtson, J.L. & Laws, R.M. 1985. Trends in age of maturity of crabeater seals: an insight into Antarctic marine interactions. In *Antarctic Nutrient Cycles and Food Webs*, ed. W.R. Siegfried, P.R. Condy & R.M. Laws, pp. 669–75. Springer-Verlag, Berlin.
Bester, M.N. 1990. Reproduction in the male sub-Antarctic fur seal *Arctocephalus tropicalis. J. Zool., Lond.* **222**, 177–85.
Bigg, M.A. 1973. Adaptations in the breeding of the harbour seal, *Phoca vitulina. J. Reprod. Fert.*, Supplement 19, 131–42.
Bonner, W.N. 1955. Reproductive organs of foetal and juvenile elephant seals. *Nature* **176**, 982–83.
Bonner, W.N. 1968. *The Fur Seal of South Georgia.* British Antarctic Survey, Scientific Reports, no. 56, 1–91.
Bonner, W.N. 1984. Lactation strategies in pinnipeds – problems for a marine mammalian group. Zoological Society of London Symposia, No. 51, 253–72.
Bowen, W.D., Oftedal, O.T. & Boness, D.J. 1985. Birth to weaning in 4 days: remarkable growth in the hooded seal, *Cystophora cristata. Can. J. Zool.* **63**, 2841–6.
Boyd, I.L. 1984. The occurrence of hilar rete glands in the ovaries of grey seals (*Halichoerus grypus*). *J. Zool.* **204**, 585–8.
Boyd, I.L. 1991. Changes in plasma progesterone and prolactin concentrations during the annual cycle and the role of prolactin in the maintenance of

lactation and luteal development in the Antarctic fur seal (*Arctocephalus gazella*). *J. Reprod. Fert.* **91**, 637–47.

Brambell, F.W.R. 1956. Ovarian changes. In *Marshall's Physiology of Reproduction*, vol. 1, part 1, 3rd ed, ed. A.S. Parkes, pp. 397–542. Longmans, Green & Company, New York.

Campbell, R.C., Dott, H.M. & Glover, T.D. 1956 Nigrosin-eosin as a stain for differentiating live and dead spermatozoa. *J. Agric. Sci.* **48**, 1.

Coulson, J.C. & Hickling, G. 1961. Variation of the secondary sex ratio of the grey seal, *Halichoerus grypus* (Fab.), during the breeding season. *Nature* **190**, 281.

DeMaster, D.P. 1978. Calculation of the average age of sexual maturity in marine mammals. *J. Fish. Res. Bd Can.* **25**, 912–15.

Eberhardt, L.L. & Siniff, D.B. 1977. Population dynamics and marine mammal management policies. *J. Fish. Res. Bd Can.* **34**, 183–90.

Elias, J.M. 1982. *Principles and Techniques in Diagnostic Histopathology and Enzyme Histochemistry*. Noyes Publications, Park Ridge, New Jersey.

Gibbney, L. 1953. Delayed implantation in the elephant seal. *Nature* **172**, 590.

Glauert, A.M. 1977. *Practical Methods in Electron Microscopy*. North Holland Publishing Company, Oxford.

Hamilton, J.E. 1939. The leopard seal, *Hydrurga leptonyx* (de Blainville). *Disc. Rep.* **18**, 239–64.

Harrison, R.J. 1969. Reproduction and reproductive organs. In *Biology of Marine Mammals*, ed. T.H. Anderson, pp. 253–348. Academic Press, London.

Harrison, R.J., Matthews, L.H. & Roberts, J.M. 1952. Reproduction in some Pinnipedia. *Trans. Zool. Soc. Lond.* **27**, 437–541.

Harrison, R.J. & Weir, B.J. 1977. Structure of the mammalian ovary. In: *The Ovary*, vol. 1, 2nd edn, ed. Lord Zuckerman & B.J. Weir, pp. 113–217. Academic Press, New York.

Hayat, M.A. 1973. *Principles and Techniques of Electron Microscopy: Biological applications*, vol. 3. Van Nostrand and Reinhold Company, New York.

Hopwood, D. 1985. Cell and tissue fixation: review. *Histochem. J.* **17**, 389–442.

Huggett, A. St. G. & Widdas, W.R. 1951. The relationship between mammalian foetal weight and conception age. *J. Physiol.* **114**, 306–17.

Ishikawa, E., Kawai, T. and Miyai, K. (Eds.) 1981. *Enzyme Immunoassay*. Igaku-shoin, Tokyo.

Jeffreys, A.J., Wilson, V. and Thein, S.L. 1985. Individual-specific 'fingerprints' of human DNA. *Nature* **316**, 76–9.

Joubert, D.M. 1963. Puberty in farm animals. *Anim. Breeding Abs.* **31**, 295.

Kirby, L.T. 1990. *DNA Fingerprinting: an Introduction*. Macmillan, London.

Laws, R.M. 1953. A new method of age determination for mammals with special reference to the elephant seal, *Mirounga leonina* Linn. *Sci. Rep. Falkld Isl. Depend. Surv.* No. 2, 1–11.

Laws, R.M. 1956a. The elephant seal (*Mirounga leonina* Linn.). II. General, social and reproductive behaviour. *Sci. Rep. Falkld Isl. Depend. Surv*, No. 13, 1–88.

Laws, R.M. 1956b. The elephant seal (*Mirounga leonina* Linn.). III. The physiology of reproduction. *Sci. Rep. Falkld Isl. Depend. Surv.* No. 15, 1–66.

264   R.M. Laws and A.A. Sinha

Laws, R.M. 1960. The southern elephant seal (*Mirounga leonina* Linn.) at South Georgia. *Norsk Hvalfangsttid* **49**(10), 466–76, (11), 520–42.
Laws, R.M. 1961. Reproduction, growth and age of southern fin whales. *Disc. Rep.* **31**, 327–486.
Laws, R.M. 1977. The significance of vertebrates in the Antarctic marine ecosystem. In *Adaptations within Antarctic Ecosystems*, ed. G.A. Llano, pp. 411–38. Houston, Gulf Publishing Company.
Laws, R.M. 1984. Seals. In *Antarctic Ecology* vol. 2. ed. R.M. Laws, pp. 621–715. Academic Press, London.
Laws, R.M., Parker, I.S.C. & Johnstone, R.C.B. 1975. *Elephants and their Habitats: the Ecology of Elephants in North Bunyoro, Uganda*. Oxford University Press, Oxford.
McCann, T.S. 1980. Population structure and social organization of southern elephant seals, *Mirounga leonina*. *Biol. J. Linn. Soc.* **14**, 133–50.
Mansfield, A.W. 1958. The breeding behaviour and reproductive cycle of the Weddell seal (*Leptonychotes weddelli* Lesson). *Sci. Rep. Falkld Isl. Depend. Surv*, No. 18, 1–41.
Øritsland, T. 1970. Sealing and seal research in the south west Atlantic pack ice, Sept. – Oct. 1964. In *Antarctic Ecology*, vol. 1. ed. M.W. Holdgate, pp. 367–76. Academic Press, London.
Perry, J.S. 1971. *The Ovarian Cycle of Mammals*. Oliver and Boyd, Edinburgh.
Polak, J.M. & Varndell, I.M. 1984. *Immunolabelling for Electron Microscopy*. Elsevier, Oxford.
Scheffer, V.B. & Slipp, 1944. The harbor seal in Washington State. *Am. Midl. Nat.* **32**, 373–416.
Sinha, A.A. & Erickson, A.W. 1972. Ultrastructure of the corpus luteum of Antarctic seals during pregnancy. *Z. Zellforsch. Mikrosk. Anat.* **117**, 35–45.
Sinha, A.A. & Erickson, A.W. 1974. Ultrastructure of the placenta of Antarctic Seals during the first third of pregnancy. *Am. J. Anat.* **141**, 263–80.
Sinha, A.A., Erickson, A.W. & Seal, U.S. 1977a. Fine structure of seminiferous tubules in Antarctic seals. *Cell Tiss. Res.* **178**, 183–8.
Sinha, A.A., Erickson, A.W. & Seal, U.S. 1977b. Fine structure of Leydig cells in crabeater, leopard and Ross seals. *J. Rep. Fert.* **49**, 51–4.
Smith, T.G. 1987. The ringed seal, *Phoca hispida*, of the Canadian Western Arctic. *Can. Bull. Fish. Aquatic Sci.* **216**, 1–81.
Snedecor, G.W. & Cochran, W.G. 1967. *Statistical Methods*. Iowa State University Press, Iowa.
Stanley, C.J., Paris, F., Plumb, A., Webb, A. and Johannson A. 1985. *Enzyme Amplification* American Biotechnology Laboratory.
Sternberger, L.A. 1979. *Immunocytochemistry*. 2nd ed. John Wiley & Sons, New York.
Stirling, I. 1971. Variation in sex ratio of newborn Weddell seals during the pupping season. *J. Mammal.* **52**, 842.

APPENDIX 12.1

The following tables (from DeMaster, 1978) based on hypothetical data, give examples of using the procedures outlined in section 'Puberty and sexual maturity' to estimate the average age of sexual maturity (Table 12.1) and the average age at first pregnancy (Table 12.2).

Table 12.1. *Calculation of average age of sexual maturity in a hypothetical population of females. Average age of sexual maturity = 4.40, var(x)= 0.032.*

Age (x)	Number of females observed $n(x)$	Number ovulated $t(x)$	P (1st ovulation at or before age x) $f(x)$	P (1st ovulation at age x) $P(x)$	$xP(x)$
0	35	0	0.0	0.0	0.00
1	30	0	0.0	0.0	0.00
2	25	0	0.0	0.0	0.00
3	20	4	0.2	0.2	0.60
4	20	12	0.6	0.4	1.60
5	15	12	0.8	0.2	1.00
6	15	15	1.0	0.2	1.20
7	10	10	1.0	0.0	0.00
8	10	10	1.0	0.0	0.00
					4.40

Table 12.2. *Calculation of average age of first pregnancy in a hypothetical population of females. Average age of first pregnancy = 3.34, var(x) = 1.21, a = 4.*

Age (x)	Number of females observed $n(x)$	Number of pregnant females $y(x)$	P (pregnant at or before age x $z(x)$	P (1st pregnant at age x) $r(x)$	$xr(x)$
0	35	0	0.00	0.0	0.00
1	30	0	0.00	0.0	0.00
2	25	6	0.24	0.24	0.48
3	20	13	0.65	0.41	1.23
4	20	16	0.80	0.15	0.60
5	15	12	0.80	0.16	0.80
6	15	12	0.80	0.032	0.19
7	10	8	0.80	0.0064	0.04
8	10	8	0.80	0.000128	—
					3.34

APPENDIX 12.2

Enzyme imunoassay is described in section 'Enzyme immunoassays'. Here is a detailed example of a method of assaying the hormones progesterone and prolactin in Antarctic fur seals, *Arctocephalus gazella* (from Boyd, 1991, courtesy Society for Reproduction and Fertility).

Hormone assays

Oestradiol-17β, progesterone and prolactin were assayed using an enzyme immunoassay developed for the measurement of these hormones in human plasma ('Serozyme', Serono Diagnostics Ltd, Woking, Surrey GU21 5JY, UK). In this study, the assays were modified to use small volumes of plasma by reducing the total volume of the assays by one quarter and running them on micro-titre plates. All reagents were used at the concentrations at which they were supplied. In the case of the steroid assays, plasma (50 μl for oestradiol and 12.5 μl for progesterone) was incubated at 37°C for 20 min with fluorescein-labelled monoclonal antibody and a standard quantity of enzyme labelled steroid. In the prolactin assay, plasma (25 μl) was incubated at 37°C for 20 min with two monoclonal antibodies specific to different parts of the prolactin molecule. One was labelled with fluorescein and the other with enzyme. Anti-fluorescein coupled to a magnetic solid phase was then added in excess and incubated at 37°C for 10 min. After magnetic separation, the supernatant was discarded and the sediment was washed with sodium phosphate buffer. Separation was repeated and the sediment was incubated for 30 min at 37°C with a standard volume of enzyme substrate before the reaction was stopped with a weak solution of sodium hydroxide as provided by the suppliers of the assay. The blank-corrected absorbence of the supernatant was then read at 550 nm on a Cambridge Life Sciences CLS962 Microplate Photometer and hormone concentrations were determined from a standard curve run in duplicate with each plate. Samples of seal plasma were assayed in duplicate. The prolactin standard was human prolactin with biological activity calibrated against the 2nd IRP for hPRL (83/562).

Pooled plasma from male fur seals was used as the blank in both the steroid assays and assay diluent was used in the prolactin assay. These gave values below the limits of detection which were the concentrations equivalent to the absorbance of the blank minus (or plus in the case of the prolactin assay) two standard deviations (20 pg/ml for oestradiol,

0·2 ng/ml for progesterone and 0·5 ng/ml for prolactin). Addition of a known mass of steroid gave mean recoveries of 80·35 ± 3·1% for oestradiol and 112·5 ± 12.5% for progesterone. A plasma sample which gave a high (40 ng/ml) value for prolactin in the assay was diluted in assay diluent at 1:1·5 to 1·25 and this gave a dilution curve parallel to the standard curve. The slope of the regression between observed and expected values was 1·3 and the regression coefficient (r^2) was 0·954. The top prolactin standard (250 ng/ml or 5000 μ i.u./ml) diluted in seal plasma also gave a dilution curve parallel to the standard curve, suggesting that plasma effects did not invalidate the assay. In the circumstances, it was not possible to validate the assay using dilution of pituitary extracts, although this procedure will be carried out in due course. The intra-assay coefficient of variation was 8·4%, 6·1% and 9·1% for oestradiol, progesterone and prolactin respectively and the inter-assay coefficient of variation was 9·4%, 6·4% and 15·8% for oestradiol, progesterone and prolactin respectively. These coefficients of variation were calculated for quality controls of 170 pg/ml, 5·4 ng/ml and 6·5 ng/ml for oestradiol progesterone and prolactin respectively. The values of quality control samples supplied with the assay kits were within the range quoted by the supplier. Cross-reactions in the oestradiol assay were only significant for oestradiol sulphate (5·2%) and oestradiol 3β-D-glucuronide (81·1%). In the progesterone assay there was 100% cross-reaction with 11α-hydroxyprogesterone and other significant cross-reactions were with pregnenolone-3-glucuronide (8·0%) and 5α-pregnanedione (14·5%). There was no significant cross-reaction with either 17α-hydroxyprogesterone or 20α-dihydroprogesterone. In the prolactin assay, there were no cross-reactions with hLH, hFSH or hTSH but there was 7% cross-reaction with hGH.

13

Diet

J.P. CROXALL

Introduction

Estimation of food consumption is usually approached by combining dietary data with information on metabolic energy requirements. Bioenergetic considerations are dealt with in chapter 14; here we consider the main methods for investigating seal diets. Knowledge of their diet is obviously fundamental to studies of the ecology of Antarctic seals and especially to any understanding of their role as predators in the Southern Ocean marine ecosystem. Unfortunately, while we have acceptable qualitative information on the diet of most species (well summarized by Øritsland, 1977), there is insufficient quantitative data to characterize adequately the diet of any species, even in summer. This chapter describes the best methods available at present for the collection, preservation and analysis (including species' identification) of various types of food sample. Suggestions are also provided for appropriate ways in which to present the results of such studies.

Nowadays, the first main objective of dietary studies of seals must be to determine the quantitative composition of the diet. There is no method for acquiring and analysing prey eaten by seals that is free from bias of one kind or another. Research to estimate the extent of these biases and to provide correction factors which may be applied to appropriate compositional data is therefore an important additional requirement.

Collection of food samples

There are three main sources of material suitable for studies of seal diets:

1. complete stomachs (and intestines) from killed seals;

2. partial samples of stomach contents obtained from live seals either by using emetics, stomach pumps or other stomach flushing techniques or by natural regurgitation;
3. collection of faecal droppings (scats).

Complete stomachs from killed seals

Although the development and improvement of non-lethal methods for acquiring seal samples is highly desirable, at present complete stomach contents are likely to provide the most detailed information and present the fewest biases in interpretation and quantification. However, only a small proportion (probably less than 10%) of stomachs of randomly shot animals contain food (or at least other than highly digested prey remains) (Fig. 13.1). If seals are to be collected for dietary studies, it is essential that they should be taken as soon after haul-out as possible. For pack ice seals the best time is likely to be in the early morning. However, for seals that may travel long distances to and from feeding areas, acquisition of adequate collections of stomach contents may require that specimens be taken at sea near the feeding grounds.

Stomachs should be removed and treated as soon as possible after a seal is killed in order to stop further digestion. Removal is easily accomplished

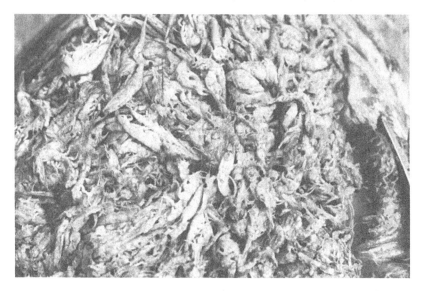

Fig. 13.1 Stomach of crabeater seal, containing Antarctic krill, *Euphausia superba*. (Photograph courtesy A.W. Erickson.)

by tying off the oesophagus immediately anterior to the stomach and at the duodenum. If both the stomach and intestines are to be taken, the entire gastrointestinal tract can be removed as a unit, in which case the end of the large intestine is tied off. It is still desirable, however, to tie off the duodenal junction, particularly if the specimen is to be examined for parasites which could move into an adjacent portion of the gastrointestinal tract.

Partial samples from live seals

There have been few, mainly unsuccessful, attempts to use emetics to induce regurgitations in anaesthetized seals. P.R. Condy (pers. comm.) was unable to effect regurgitation in Sernylan-anaesthetized crabeater and Ross seals. G.A. Antonelis (pers. comm.) was also unable to induce regurgitation in northern fur seals by tube-feeding three physically restrained adult females a single dose of either 237 ml of 1% H_2O_2, 474 ml of 1% H_2O_2 or 237 ml of a 2.0 M solution of LiCl. Each female was captured within an hour after returning from feeding at sea and remained in holding cases for three hours after the emetic was administered. Green & Burton (1987) had limited success using apomorphine hydrochloride injected into Weddell seals. At one site they obtained no results 20 min after dosing three animals with 0.33 to 0.40 mg/kg, despite the animals having full stomachs. At a second site, using dosages of 0.54 to 0.65 mg/kg and 0.22 to 0.39 mg/kg, one of three seals in each case vomited stomach contents (after 27 and 55 min) but the rest showed no response after up to 120 min.

Stomach-flushing techniques, however, have been more successful, though tried on only three species so far. Antonelis *et al.* (1987) lavaged the stomachs of 60 northern elephant seals choosing individuals which had recently hauled out. After capture and restraint, seals were immobolized by intramuscular injection of ketamine hydrochloride, at dose rates of 2.5 to 3.0 mg/kg for males (>250 cm in length) and 2.5 to 3.5 mg/kg for females (210–283 cm in length) and young males (<250 cm in length). Injections were administered using a 35 ml syringe and a human spinal needle (8.9 cm × 18 gauge). A clear tube of flexible PVC (2.54 cm inside diameter, 0.48 cm wall thickness, 3 m length) was inserted into the animal's mouth and gently pushed past the oesophagus into the stomach. Passage of the tube was facilitated by rounding the edges of the end of the tube and coating it with surgical lubricant. In most cases the tube was passed through a 3.8 cm hole in a block of wood (38 × 3.8 ×

7.6 cm) which was placed in the animal's mouth. Once in place, the tube was connected to the discharge fitting of a manually operated suction pump and approximately three to four litres of water were pumped into the animal's stomach. The suction fitting of the pump was then connected to one of two hose fittings on a 10 L collecting bottle while the other fitting was attached to the lavage tube. A vacuum was created in the collecting bottle and the slurry of water and undigested food parts were evacuated into the bottle where a screen bag (1 mm^2 mesh) separated food items from the liquid. The remaining liquid was passed through a second screen (0.5 mm^{-2} mesh) to ensure that all of the identifiable parts of prey had been recovered. Most seals recovered from the anaesthetic within one hour of injection; all but three contained food remains.

Using the lavage technique (Antonelis *et al.*, 1987) and the immobilization methods (McCann, Fedak & Harwood, 1989, and Baker *et al.*, 1990), samples of stomach contents were obtained from 46 of 50 (92%) southern elephant seals recently hauled out to moult (Rodhouse *et al.*, 1992). The only modification to the lavage technique was that instead of using a pump, water was poured manually into a PVC-tube inserted into the stomach. The diluted stomach contents were emptied by lowering the PVC-hose below the seal's head and the sample collected in a fine-mesh net.

A similar technique was used by J.P. Croxall and co-workers (unpub. data) on female Antarctic fur seals. Recently hauled out animals were selected, captured, placed on a restraint board (Gentry & Holt, 1982) on a bench 1 m above the ground and injected intravenously with Saffan (45 mg alphaxalone; 15 mg alphadolone) at dose rates of 0.10 to 0.15 ml/kg. An opaque semi-flexible PVC tube (2.7 cm external diameter, 0.8 cm wall thickness, 2 m length) was carefully inserted into the stomach and 2–3 L of seawater passed (by gravity) into the stomach. The free end of the tube was then lowered and water and food remains flowed out into a container. About 60% of animals treated provided samples of food. Using Saffan, animals were immobilized within one minute and recovered full mobility within 45 minutes of the original injection. All studies provided good samples of small prey items (krill from fur seals; pelagic crustaceans, fish otoliths, lenses and bones and squid beaks from elephant seals) but items larger than the tube diameter will not be recovered.

Natural regurgitation of krill, fish and boluses of squid beaks has been observed for several species of fur seals and sea lions, but only for Weddell seals among phocid seals. Cape fur seals in captivity regurgitated 95% of 700 otoliths fed to them in the form of whole *Pagellus natalensis*, the remaining 5% appearing in the faeces (G.J.B. Ross, pers. comm.). Such regurgitations by otariid seals may be a useful source of information on their diet, but the habit will complicate studies of diet based on faecal samples.

Faecal droppings

Faecal material from Antarctic seals has been collected systematically only from fur seals (North, Croxall & Doidge 1983; Green, Burton & Williams, 1989), and Weddell seals (Testa *et al.*, 1985; Green & Burton, 1987). Ideally scats should be collected fresh; they are usually produced by animals that have fairly recently hauled out. Faecal sample analysis depends on using hard parts that have resisted digestion. It is therefore most appropriate for seals which eat fish, especially larger individuals or species, because small otoliths can be completely digested fairly rapidly (see later). Squid beaks appear often to be retained in the stomach; invertebrates are usually highly digested but fresh scats may contain recognizable exoskeletons. The development of correction factors to allow for the loss of hard parts (e.g. Murie, 1987), will greatly enhance the value of faecal samples, but considerable work will be needed to achieve this for Antarctic seal species.

Preservation of food samples

At the time of collection, the date, time and place should be recorded, together with the sex of the animal and its specimen number. Ideally, numbered tags should be attached (preferably with strong string) to the main stomach sample and good quality paper tags or labels used with other samples (including elements removed from the main stomach sample). Information on these tags should be inscribed with a soft lead pencil.

Neutralized 4% formaldehyde in seawater is a convenient fixative (with recommended ratio of material to fixative of 1:9) for the majority of seal prey items. (Excess calcium carbonate should be used in the stock solution to neutralize the acidity of the formaldehyde fixing and preserving solutions.) Propylene phenoxytol and propylene glycol added to this solution give good bactericidal, fungicidal and softening properties, especially with crustacean material. Fish otoliths (and squid statoliths), however, are dissolved by formalin, they are also affected to some extent by alcohol and are best preserved dry in gelatin capsules. This means that samples including fish material (and whole squid if statoliths are required) should be processed at the time of collection before preservation of the bulk of the sample.

Alternatively, samples can be frozen in their entirety. This may produce problems of storage and, later on, of analysis because specimens that are thawed out are usually less easy to handle for measurement and critical identification. However, if time is insufficient to extract otoliths from fish specimens (or statoliths from squid), it may be essential to freeze these for subsequent processing.

Sorting and identification of food material

General

The ultimate aim of the processing operation is to record or estimate the weight (and/or volume) and number of individuals present of each species of prey, together with as much information as possible on their size, age and reproductive status. For faecal samples, extraction, identification and measurement of hard parts is the most that is usually possible. The main problems in sorting are the recovery of small items, such as otoliths and squid beaks that are often critical to prey identification, and inadequate experience in identifying material from often fragmentary remains.

Ideally samples should be processed in as much detail as possible as soon after collection as practicable. This is especially critical in the case of samples containing fish (and intact squid, if statoliths are required). Samples, or parts thereof, that have been preserved in formalin will need to be placed in a water bath for several hours (e.g. overnight) in order to remove the bulk of the preservative before sorting can start.

Sorting

Complete fresh or frozen stomachs should be weighed, emptied into a large tray, rinsed thoroughly and reweighed. Preliminary sorting, that is extracting fairly intact specimens and classifying into major prey types (e.g. fish, squid, crustacean, etc.), is often conveniently done at this stage. When this is completed, the next requirement is to extract all otoliths, squid beaks and statoliths. Some of these are very small but all are usually dense enough to sink in water and effective retrieval techniques depend on sieving prey remains and allowing dense material to sink in a receptacle whence it can be collected. A relatively simple technique involving three sieves (5.6 mm, 1.0 mm and 0.55 mm mesh diameter) and standing

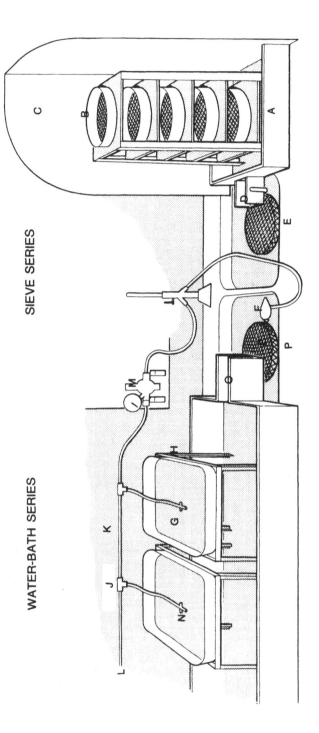

Fig. 13.2 Apparatus for sorting seal stomach contents to retain otoliths (from Murie & Lavigne, 1985).

Key:
A Sieve stand
B Sieves
C Splashguard
D Outflow channel

E Sieve
F High pressure nozzle
G Pan
H Pan carrier

I Hose
J T-connector
K Water-pipe
L Water source

M Pressure reducing valve
N Hose clamps
O Outflow channel
P Draining sieve

material in water-filled beakers is described by Treacy & Crawford (1981).
A more sophisticated apparatus (Fig. 13.2) using five sieves (4.7 mm,
2.36 mm, 1.40 mm, 0.85 mm, 0.425 mm) and five inclined waterbaths,
with the rate of water flow adjusted for each bath so that otoliths are
retained and flesh fragments carried over the lip, is described by Murie
& Lavigne (1985a). They also tested their method for precision and
accuracy; errors were very small and the technique is clearly highly
efficacious. Bigg & Olesiuk (1990) have also developed a complex and
efficient apparatus for processing scats quickly, with minimal exposure to
odour and little damage to skeletal structures.

   Where it is necessary rapidly to sort large volumes of stomach material
in the field (or on board ship) it may be preferable to use large black-
bottomed trays. Otoliths are clearly visible against this background and
the method also avoids the problem of small otoliths becoming stuck in
the mesh of sieves.

   With faecal samples, extraction of identifiable hard parts of prey is
facilitated by suspension in a mixture of water (three parts) 95% ethyl
alcohol (10 parts) and 0.4% carboxymethylcellulose solution (one part)
(Treacy & Crawford, 1981). A cheaper and equally effective method is to
use a mixture of one part liquid dish-washing detergent to 100 parts water
(Antonelis, Fiscus & DeLong, 1984).

   The final sorting of readily identifiable material can proceed concurrently with the extraction of otoliths, squid beaks and statoliths. Material
remaining after these extractions will be highly digested and/or very
fragmented. If the amount of such material is small relative to the total
contents, it is usually sufficient to weigh it and assume that its composition is identical to that of the rest of the sample. If there is a fair amount
of fragmentary material, or it is obvious that not all prey types are represented therein, then it is probably necessary to remove one or more
subsamples for detailed sorting and analysis.

Identification and analysis

By this stage all potentially identifiable material should have been sorted
at least into the major prey classes. We can now consider procedures relevant to the identification and analysis of each of these.

   A common element to much of what follows is the need to identify
intact and/or fragmentary material, especially including certain potentially diagnostic hard parts. Workers undertaking a considerable amount
of such work will, necessarily, build up fairly comprehensive reference

collections. In all cases, however, it is vital that full sets of voucher specimens are retained and, wherever possible, that the identification of these is checked by a specialist in the appropriate field.

In addition, there is increasing use of measurements of the eyes, otoliths and beaks of krill, fish and squid to estimate the original length and mass of ingested food items (see below). Many (if not most) of the relationships enabling this to be done need improving. This can be achieved by forwarding material to specialists who have already developed such relationships, or by publishing appropriate measurements, or by providing new length-mass (or similar) relationships. These relationships usually take the general form of $Y = aXb$. In order to estimate a and b from a linear regression of $\log_e Y$ (W) on $\log_e X$ (V), the minimum information required is ΣWV, $\Sigma\Psi^2$, ΣV^2, ΣW, ΣV, and the sample size (n), which allows confidence limits to be calculated and the relationship to be expressed in reverse form. Power curves are not necessarily the best for all data sets, particularly where larger and older fish are concerned and workers are advised to check the fit of curves to their raw data.

Fish

Relatively intact material should be fairly readily identifiable using standard reference works and keys. The most important of these for Antarctic species are: General: Norman (1937, 1938); DeWitt (1964); Marshall (1955); Nototheniidae: Andersen (1984); Fischer & Hureau (1985); Myctophidae: Hulley (1981).

Some skeletal material is also diagnostic (Iwami, 1985) and information relevant to the identification of fish from skeletal structures is accumulating steadily (e.g. Eakin, 1981; Andersen, 1984). However, at present, access to appropriate specialists and/or reference collections is essential. Sagittal otoliths are becoming widely used in species identification of digested material (especially from scats), although some closely related species cannot be distinguished at present on the basis of otolith structure. When otoliths are still retained within the skulls of fish they should be removed (see Williams & Bedford, 1974) and kept together as a 'pair' (i.e. not mixed with collections of loose otoliths). Ideally the specimen from which they were removed, if it is relatively intact and still likely to possess diagnostic characters, should be preserved, as it may help to resolve problems of identification between species having similar otoliths.

Otoliths can also provide valuable information on the age and size (length) of fish and allow estimates to be made of mass, using standard length–mass relationships. The most usual measurement made on otoliths is maximum length; maximum breadth is often recorded (Fig. 13.3) and sometimes also thickness and weight. Otoliths from prey samples (especially faecal ones) may be partially digested and it is obviously desirable to correct for this (or to exclude such otoliths from analysis) before estimates of fish length are derived. It is possible that digestion of the otolith surface progresses at a more constant rate than that of the perimeter (which can be unevenly eroded or even broken). Prime (1979) suggested that otolith thickness might be the best measurement to make and used this in extensive work deriving digestion ratios (difference in thickness between undigested and digested otoliths) and relationships between undigested otolith thickness and fish weight (as a cube root), in studies with grey seals (Prime & Hammond, 1985, 1987).

Other measurements potentially very relevant to digestion studies are otolith weight and circumference. There is obviously a need for a critical study of otolith digestion to determine which measurements are the most

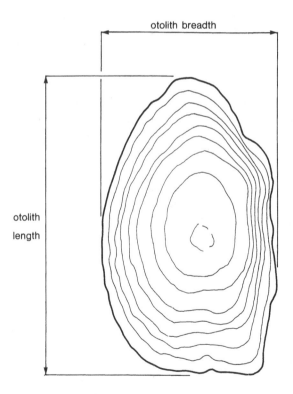

Fig. 13.3 Fish otolith, showing standard overall length and breadth measurements, for use in relating dimensions to fish length and weight.

278   *J.P. Croxall*

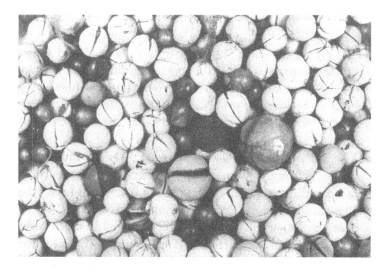

Fig. 13.4 Eyeballs from the stomach of a Weddell seal. The eyeball just
below the centre is from a cephalopod; the rest are from fish, almost all
Pleuragramma and are about 2–3 mm in diameter; the very largest is
from *Dissostichus mawsoni*. (Photograph courtesy J. Plötz; twice
natural size.)

appropriate. In the meantime it is recommended that with otoliths from
predator stomachs, digested (i.e. loose) and undigested (i.e. removed
from fish heads) ones are kept separate and that measurements be made
of overall length, breadth, thickness and weight.

An atlas to the identification of otoliths of Southern Ocean fish species
(Hecht, 1987) incorporates appropriate mathematical relationships for
undigested otoliths. Additional relationships of Antarctic fish are published in Adams & Klages (1987), Croxall, North & Prince (1988) and
Hindell (1988).

In some cases the number and size of fish (and squid) eyeballs may be
a useful index of the number of prey ingested and even of the range of
sizes consumed, particularly in cases where otoliths are badly eroded or
may have been completely digested (J. Plötz pers. comm.; see Fig. 13.4).

Squid

Even relatively intact squid are difficult to identify, except by specialists.
Okutani & Clarke (1985) provide a very useful guide to the identification
of 23 species of common Antarctic squid and 31 species are treated in full
in Fischer & Hureau (1985). A number of other species also occur in the

region, however, including taxa whose taxonomic status is still uncertain. The benthic octopods are particularly poorly understood.

In the last two decades much progress has been made on identification of squid from their keratinous beaks (Clarke, 1962, 1980) and a recent handbook (Clarke, 1986) synthesizes all pertinent available information. Nevertheless, identification of squid beaks is still difficult and requires much practice and experience. It is vital that workers making their own identifications should have these checked by specialists and that all material should be preserved in case future work permits more critical re-identification. This is particularly so for groups like octopods and Cranchiidae where, with beak material, distinguishing even between genera can be extremely difficult, if not impossible, at present. A typical octopod and decapod lower beak are shown in Fig. 13.5.

Clarke (1986) also provides extensive information on the removal of beaks from specimens (especially from buccal masses; see Fig. 13.6) and of all aspects of the preservation, examination and documentation of beak material. This volume also includes the relationships enabling squid mass to be estimated from the rostral length of lower beaks. Additional relationships for Antarctic squid are published in Rodhouse (1989),

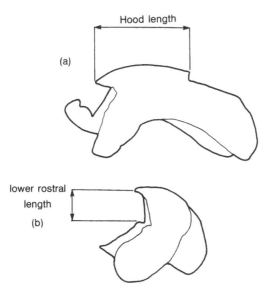

Fig. 13.5 Lower beaks from (a) octopod and (b) decapod cephalopods, showing the standard measurements of lower rostral length and hood length.

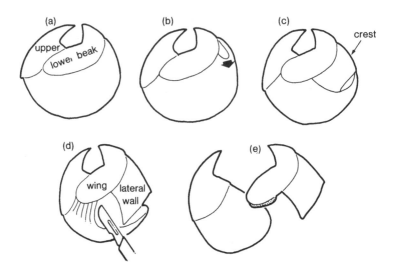

Fig. 13.6 Successive stages in the removal of squid beaks from the muscle of the buccal mass (after Clarke 1986).

Rodhouse *et al.* (1990), Rodhouse & Yeatman (1990) and Rodhouse *et al.*, 1992.

The potential use of statoliths in providing data on the age of squid has been indicated (Clarke *et al.*, 1980) but considerable validation work needs to be undertaken before statoliths can be used routinely in this fashion.

Crustaceans

Intact crustaceans should be identifiable using standard keys to the main groups: Euphausiacea: Kirkwood, 1982; Mysidacea: Kirkwood, in press; Decapoda: Kirkwood, 1983; Amphipoda: Barnard, 1932; Barnard 1969; Bellan-Santini, 1972a, b; Bowman, 1973; Bowman & Gruner, 1973; Thurston, 1972, 1974. Crustaceans belonging to other groups are rare in stomach contents of Antarctic seals. Digested material is often identifiable by reference to intact material in the same sample. Particularly with euphausiids it is important to record additional information on the size, sex and reproductive status of at least a sample (ideally randomly selected) of the animals in each stomach. Body length measurements of krill have been reviewed by Mauchline (1981) and Morris *et al.* (1988). The best standard measurement is from the anterior edge of the eye to the tip of the telson, excluding setae (Fig. 13.7). In digested samples (and especially

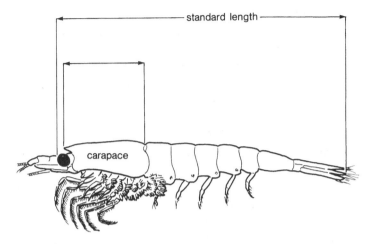

Fig. 13.7 Recommended measurements for standard length and carapace length of Antarctic krill *Euphausia superba*.

in scats), however, it may be impossible to obtain an adequate sample of individuals in sufficiently good condition to make this measurement. In such cases the length of the removed carapace should be measured (Hill, 1990), from the tip of the rostrum to the mid-dorsal posterior edge (Fig. 13.7). This can be converted to total length using the relationships given in Hill (1990). In making this conversion much greater accuracy is achieved if the sex and stage of the krill are known.

In extreme cases, the only undigested material of krill remaining may be the eyes, the diameters of which can provide an estimate of total body length (Nemoto, Okiyama & Takahashi, 1985) but the mathematical relationship for doing this has not yet been published.

The sexual maturity stages of krill are classified by Makarov & Denys (1981). Determination of reproductive status can only be made on virtually intact individuals.

Other groups

In general, other invertebrate groups provide little food for seals. Weddell seal stomachs have contained isopods (Dearborn, 1965), lamellibranchs and holothurians (Wilton, Pirie & Brown, 1908), hydrozoans, bryozoans and ascidians (Green & Burton, 1987). Lamellibranchs (Wilton *et al.*, 1908) and ascidians (Hamilton, 1939) are recorded for crabeater seals.

In contrast, leopard seals frequently prey on seabirds and other species of seal; feather and skin remains are often found in their stomachs.

Portions of seal skin with hairs attached can be identified with practice, using Scheffer (1964) as an initial guide. Penguin feathers, especially rectrices, are very distinctive but species discrimination is not easy. Identification of skulls of smaller seabirds is relatively straightforward but identification of feather remains is always difficult, even if excellent reference material is available.

Biochemical techniques

Flesh remains, especially in highly digested samples, are often difficult to identify to prey category, let alone species. Immunological techniques have been used to identify invertebrate tissue in predator–prey studies (Pickavance, 1970; Calver, 1984; Grisley & Boyle, 1985). These approaches, including the recent use of enzyme-linked immunosorbent assay (ELISA) (Walter, O'Neill & Kirby, 1986), depend on raising antibodies against the proteins of specific prey species. The technique is, therefore, more suitable for predators which exploit only a low diversity of prey species and preferably those of which fresh material (for raising antibodies) is readily available. The relative complexity of the basic techniques is likely to restrict the use of such approaches to very detailed dietary studies, unless some organization is prepared to establish and maintain a reference bank of material and run assays on a service basis.

Interpretation and presentation of results

Quantifying and interpreting the results of diet composition studies is not straightforward. Important biases derive from the facts that:

1. seal stomachs will have been collected at various times after consumption of the last meal (i.e. there is usually appreciable heterogeneity between samples in the degree to which digestion has proceeded);
2. not all the food in a stomach was ingested at the same time (i.e. material comes from an unknown number of meals ingested over an unknown period of time);
3. different types of food are digested at different rates. Generally, squid flesh is digested faster than fish (Bigg & Fawcett, 1985).

Additional problems attend the ways in which the results of dietary analyses are expressed. A common method is to give frequency of occurrence (i.e. the proportion of samples in which a particular taxon is represented). This gives equal weight to abundant and rare constituents and may thus present a rather misleading picture of the basic diet, even when species occurring just as 'traces' are excluded. In addition, hard parts of some groups (e.g. squid beaks) may be retained in a stomach for much longer than the remains of other prey groups. Their importance will, therefore, be considerably over-estimated by frequency of occurrence analysis. Presentation of the results of dietary analyses solely in terms of frequency of occurrence is not recommended when additional information can be presented from actual measurements of prey or from estimates derived from known relationships between prey size and specific hard-parts (e.g. squid beaks and fish otoliths).

For most purposes, and especially when prey stock assessment and management is involved, more useful measures are composition of prey species by volume or biomass. The main problem here is that this does not take into account differential rates of digestion of various types of prey.

Theoretically, a way of compensating for this is to estimate the original mass of each prey item ingested, using hard parts (squid beaks, otoliths) which are retained in the stomach. This depends on the quality of the relevant relationships between hard part dimension and total prey size and mass. It also requires that all types of prey possess hard part remains equally suitable for constructing such relationships, and that these structures are not selectively rejected (see Boulva, 1973 and Pitcher, 1980 for suggestions that seals may frequently discard fish heads). Most important, it assumes that there is no loss, (or significant change in size), of hard parts during digestion.

There are few relevant data for squid beaks, but very small beaks and those of young squid where little 'darkening' has taken place, are probably liable to digestion if retained in the gastrointestinal tract for more than one to two days. Although squid beaks can appear in faeces within 7 h of ingestion of buccal masses (Bigg & Fawcett, 1985), their irregular shape may make them prone to retention in the rugae of the stomach. Large beaks are liable to abrasion, which, when the tip is damaged, may preclude accurate estimation of rostral length and hence of original mass.

Four recent studies, in addition to the review by Jobling & Breiby (1986), have shown the biases that need to be recognized in using otoliths to reconstruct fish diets of seals. Murie & Lavigne (1985b, 1986) fed known quantities of intact Atlantic herring *Clupea harengus* (averaging 25 cm standard length, 214 g mass and with sagittal otolith lengths of 4.4 to 5.2 mm) to grey seals. After 2.8 h, 100% of the otoliths were recovered from seal stomachs; by 5.7 h, however, 70% of otoliths were in the stomach, 2% in the intestines and 28% had been destroyed by digestion;

after 18 h no otoliths remained in the stomach, 39% were recovered from the intestines and the rest had disappeared. The digestion of the recovered otoliths was very variable. Otoliths remaining inside fish heads (13–93% of otoliths after 3 h) were unaffected; 40% of loose otoliths showed signs of digestion after 6 h.

Silva & Neilson (1985) fed herring of various sizes (from which one otolith had been removed as a control) to harbour seals and collected scats over the next 48–72 h. Only two otoliths were recovered, both from herring of the largest size (30–35 cm standard length), representing 4% of otoliths ingested from fish of this size. These two otoliths were 26% and 13% shorter and 32% and 25% narrower than the controls. Annulus counts for age determination were also more difficult. Small fish, or fish with small otoliths, may be disproportionately affected by otolith digestion because while Prime (1979) lost only 14% of gadoid otoliths fed to a common (harbour) seal, he noted that there were few signs of otoliths from the herring used to carry the gadoid otoliths.

Treacy (1986) showed that the heads of adult salmonid fish were often discarded by harbour seals. Scats were collected from seals that had been fed on salmonid smolts. With coho salmon *Oncorhynchus kisutch* (fork length 140 mm), 70% of otoliths were recovered; with chinook salmon *O. tschawytscha* (fork length 80 mm), 33% were recovered; with steelhead trout *Salmo gairdneri* (fork length 50 mm, otolith length 1 mm), 42.5% were retrieved. The passage times of coloured beads inserted into smolt and Atlantic herring were also monitored. The mean times for the first bead to appear in the faeces was 9.0 h (range 5.7 to 12.3 h) and for the last bead it was 29.6 h (range 25.2 to 33.9 h).

Dellinger & Trillmich (1988) found that only 40% of otoliths in intact herring and sprats fed to California sea lions and South American fur seals were found in the scats. Partial digestion of the otoliths recovered would have resulted in a 16% underestimation of fish length and a 35% underestimation of fish mass. However the relative frequencies of the otoliths of the two species in the scats were only slightly different from the relative frequencies of the fish actually ingested.

Murie (1987) quantified the relationships between the degree of digestion of stomach contents and the time elapsed since feeding and between the latter and the proportion of otoliths recovered, mainly using grey seals fed on Atlantic herring. This allows corrections to be applied, according to the state of digestion of stomach contents, for the proportion of otoliths lost in future studies on this species. However, the method will not allow for the loss of disproportionate numbers of small otoliths and it will require separate relationships to be established for each seal species (and even, perhaps, different times of year) and especially for different prey species, which are likely to be digested at different rates. Seals with varied diets will, therefore, pose special problems. Nevertheless, this is clearly a very fruitful line of investigation and much more experimental work on otolith retention and digestion is needed.

Because of the difficulties experienced in using any single statistic to portray seal diet composition, attempts have been made to use compound indices, in which it is hoped the biases of each component might tend to cancel each other out. Pinkas, Oliphant & Iverson (1971) derived an index of relative importance (IRI), whereby the percentage frequency of occurrence of each prey species was multiplied by the percentage volume and then by the percentage count of individuals. Bigg & Perez (1985) describe a modified volume index for use with species which have a mixed diet of fish and squid. First, the proportion of total fish and total squid in the diet, using non-trace frequency of occurrence data, is calculated. Secondly, the ratio of each species within only fish and within only squid is determined by volume. Thirdly, the volumetric ratios for squid and for fish species are adjusted to sum, respectively, to the total proportion of squid and fish in the diet. Finally, all values are readjusted to total 100%.

Both these indices appear to be useful compromises. Bigg & Perez (1985) compared the results of analysing their data using these and a variety of other, single index, methods. Values for the squid component of the diet ranged from 48% (simple frequency of occurrence) to 8% (non-trace IRI), with composition by volume (17%) and modified volume (26%) being roughly intermediate.

It should be clear by now that there is no single ideal way in which to present the results of seal dietary studies. It is obviously preferable to produce a synthesis of the original data in as many ways as possible. It is recommended that these should include composition by mass (or volume), by frequency of occurrence and by number of individuals. Where appropriate the presentation should also document the result of excluding hard parts (i.e. loose otoliths and squid beaks) which may have been retained for some time. Composite indices may be useful for comparative purposes and non-trace IRI and modified volume appear to be the most straightforward ones at present.

Acknowledgements

I am most grateful to G.A. Antonelis, Jr., A.W. Erickson, P. Hammond, H.J. Hill, J. Plötz, J.H. Prime and G.J.B. Ross for helpful comments on the manuscript and to P.R. Condy and G.J.B. Ross for allowing me to use their unpublished data.

References

Adams, N.J. & Klages, N.T. 1987. Seasonal variation in the diet of the king penguin (*Aptenodytes patagonicus*) at sub-Antarctic Marion Island. *J. Zool. Lond.* **212**, 303-24.

Andersen, N.C. 1984. Genera and subfamilies of the family Nototheniidae (Pisces, Perciformes) from the Antarctic and Subantarctic. *Steenstrupia* **10**, 1-34.

Antonelis, G.A., Jr., Fiscus, C.H. & DeLong, R.L. 1984. Spring and summer prey of California sea lions, *Zalophus californianus*, at San Miguel Island, California 1978-79. *Fish. Bull.* **82**, 67-76.

Antonelis, G.A., Jr., Lowry, M.S., DeMaster, D.P., & Fiscus, C.H. 1987. Assessing northern elephant seal feeding habits by stomach lavage. *Mar. Mammal Sci.* **3**, 308-22.

Baker, J.R., Fedak, M.A., Anderson, S.S., Arnbom, T. & Baker, D.R. 1990. Use of a tiletamine-zolazepam mixture to immobilise wild grey seals and southern elephant seals. *Vet. Rec.* **126**, 75-7.

Barnard, J.L. 1969. The families and genera of marine gammaridean Amphipoda. *Bull. U.S. Natl. Mus.* **271**, 1-535.

Barnard, K.H. 1932. Amphipoda. *Discovery Rep.* **5**, 1-326.

Bellan-Santini, D. 1972a. Invertébrés marins des XIIeme et XVeme Expéditions Antarctiques Françaises en Terre Adélie. 10. Amphipodes gammariens. *Tethys* Suppl. **4**: 157-238.

Bellan-Santini, D. 1972b. Amphipodes provenant des contenus stomacaux de trois espèces de poissons Nototheniidae recoltes en Terre Adélie (Antarctique). *Tethys* **4**, 683-702.

Bigg, M.A. & Fawcett, I. 1985. Two biases in diet determination of northern fur seals *Callorhinus ursinus*. In *Marine Mammal - Fishery Interactions*, ed. J. Beddington, R.J.H. Beverton, & D.M. Lavigne, pp. 284-91. Allen & Unwin, London.

Bigg, M.A. & Olesiuk, P.F. 1990. An enclosed elutriator for processing marine mammal scats. *Mar. Mammal Sci.* **6**, 350-5.

Bigg, M.A. & Perez, M.A. 1985. Modified volume: a frequency volume method to assess marine mammal food habits. In *Marine Mammal - Fishery Interactions*, ed. J. Beddington, R.J.H. Beverton, & D.M. Lavigne, pp. 278-83. Allen & Unwin, London.

Boulva, J. 1973. *The harbor seal* Phoca vitulina concolor *in eastern Canada*. PhD thesis: Dalhousie University, Halifax.

Bowman, T.E. 1973. Pelagic amphipods of the genus *Hyperia* and closely related genera (Hyperiidea: Hyperidae). *Smithson. Contrib. Zool.* **136**, 1-76.

Bowman, T.E. & Gruner, H.E. 1973. The families and genera of Hyperiidea (Crustacea: Amphipoda). *Smithson. Contrib. Zool.* **146**, 1-64.

Calver, M.E. 1984. A review of ecological applications of immunological techniques for diet analysis. *Austr. J. Ecol.* **9**, 19–25.

Clarke, M.R. 1962. The identification of cephalopod 'beaks' and the relationship between beak size and total body weight. *Bull. Br. Mus. (Nat. Hist.) Zool.* **8**, 419–80.

Clarke, M.R. 1980. Cephalopoda in the diet of sperm whales of the southern hemisphere and their bearing on sperm whale biology. *'Discovery' Rep.* **37**, 1–324.

Clarke, M.R. (Ed.) 1986. *A Handbook for the Identification of Cephalopod Beaks*. Clarendon Press. Oxford.

Clarke, M.R., Fitch, J.E., Kristensen, T., Kubodera, T. & Maddock, L. 1980. Statoliths of one fossil and four living squids (Gonatidae: Cephalopoda). *J. Mar. Biol. Ass. U.K.* **60**, 329–47.

Croxall, J.P., North, A.W. & Prince, P.A. 1988. Fish prey of the wandering albatross *Diomedea exulans* at South Georgia. *Polar Biol.* **9**, 9–16.

Dearborn, J.H. 1965. Food of Weddell seals at McMurdo South, Antarctica. *J. Mammal.* **46**, 37–43.

Dellinger, T. & Trillmich, F. 1988. Estimating diet composition from scat analysis in otariid seals (Otariidae): is it reliable? *Can. J. Zool.* **66**, 1865–90.

DeWitt, H.H. 1964. A redescription of *Pagothenia antarctica*, with remarks on the genus *Trematomus* (Pisces, Nototheniidae). *Copeia* 4, 683–686.

Eakin, R.R. 1981. Osteology and relationships of the fishes of the Antarctic family *Harpagiferidae* (Pisces, Notothenioidei) *Antarct. Res. Ser.* **31**, 81–147.

Fischer, W. & Hureau, J.C. 1985. *FAO Species Identification Sheets for Fishery Purposes: Southern Ocean*. 2 vols. FAO, Rome.

Gentry, R.L. & Holt, J.R. 1982. *Equipment and Techniques for Handling Northern Fur Seals*. NOAA Tech. Rep. NMFS SSRF-758,– 15.

Green, K. & Burton, H.R. 1987. Seasonal and geographical variation in the food of Weddell seals *Leptonychotes weddellii* in Antarctica. *Aust. Wildl. Res.* **314**, 475–89.

Green, K., Burton, H.R. & Williams, R. 1989. The diet of Antarctic fur seals *Arctocephalus gazella* (Peters) during the breeding season at Heard Island. *Antarct. Sci.* **1**, 317–24.

Grisley, M.S. & Boyle, P.R. 1985. A new application of serological techniques to gut content analysis. *J. Exp. Mar. Biol. Ecol.* **90**, 1–9.

Hamilton, J.E. 1939. The leopard seal *Hydrurga leptonyx* (de Blainville). *'Discovery' Rep.* **18**, 239–64.

Hecht, T. 1987. A guide to the otoliths of Southern Ocean fishes. *S. Afr. J. Antarct. Res.* **17**, 1–87.

Hill, H.J. 1990. A new method for the measurement of Antarctic krill *Euphausia superba* Dana from predator food samples. *Polar Biol.* **10**, 317–20.

Hindell, M.A. 1988. The diet of the king penguin *Aptenodytes patagonicus* at Macquarie Island. *Ibis* **130**, 193–203.

Hulley, P.A. 1981. Results of the research cruises of FRV 'Walter Herwig' to South America. LVIII Family Myctophidae (Osteichthyes, Myctophiformes). *Arch. FischWiss.* **31**, 1–300.

Iwami, T. 1985. Osteology and relationships of the family Channichthyidae. *Mem. Natl. Inst. Polar Res.* Ser. E. **36**, 1–69.

Jobling, M. & Breiby, A. 1986. The use and abuse of fish otoliths in studies of feeding habits of marine piscivores. *Sarsia* **71**, 265–74.

Kirkwood, J.M. 1982. *A Guide to the Euphausiacea of the Southern Ocean.* ANARE Res. Notes, No. 1.

Kirkwood, J.M. 1983. *A Guide to the Decapoda of the Southern Ocean.* ANARE Res. Notes, No. 11.

Kirkwood, J.M. *A Guide to the Mysidacea of the Southern Ocean.* ANARE Res. Notes, No. 12. (In press.)

Makarov & Denys, G.J. 1981. *Stages of Sexual Maturity of* Euphausia superba *Dana.* BIOMASS Handbook 11.

Marshall, N.B. 1955. Studies of alepisauroid fishes. *'Discovery' Rep.* **27**, 303–36.

Mauchline, J.D. 1981. *Measurement of Body Length of* Euphausia superba *Dana.* BIOMASS Handbook 4.

McCann, T.S., Fedak, M.A. & Harwood, J. 1989. Parental investment in southern elephant seals, *Mirounga leonina. Behav. Ecol. Sociobiol.* **25**, 81–7.

Morris, D.J., Watkins, J.L., Ricketts, C., Buchholz, R. & Priddle, J. 1988. An assessment of the merits of length and weight measurements of Antarctic krill *Euphausia superba. Bull. Br. Antarct. Surv.* **79**, 27–50.

Murie, D.J. 1987. Experimental approaches to stomach content analyses of piscivorous marine mammals. In *Approaches to Marine Mammal Bioenergetics,* ed. A.C. Huntley, D.P. D.P., Costa, G.A.J. Worthy, & M.A. Castellini, pp. 147–63. Society for Marine Mammalogy, Lawrence, Kansas.

Murie, D.J. & Lavigne, D.M. 1985a. A technique for the recovery of otoliths from stomach contents of piscivorous pinnipeds. *J. Wildl. Manag.* **49**, 910–12.

Murie, D.J. & Lavigne, D.M. 1985b. Digestion and retention of Atlantic herring otoliths in the stomachs of grey seals. In *Marine Mammal – Fishery Interactions,* ed. J. Beddington, R.J.H. Beverton, & D.M. Lavigne, pp. 292–9. Allen & Unwin, London.

Murie, D.J. & Lavigne, D.M. 1986. Interpretation of otoliths in stomach content analyses of phocid seals: quantifying fish consumption. *Can. J. Zool.* **64**, 1152–7.

Nemoto, T., Okiyama, M. & Takahashi, M. 1985. Aspects of the roles of squid in food chains of marine Antarctic ecosystems. In *Antarctic Nutrient Cycles and Food Webs,* ed. W.R. Siegfried, P.R. Condy, & R.M. Laws, pp. 415–20. Springer-Verlag, Berlin.

Norman, J.R. 1937. Coast fishes Part II. The Patagonian Region. *'Discovery' Rep.* **16**, 1–150.

Norman, J.R. 1938. Coast fishes Part III. The Antarctic Zone. *'Discovery' Rep.* **18**, 1–104.

North, A.W., Croxall, J.P. & Doidge, D.W. 1983. Fish prey of the Antarctic fur seal *Arctocephalus gazella* at South Georgia: methods and results of otolith examination. *Bull. Br. Antarct. Surv.* **61**, 27–38.

Okutani, T. & Clarke, M.R. 1985. *Identification Key and Species Descriptions for Antarctic Squids.* BIOMASS Handbook 21.

Øritsland, T. 1977. Food consumption of seals in the Antarctic pack ice. In *Adaptations within Antarctic Ecosystems,* ed. G.A. Llano, pp. 749–68. Smithsonian Institution, Washington D.C.

Pickavance, R. 1970. A new approach to the immunological analysis of invertebrate diets. *J. Anim. Ecol.* **39**, 715-24.

Pinkas, L., Oliphant, M.S. & Iverson, I.L.K. 1971. Food habits of albacore, bluefin tuna and bonito in California waters. *Calif. Dept. Fish Game, Fish. Bull.* **152**.

Pitcher, K.W. 1980. Stomach contents and faeces as indicators of harbor seal, *Phoca vitulina*, diets in the Gulf of Alaska. *Fish Bull.* **78**, 797-8.

Prime, J.H. 1979. *Observations on the Digestion of Some Gadoid Fish Otoliths by a Young Common Seal.* Int. Counc. Expl. Sea CM 1979/N 14.

Prime, J.H. & Hammond, P. 1985. Estimating fish weight from size of otoliths from faecal remains. In: *The Impact of Grey and Common Seals on North Sea Resources*, ed. P.S. Hammond, & J. Harwood, pp. 59-83. Contract report ENV 665 UK (H) to the Commission of the European Communities.

Prime, J.H. & Hammond, P.S. 1987. Quantitative assessment of grey seal diet from fecal analysis. In *Marine Mammal Energetics*, ed. A.C. Huntley, D.P. Costa, G.A.J. Worthy, & M.A. Castellini, 165-81. Society for Marine Mammalogy, Lawrence, Kansas.

Rodhouse, P.G. 1989. Pelagic cephalopods caught by nets during the Antarctic research cruises of the 'Polarstern' and 'Walter Herwig', 1985-1987. *Arch FischWiss.* **39**, 111-21.

Rodhouse, P.G. & Yeatman, J. 1990. Redescription of *Martialia hyadesi* Rochbrune and Mabille, 1889 (Mollusca: Cephalopoda) from the Southern Ocean. *Bull. Br. Mus. Nat. Hist. (Zool.)* **56**, 135-43.

Rodhouse, P.G., Prince, P.A., Clarke, M.R. & Murray, A.W.A. 1990. Cephalopod prey of the grey-headed albatross *Diomedea chrysostoma*. *Mar. Biol.* **104**, 353-62.

Rodhouse, P.G., Arnbom, T.R., Fedak, M.A., Yeatman, J. & Murray, A.W.A. 1992. Cephalopod prey of the southern elephant seal, *Mirounga leonina. Can. J. Zool.* **70**, 1007-15.

Scheffer, V.B. 1964. Hair patterns in seals (Pinnipedia). *J. Morph.* **115**, 1-304.

Silva, J. da & Neilson, J.D. 1985. Limitations of using otoliths recovered in scats to estimate prey consumption in seals. *Can. J. Fish. Aquat. Sci.* **42**, 1439-42.

Testa, J.W., Siniff, D.B., Ross, M.J. & Winter, J.D. 1985. Weddell seal - Antarctic cod interactions in McMurdo Sound, Antarctica. In *Antarctic Nutrient Cycles and Food Webs*, ed. W.R. Siegfried, P.R. Condy, & R.M. Laws, pp. 561-5. Springer-Verlag, Berlin.

Thurston, M.H. 1972. The Crustacea Amphipoda of Signy Island, South Orkney Islands. *Sci. Rep. Br. Antarct. Surv.* **71**, 1-127.

Thurston, M.H. 1974. Crustacea Amphipoda from Graham Land and the Scotia Arc collected by Operation Tabarin and the Falkland Islands Dependencies Survey 1944-59. *Sci. Rep. Br. Antarct. Surv.* **85**, 1-89.

Treacy, S.D. 1986. *Ingestion of Salmonids and Gastrointestinal Passage in Captive Harbor Seals Phoca vitulina.* Final report to Marine Mammal Commission, Contract No. MM 2079357-5.

Treacy, S.D. & Crawford, T.W. 1981. Retrieval of otoliths and statoliths from gastrointestinal contents and scats of marine mammals. *J. Wildl. Manage.* **45**, 990-3.

Walter, C.B., O'Neill, E. & Kirby, R. 1986. 'Elisa' as an aid in the
    identification of fish and molluscan prey of birds in marine ecosystems.
    J. Exp. Mar. Biol. Ecol. **96**, 97–102.
Williams, T. & Bedford, B.C. 1974. The use of otoliths for age determination.
    In *The Ageing of Fish*, ed. T.B. Bagenal, pp. 114–23. Old Woking,
    Surrey: The Gresham Press.
Wilton, D.W., Pirie, J.H.H. & Brown, R.N.R. 1908. Zoological log. *Rep.
    Sci. Results Scot. Natl. Antarct. Exped.* **4** (Zool), 1–105.

14

Bioenergetics

S.S. ANDERSON, D. COSTA AND M.A. FEDAK

Introduction

The study of bioenergetics offers powerful methods for quantifying many aspects of animal life histories and, more importantly, provides a common currency in the form of energy units for making comparisons across species. Seals provide excellent opportunities for examining specific topics, such as the costs of reproduction, parental investment and the energetics of foraging, although there are logistical problems which must be considered when studying large animals in remote places. The separation of feeding and breeding which characterizes most phocid seal life histories makes the estimation of energy costs of reproduction relatively simple. The animals fuel their requirements from stored body resources and thus it is necessary to measure only energy output. The pupping and mating seasons are discrete and well-synchronized, and the animals are relatively accessible because they spend all or most of their time out of the water. Parental investment is provided solely by the female and is terminated abruptly by her departure. In otariid seals, care of the young is similarly the prerogative of the female, although the estimation of reproductive costs in this group is complicated by the alternation of feeding and lactation bouts. There is the compensatory advantage that the predictable return of females to their young enables studies to be made of foraging costs.

In this chapter we propose to list some of the questions that might be asked in bioenergetic studies of seals and to cover in detail the techniques which can be used to address them. These techniques fall into three main categories: changes in the weight and composition of the body over time, changes in the rate at which substances turn over within the body and direct measurements of the oxygen consumed and carbon dioxide produced by aerobic metabolism. Additional techniques are described in section 'Measurement of correlates of energy expenditure' which are supplementary to the main methods. They are concerned with correlates of energy expenditure which can be used as substitutes for direct measurements when these are inappropriate or impossible to make. They can also be important in furthering understanding of how and why animals use energy in various ways. The limitations of the methods described will be discussed.

Definitions of terms

An extensive literature exists on the energy and nutritional requirements of domestic animals and man (Brody, 1945; Kleiber, 1975; Baldwin & Bywater, 1984). The applicability of this literature to pinnipeds was largely ignored until the extensive review of pinniped bioenergetics completed by Lavigne, *et al.* (1982). This review points out the importance of using widely accepted units and definitions that have been established by a number of commissions formed to summarize these data for practical use (Agricultural Research Council, 1965, 1968; National Research Council, 1976, 1981a, 1981b). In this chapter we have adopted the most recent definitions of the National Research Council (1981b) of the USA.

The energy requirements of animals vary as a function of body mass, sex, physiological state, activity, environment, reproductive or growth status and nutritional value of the food (Baldwin & Bywater, 1984). This requires knowledge of the partitioning of energy flow in and out of the animal. Energy utilization can be thought of as a series of chemical reactions or processes that occur within a series of black boxes. A schematic representation of energy flow in animals is given in Fig. 14.1. The intake of energy in food (IE) is equivalent to the net chemical energy liberated as heat, if food is completely oxidized. Food energy remaining after digestion and elimination of faecal energy (FE) is known as the digestible energy (DE). Chemical energy lost as the end product of catabolism is known as urinary energy (UE). The most significant component of UE results from urea formation, the end product of protein catabolism. Essentially no UE results from catabolism of fat or carbohydrate. Metabolizable energy (ME) is the net energy remaining after faecal and urinary energy losses and is that available for work, heat production, and/or storage. Stored or recovered energy (RE) is that used for growth and reproduction (National Research Council, 1981b).

Modifications in food energy intake or diet composition can be compensated for by changes in any of the above parameters. For example,

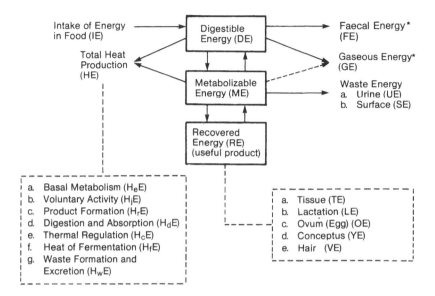

*Under some circumstances the energy contained could be considered to be a useful product.

Fig. 14.1 Schematic diagram of a conceptual model of energy flow through a mammal. (National Research Council, USA, 1981.)

changes in the protein content of the diet will result in modifications of UE or H_dE, which is the heat produced as a result of the physical and biochemical processes associated with digestion, absorption and assimilation. The metabolic efficiency (the ratio of ME:IE) may also change as a function of the diet or quantity of food consumed. Alternatively, the amount of energy partitioned to activity, maintenance or storage (growth or reproduction) may also change. The capability of marine mammals to alter or modify the partitioning of energy between these different pathways in response to changes in diet or to over or under-nutrition has not been examined.

Previous studies of FE and UE have shown that they are relatively predictable within a species, but that they vary following changes in the proximate composition of the diet (Woods, 1982; Costa, 1982; Ashwell-Erickson & Elsner, 1982; Ronald, *et al.*, 1984; Keiver, Ronald & Beamish, 1984). Measurements of DE show that in northern fur seals this changes depending on whether the diet consists predominantly of capelin (88%), herring (93%), pollock (90%), or squid (88%) (Miller, 1978). In California sea lions ME changes with diet mainly of herring (88.2%), anchovy (91.6%), mackerel (91.4%) or squid (78.3%) (Costa, 1986). These data show that the energy intake of seals varies widely with the prey species taken.

After faecal and urinary losses, the remaining energy must be sufficient to meet the costs of homeostasis. This includes basal metabolism, heat liberated as a result of voluntary activity, heat used for thermoregulation outside the zone of thermoneutrality, and the $H_d E$. When energy intake is greater than that required for maintenance, the remaining energy is stored as RE. If food energy is insufficient, the previously stored RE is then utilized. If energy requirements are increased because of thermoregulatory needs, animals may experience a rapid weight loss because they need to use their insulating blubber layer to sustain their metabolism. When energy intake exceeds maintenance requirements, mature animals deposit the excess chemical energy as fat. Most animals maintain a stable body composition at or near the species norm even when food is available *ad libitum*.

Questions which may be addressed by bioenergetic studies

The purpose of listing possible areas for investigation is to present some indication of the range of biological topics which can be tackled through a study of bioenergetics. It is not an exhaustive list, and the topics are noted below as headings, which would have to be expanded into detail before being applied to a particular species.

1. Resting metabolism of individuals of different ages and sexes
2. Costs of reproduction
   Females,
   - heat increment of gestation plus tissue growth of foetus;
   - cost of lacation, that is, milk plus maternal metabolic costs;
   - efficiency of lactation;
   - additional costs of parental investment (e.g. pup protection);
   - effect of sex of pup and maternal age and body condition on level of parental investment.
   Males,
   - costs of resource defence;
   - costs of alternative reproductive strategies;
   - effect of male age on energetic investment in reproduction.
3. Foraging energetics
   - costs of transport to feeding grounds (swimming energetics);

- cost of catching prey;
- activity budget while foraging;
- assimilation efficiency (to produce metabolizable and stored energy);
- overall foraging efficiency.
4. Year to year changes in energy availability and use
   - effect of environmental perturbations on food/energy supplies;
   - changes in body condition associated with reduced level of parental investment (phocids) or reduced foraging efficiency (otariids).

Measuring changes in body weight and condition with time

Where a biological activity is accompanied by starvation, weight changes of animals can be used to obtain estimates of energy use. This method requires that information is gathered on the energy equivalent of the weight change, for instance, by carcase analysis or by isotope dilution (section 'Assigning energy densities to weight changes'). The most direct method of assessing weight change is to weigh the animals during the period of weight loss, but where the logistics of handling large animals rules this out, it is possible to make physical measurements which give estimates of body weight and condition. Both approaches are discussed below.

Weight changes and physical measurements

Weight measurements

The prerequisite of this study is first to capture the seal (see chapter 3). The next requirement is to get the animal off the ground, which is a simple matter when it can be lifted by two people. A sling can both aid lifting and act as a restraint; a useful type consists of two wooden or alloy poles joined together by netting whose width is just greater than the compressed circumference of the animal. Once the animal is positioned, the poles can be clipped together and the netting closed by a drawstring at each end. The seal is then suspended from weighing scales hung on a pole supported by two people. For larger animals, a simple support of two light poles, formed into an A-frame, can increase lifting capacity, a third pole being used as a horizontal member from the A-frame to the shoulder of an assistant (preferably tall!). The weighing device and seal are suspended midway along this pole. This arrangement has been used to weigh animals of up to 200 kg without the need for any lifting gear if four people are available – one stabilizing the A-frame, two aiding the initial lift and the fourth providing the shoulder.

Heavier animals obviously require lifting devices and mechanical supports. Animals up to 1000 kg have been weighed on a lightweight tripod of aluminium poles using a small mechanical hoist (e.g. Minilift Hoist, Model J, Didsbury Engineering Co. Ltd, UK). The constraining factor in dealing with large animals is likely to be the cost of the drug needed to restrain them, rather than the logistics of weighing.

The weight measurement itself can be made using either mechanical (spring or balance) scales or electronic weighing devices. The mechanical devices tend to be heavy and bulky, especially for heavier loads but may be more robust in environments hostile to electronic devices. The strain-gauge or tensile link device, which is based on a load cell combined with a transducer, can be much more accurate and is often very small and light (e.g. 'Mini Weigh', Offshore Monitoring Services, Southampton, UK). Accuracies of up to 0.1% are obtainable which is usually far better than that required.

The weighing techniques described above are appropriate for use on free living, wild seals. Examples of the use of these methods on grey seals (*Halichoerus grypus*) can be found in published work (Fedak & Anderson, 1982; Anderson & Fedak, 1985, 1987a, b). Similar studies have been carried out recently on northern (*Mirounga angustirostris*) and southern elephant seals (*M. leonina*) (Costa *et al.*, 1986; McCann, Fedak & Harwood, 1989). When it is not possible to follow individual seals throughout their period of starvation, an alternative technique is to shoot a sample of seals of known status. Detailed sampling can then be undertaken, including assessment of the body condition of individual seals. Such a study was undertaken on harp seals (*Phoca groenlandica*) by Stewart & Lavigne (1984) who measured body core and sculp (skin plus blubber) weights in lactating females. The stage of lactation was estimated from pup age.

Body dimensions and condition

The relationship between various linear measurements, mainly girth and length, and the weight of an animal can be calculated. These measurements can also be used to calculate condition indices, which give a relative measure of the fatness and thus the energy stores available to the animal. A condition index of maximum girth × 100/body length has been used to compare the degree of fatness between species of seal and between individuals of the same species (Smirnov, 1924; Laws, 1953; Smith, 1966; Bryden *et al.*, 1984). Measurements of body condition have recently been improved by the use of a portable ultrasonic scanner for measuring blubber thickness. This equipment was developed for measuring fat depth and detecting pregnancy in pigs (e.g. 'Scanoprobe 731A', Agro-process Ltd., Framlingham, Woodbridge, Suffolk, IP13 9PT, UK). A small transducer is placed against the skin with oil or bubble-free water being used to facilitate impedance matching. The sonic signal is reflected back to the transducer from discontinuity layers, such as the blubber/muscle junction. The time taken for the signal to travel is converted to a digital readout of the depth. Some practice in the use of the instrument is required, and, ideally, validation trials on dead animals should be carried out. The technique has been used on harp seals (Worthy, 1982), grey seals (Øritsland *et al.*, 1985; M.A. Fedak & S.S. Anderson, pers. comm.) and on northern and southern elephant seals (P. Morris, D.P. Costa and B.J. Le Boeuf, pers. comm.; Gales & Burton, 1987).

Assigning energy densities to weight changes

Assumptions can be made about the components of the body which are depleted during starvation, and calculations made of the known energy equivalents of these components. For instance, Stewart & Lavigne (1984) assumed that only fat was used when they calculated the energy costs of lactation in harp seals from weight differences. However, a more accurate conversion from weight to energy values can be made using one or more of the methods noted below.

Carcase analysis

A sample of seals, representative of the species under study, is taken at the beginning and end of the period of study during which no feeding occurs. Each carcase is weighed and analysed for total energy content (in joules or calories). The change in total joules or calories can then be divided by the weight change to provide an energy value per unit weight change.

That is:

$$ED = (W_1C_1 \times W_2C_2)/W_1 - W_2$$

Where ED is the energy density of the weight change, W_1 and W_2 are the weights at the start and end of the period and C_1 and C_2 are the appropiate energy densities of each carcase.

The assumption is then made that this value is roughly constant for the population at large. This method has been used to assign energy values associated with weight changes during lactation in grey seals (Fedak & Anderson, 1982), but the principles apply to weight change studies in general. The assumption concerning the applicability of the values calculated will only hold if the samples chosen are representative of the population as a whole and there is no systematic variation of energy density of weight change with variables, such as animal size, age or condition. The method can be improved by obtaining a sample of seals for carcase analysis which includes a range of weights and stages of lactation. The energy density of each can be regressed on weight and day of lactation to form an empirical predictive model. Subsequent studies of weight changes could then use the model to predict the calorific density of each seal for its weight and day of lactation, and hence the energy used by each animal could be derived. By using the actual weight changes of individual animals, rather than the average difference in weight between those samples collected for the carcase analysis, at least some of the assumptions mentioned above will be avoided.

Carcase analysis presents a variety of problems to the biologist operating in remote conditions. Processing large quantities of animal tissue requires either ingenuity or specialized equipment, such as an industrial mincer. For example, prior to mincing it is often necessary to separate the blubber from the rest of the carcase. Care must be taken that the tissues are fully homogenized prior to sub-sampling. Laboratory facilities are required for accurate weighing, oven drying, bomb calorimetry and for lipid and protein analysis, though not necessarily in the field as these can be done from frozen samples.

Bomb calorimetry

The energy content of tissues can be calculated from the lipid, protein and carbohydrate content assuming standard energy values for these components, or one can determine energy content by bomb calorimetry. In this method the energy content is determined by the energy released after combustion in a high pressure oxygen atmosphere (30 atm). Although a straightforward procedure, this requires access to specialized equipment such as a drying oven, an oxygen bomb calorimeter, a homogenizing mill and a press to form the homogenized sample into pellets prior to combustion. These items can be found at most universities, and are particularly common in agricultural or nutrition departments. There is a variety of instruments available, from the micro-bomb calorimeter, capable of determining energy content of 50 mg samples, to the simple manual non-adiabatic and automatic adiabatic calorimeters that require 0.5 to 1.0 g of material. Although the micro-bombs require less material, it is easier to assure sample homogeneity when larger sample sizes are used. Specific methods vary with the type of calorimetry employed and most equipment comes with detailed instructions. Difficulties can be encountered during the drying and homogenization of samples because of the high lipid contents of seals and their prey. A more detailed discussion of the techniques and problems associated with bomb calorimetry can be found in Paine (1971) and Leith (1975).

Isotope-dilution

This technique can provide an indirect form of carcase analysis by measuring the volume of water within an animal's body and calculating the mass of the component tissues from a knowledge of the water contents of adipose and lean tissue. The technique is described in detail in the next section. It has the advantage that it does not require sacrificing the animals involved in the weight change study.

Ideally this indirect technique requires that carcase analysis is performed to calibrate the relationship between water and energy content. This has been done in grey seals by Reilly & Fedak (1990) who have shown that a similar relationship between water and chemical composition holds for this species as for other animals. It needs to be emphasized that this technique estimates the amount of water, fat and protein in the body, not the amount of any dissectible or anatomically defined tissue.

Radioisotope methods

The injection of radioisotopes into animals, which can be recaptured at a later date, provides not only information on body composition but also a way of measuring the flux of various components with time (Lifson & McClintock, 1966). When known amounts of tritium or deuterium and oxygen-18 labelled water are injected into an animal, the oxygen-18 water freely equilibrates with the animal's CO_2 and water pools, and dilutes as a function of water influx and CO_2 production (Fig. 14.2). In contrast, tritiated water equilibrates only with the water pool and dilutes as a function of water influx. The initial dilution of these isotopes (determined by measuring their specific activity in a blood sample), after an appropriate equilibration period, allows the determination of the total body water (TBW) volume (Foy & Schneiden, 1960; Schoeller *et al.*, 1980). Equilibration time varies with the size and feeding state of the animal. Larger animals require longer equilibration times than smaller animals (Costa, 1987). In special circumstances, such as milk intake studies, equilibration may not be complete due to the slow exchange of water from the volume of lipid-rich milk contained in the stomach. In pinnipeds of up to 90 kg, equilibration is complete in 90 minutes, whereas in 500 kg elephant seals, three hours are required. After equilibration, the dilution of the isotopes is a function of their turnover rate and is monitored by taking serial blood samples through time. The measurement interval is then completed by recapturing the animals and taking a final blood sample. If the animal undergoes changes in body mass, it may be necessary to determine the new total body water by an additional injection of isotopic water, followed by a post equilibration blood sample. The obvious limitation of this technique is the need to recapture the individual. Measuring food intake and energy consumption during key periods are two of the most important applications of these techniques, although they are also useful to estimate body composition which does not require the second capture of animals.

Determination of body composition

Stable or radioisotopes of water can be used to measure the water content of a living animal and then the body composition can be estimated from that value. The procedure is based on the differences in the water content of adipose tissue (10%) and lean tissue (73%) (Pace & Rathbun, 1945). Any ratio of total body water between 10% and 73% is made up of some percentage of adipose and lean tissue. This ratio can be calculated and the body composition of the animal can be determined (Ortiz, Costa & Le Boeuf, 1978). This technique has been used to estimate body composition in several animals, including man (Stock & Rothwell, 1982). In pinnipeds, it has been used to study changes in fat content of northern elephant seal females, pups and weanlings (Costa, Le Boeuf *et al.*, 1986; Ortiz *et al.*, 1978; Ortiz, Le Boeuf & Costa, 1984), in California sea lions (*Zalophus californianus*) (Costa, 1986; Oftedal, Iverson & Boness, 1987) and in two species of fur seal (Costa & Trillmich, 1988; Costa, Croxall & Duck, 1989).

Food intake

If the diet is known and it can be assumed that the animal does not have an exogenous source of water (i.e. snow or seawater) measurements of water influx derived from isotopic tracer studies can be used to estimate food intake. Since metabolic water production and food intake are related, a relationship can be derived between food intake and water influx if the following parameters are known: the metabolic efficiency of the diet (the ratio of the total energy contained in the food to that available to the animal), the amount of metabolic water produced per gram of food consumed, and the energy and water contents of the prey (Shoemaker, Nagy & Costa, 1976). The most likely source of error in this method is the assumption that the animals do not drink. If animals were to drink sea or fresh water this method would overestimate prey intake. However, studies have shown that phocid seals either do not or cannot drink seawater (Depocas, Hart & Fisher, 1971; Tarasoff & Toews, 1972; Ortiz *et al.*, 1978). The validity of this method for estimating food intake of pinnipeds was assessed by comparing the measured food intake to that derived from water influx measurements. Such studies on northern fur seals, California sea lions, and harbour seals (*Phoca vitulina*) indicate that seawater ingestion does not occur, and food consumption estimated from water influx was within 1.2% to 2.3% of the measured food intake (Depocas, *et al.*, 1971; Costa, 1987). This method has been used to estimate prey consumption of king penguins (*Aptenodytes patagonicus*) (Kooyman *et al.*, 1982), macaroni (*Eudyptes chrysolophus*) and gentoo penguins (*Pygoscelis papua*) (Davis, Kooyman & Croxall, 1983), little penguins (*Eudyptula minor*) (Costa, Dann & Disher, 1986); albatrosses (Costa & Prince, 1987) northern fur seals (Costa & Gentry, 1986); California sea lions (Costa, 1986) and Antarctic fur seals (*Arctocephalus gazella*) (Costa *et al.*, 1989).

Energy consumption

The energy expended by lactating and foraging animals can be measured using the oxygen-18 doubly-labelled water method. This method was developed by Lifson, Gordon & McClintock (1955) and is based on the observation that oxygen in respired CO_2 equilibrates with oxygen in the body water via the action of carbonic anhydrase in blood (Lifson *et al.*, 1949). As CO_2 is produced by metabolism, only oxygen-18 labelled water is diluted. Therefore, the difference between oxygen-18 turnover and tritiated water turnover is a measure of the animal's CO_2 production (Fig. 14.2). These radioisotopic methods of measuring water and energy consumption have been discussed by Nagy (1980), Nagy & Costa (1980), Schoeller & van Santen (1982), Schoeller *et al.* (1986) and Costa (1987).

Sample analysis

Measurements of total body water for composition studies or water turn-over measurements, used for determination of fasting metabolic rates, milk intake or feeding rates, require the injection of only a single isotope

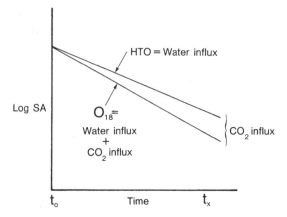

Fig. 14.2 Differential decline in the specific activity of tritium, or deuterium labelled water and the O_{18} labelled water provide a measurement of the CO_2 production of free-ranging animals. (After Costa, 1987.)

of water. For these studies the hydrogen atom of the water molecule is labelled with either tritium or deuterium. The advantages of using tritium labelled water is that only small quantities of isotope are required for injection and analysis, using a liquid scintillation counter, is straightforward. Such a counter, although expensive, is common in university laboratories. Tritium specific activity can be determined from whole plasma as long as variations due to quenching (resulting from haemolysis) and plasma water content are corrected for. Most researchers tend to avoid these problems by counting the tritium activity in pure water distilled from the serum samples (Ortiz, et al., 1978). However, a severe disadvantage of tritiated water is that, being a radioisotope, its use requires permits and other licensing that varies from country to country.

In situations where the use of radioisotope is precluded, deuterium labelled water is used. Its use is essentially the same as that of tritiated water except that, depending on the analysis used, the amount of material injected may need to be quite large. Two forms of analysis are available. One method measures the infrared absorbance of pure water samples distilled from serum, plasma or blood (Oftedal & Iverson, 1987). However, due to the poor precision of infrared spectrophotometers at low isotope concentrations, this technique requires injection of large volumes of deuterated water in order to achieve high blood isotope levels. The second analytical method uses a mass-ratio spectrometer (Mook &

Grootes, 1973; Terwilliger, 1977). This technique offers higher precision at low isotope enrichments, but requires highly sophisticated analytical equipment that is generally not available to university or government researchers. There are a few commercial laboratories that can carry out this analysis, but at a relatively high price.

Metabolic rate can be measured directly using the doubly-labelled water method. This requires the simultaneous use of oxygen-18 labelled water and either tritium or deuterium labelled water. Typically three measurements are needed, one for background, a post-injection equilibration and a final recapture sample. Although this method provides a very precise measure of metabolism, its major drawback is that both the isotope and its analysis are expensive. Oxygen-18 water is available in a variety of concentrations, but it is most commonly used in the 5–15% or 90–95% concentrations. Most studies require approximately 0.15 to 0.2 g of oxygen-18 isotope per kg of animal body mass (see Schoeller & Webb, 1984, for discussion of dosage). For this enrichment, analysis can only be done with a mass-ratio spectrometer.

Measuring energy use via oxygen consumption by indirect calorimetry

As early as 1780 Lavoisier showed that oxygen consumption is directly related to energy metabolism. The rate of use of energy by seals can readily be measured in the laboratory by monitoring oxygen consumption and carbon dioxide production. Because the techniques for accomplishing this are well established and regarded as very reliable, such measurements are particularly useful for validating other more indirect or less well established techniques as well as being useful in their own right. In practical terms, measurement of gas exchange can only be managed for periods of hours or days because of problems of maintaining animals in the constraining situations required by the gas analysis system. It is possible to extend this period if special arrangements can be made and the animals will tolerate long-term confinement. The most difficult problem is to ensure that the state of the animal in the laboratory is representative of that in the wild. It is difficult to avoid some stress such that normal activities may be suspended by the animal. Some of the problems of close confinement can be avoided in water by providing animals with a large tank which permits free swimming, but only allows them to breath in a specific spot at which gas exchange can be monitored. On land, metabolic chambers are often much more restrictive and, therefore, more likely to cause problems.

Metabolic rate is most conveniently measured using an open-circuit gas analysis system (Fig. 14.3). The essential element of such systems is that all of the animal's expired air must be captured. Ambient air is pulled through a chamber or breathing box at flow rates such that all of the animal's expired air is drawn into the analysis system and that excurrent O_2 concentrations are within 1% of ambient even at the highest metabolic levels likely to be encountered. An aliquot of the excurrent air is pushed by a pump to CO_2 absorbing and drying columns and then into the O_2 analyser. In some situations it is desirable to monitor CO_2 production simultaneously. Output from the gas analysers can be digitized and/or recorded continuously. The system can be calibrated *in situ* using the nitrogen dilution technique described by Fedak, Rome & Seeherman (1981). This procedure simply involves introducing N_2 (and perhaps CO_2) to the air stream being drawn through the box or chamber. Calibrated flow meters are used to measure the N_2 and CO_2 introduced to the system. The system can be checked for expired air loss by introducing N_2 or CO_2 at a constant known rate into the system, first into the breathing box and then, with no other changes, into the excurrent tube. It is necessary to ensure that the readings obtained are the same under all conditions. With these calibration techniques such systems can be very reliable and their operation is easily made routine. We estimate that oxygen consumption can be measured to an accuracy and precision of better than ±5% (Ku, 1969).

Other techniques can be calibrated by simultaneously measuring the animal's metabolic rate by indirect calorimetry (e.g. combining O_2 measurements with isotope techniques, see section 'Radioisotope methods'). Observations on parameters correlated with metabolic rate changes such as heart rate, breathing rate and fraction of time at the surface can also be combined with indirect calorimetry in order to establish how such indirect measures change as a function of metabolism (Fedak, 1986; Fedak, Pullen & Kanwisher, 1988). These may then be used in the field to estimate energy expenditure in nature (see next section). Several workers have used large water flumes to allow animals to swim at various speeds while monitoring gas exchange, thus allowing construction of power curves relating swimming speed to metabolic rate and allowing estimation of the energy cost of travel at different speeds (Davis, Williams & Kooyman, 1985; Fedak, 1986; Feldkamp, 1988; Williams & Kooyman, 1985).

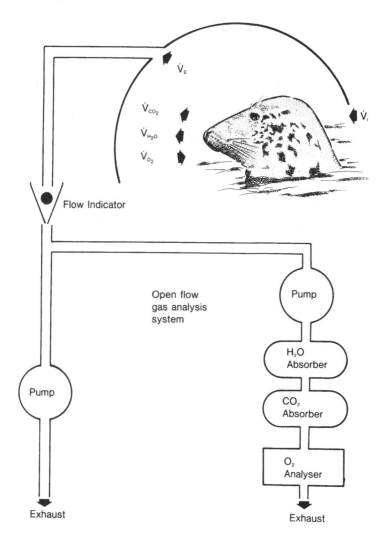

Fig. 14.3 Diagram of an open circuit VO_2 measuring system. Air is drawn through the system by a pump which exhausts to the outside. Main flow through the system is not dried, nor is CO_2 absorbed from it. A small aliquot of gas is pumped from the main flow, dried and pushed through an O_2 analyser.

$\dot{V}_E$ represents flow out of the mask or chamber. The remaining $\dot{V}$ terms refer to flows of the indicated gases in or out of the animal. (After Fedak *et al.*, 1981.)

Measurement of correlates of energy expenditure

It is not always possible to measure energy consumption by the relatively direct methods described above. There are circumstances in which it is either not appropriate to disturb animals in the field by going in to weigh, inject or measure, or when animals are not accessible, a particular problem with seals at sea. Methods which record a correlate of energy expenditure using a non-invasive or limited disturbance approach can therefore be very useful. During the 1970s a variety of recording devices, sufficiently small to attach to marine mammals, were developed to measure biological parameters. Mechanical recorders that measure both the time spent under water and the pressure have given a wealth of information on the diving behaviour of otariid seals (Gentry & Kooyman, 1986). Radio-telemetry has now been used extensively to measure the duration of various elements of behavioural repertoires of animals and to measure various aspects of physiology, in particular, heart rate, which is a useful correlate of energy expenditure (Fedak *et al.*, 1988). The advantage of using recording equipment of the types described above is that, after the initial stress of handling to attach the device, the animal can be released to continue its normal activities until recaptured to retrieve the recorder, or in the case of radio-telemetry, while recording proceeds at a discrete distance. It is, however, essential in the case of heart rate telemetry, and indeed any correlate measure, that laboratory studies on captive animals be also undertaken so that the relationship between heart rate and energy expenditure can be validated.

In addition to being useful as correlates of energy expenditure, behavioural studies can contribute to an understanding of energy flux. Gross measurements of energy expenditure are of limited value unless they are combined with measurements of biological activities which require fuelling. It is particularly important to look at differences in the behaviour of individual animals in order to interpret variations in energy use.

Heart rate as a correlate of metabolism

Heart rate is likely to be related in a predictable way to metabolic rate. Further, it is susceptible to electronic measurement and thus suited to measurement by telemetry. Correlation between heart rate and metabolism in diving grey seals has been reported by Fedak (1986) and related to free ranging dive activity in common seals (Fedak *et al.*, 1988) and in Weddell seals (*Leptonychotes weddellii*) (Hill, 1986). In grey seals, the heart rate of individuals at the surface and the rate while diving are poor predictors of metabolism, while heart rate averaged over entire dive cycles is highly correlated with metabolic rate (Fig. 14.4). However, care has to be taken in interpreting heart rate data because other influences, such as an adrenalin surge due to fright, can also affect the relationship. Heart rate might best be considered as a clue to activity level rather than as a precise predictor of metabolism.

Electrocardiogram signals can be acquired from seals swimming in salt water using electrodes mounted on the surface of the skin. The most diffi-

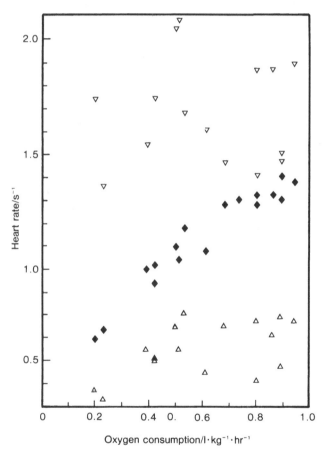

Fig. 14.4 Heart rate as a function of metabolic rate in a grey seal. Upright triangles represent average rates during dives, while downward pointing triangles represent heart rate during breathing episodes. Filled symbols represent average rates over complete dive/breathing cycles, and are comparable to heart rate curves for typical mammals. (After Fedak *et al.*, 1988.)

cult problems to overcome are: (a) isolation of the electrode from sea water and (b) the provision of a robust connection between the electrodes and the amplifier/transmitter sending the signal. This can be accomplished using readily available materials without any specialized equipment from a copper disc, woven rope and quick setting epoxy resin (Fedak *et al.*, 1988). Ultrasonic telemetry or a combined data storage and radio transmission device are obvious choices for sending the information because radio signals are blocked by sea water. Both techniques have been successfully used on seals in the wild (Hill, 1986; Fedak *et al.*, 1988).

Breathing parameters and dive duration

Breathing rate and the proportion of time spent at the surface are both possible correlates of exercise level and metabolic activity. In grey seals, the fraction of time spent at the surface increases with exercise and metabolic rate, as would be expected as the rate that O_2 and CO_2 can be exchanged is approached by the rate of use and production in the tissues. In common seals, nearly 80% of the variation in mean heart rate could be explained by considering only percent dive time and dive duration (Fedak *et al.*, 1988). These variables are likely to be good predictors of activity level in diving seals and have the advantage that they can be monitored without specialized transducers. Simple sonic pingers or depth transmitters can provide this information, although, in certain situations, direct observation might be sufficient.

Swimming speed and distance travelled

Work must be done to move through the water and, therefore, speed and distance travelled may provide an index of energy expenditure and simple tracking transmitters or swim speed transmitters can provide this information. The power requirements of swimming have been established for seals swimming in water flumes and in tanks, but direct application of these results to other species and seals swimming in the wild is problematic for two reasons. First, the flow characteristics in swimming flumes may not model those encountered in nature so that direct transfer of energy requirements from one to the other may not be possible. Secondly, the relationship between metabolic power and speed is not linear, both on theoretical and experimental grounds, so that the energy cost required to travel a given distance will vary with swimming speed and is difficult to estimate. Nonetheless, both measures can provide comparative indices of energy expenditure in some situations. Monitoring of swimming speed could also give some idea of how prey is captured and of general activity patterns.

Behavioural correlates

The way in which animals partition their time reflects in some way the partitioning of energy expenditure. This can be measured by radio-telemetry, as has been done in sea otters (Loughlin, 1979) and female fishers (Powell & Leonard, 1983), or by direct observation to determine partitioning of time into behavioural categories such as resting, suckling and foraging (Anderson & Harwood, 1985). Suitable techniques for behavioural observations are the scan sampling technique or the method of recording all occurrences of certain behaviours (see chapter 6).

Where the energy costs of these separate activities have been measured, they can then be summed in proportion to the partitioned time to derive the total energy budget (Costa, 1978). In addition, when the total energy budget has been measured, for example by mass change, but only some of the behavioural components have measured energy costs, then the energy costs of the missing elements of the behavioural repertoire can be derived by difference (e.g. flight energetics in terns, Flint & Nagy, 1984; flight energetics in albatrosses, Costa & Prince, 1987; diving energetics, Costa, 1988).

The combination of behavioural observations and measurement of energy expenditure in marked individuals has proved particularly fruitful in furthering understanding of the determinants of reproductive success in grey seals (Anderson & Fedak, 1985, 1987a). Energy can be regarded as a biological 'currency', and, like money, can be used to compare a wide range of processes and behavioural activities. For instance, the cost of investment by females in raising pups to weaning can be compared with the costs to males of completing a successful breeding season. Marked variation in energy expenditure between individual males has been positively correlated with sexual success (Anderson & Fedak, 1985). Large females are capable of investing more energy resources in their offspring, and show a tendency to have more male than female young (Anderson & Fedak, 1987a). Similar studies combining physiological and behavioural measures of maternal investment have been made in Steller's sea lion (*Eumetopias jubatus*) (Higgins *et al.*, 1988).

Conclusions

Marine mammals can present formidable problems to researchers, while at the same time being very suitable subjects for particular studies. A large part of their lifetime is spent at sea which makes them unamenable to direct observation. They are large animals – the smallest fur seals weigh about 30 kg, whereas the largest of the pinnipeds, the southern elephant seal, may be as much as 4 tonnes. Physical restraint and handling (see chapter 3) require manpower, equipment and ingenuity. The cost of immobilizing drugs and some isotopes may be prohibitively high for the larger species. On the positive side, the partitioning of their activities into offshore and terrestrial phases allows separate study of processes such as foraging and lactation. Once ashore, animals are continuously available as in the case of phocids, or predictably regular in their arrival pattern as in the case of otariids. They have relatively limited mobility while on land due to their aquatic adaptations, a feature which can contribute to their catchability even if they are hard to handle when caught.

The fact that many species of pinnipeds fast during various stages of pupping, lactation and reproduction means that they show marked changes in body size, condition and composition during the annual cycle. This characteristic, most marked in phocid seals, makes the technique of estimating energy use by weight change particularly appropriate. The advantages are that it is a simple method to use once the logistics of animal handling have been worked out, and it gives a direct measure of resources utilized. The drawbacks of this technique are that it gives a gross measurement of energy used over the period measured, without partitioning into components (such as metabolism, activity, milk production), and that it requires information on the energy density of the weight changes. It is clear from our review of the techniques available for assigning energy density values to weight change information that such techniques are labour intensive. One further drawback of the weight change technique is the assumption, that is usually made, that the components of the body which contribute to the weight loss are used in a similar manner in all age and size classes of seals, an assumption which may not be valid. In summary, the use of weight changes to estimate energy loss or gain is a straightforward method, useful in its own right and particularly valuable where conversion data are available. It is an essential adjunct to other more complicated approaches.

When animals feed during the lactation and reproductive periods, weight change cannot be used to measure energy lost or gained, but isotope injection may be the method of choice, as it offers several advantages in the study of marine mammal energetics. A direct measure of body composition can be gained from a single injection of isotope. The components of energy expenditure, such as foraging energetics and milk production, can be measured. Between the first and second capture the animal is completely undisturbed, so that an accurate measure of natural, free-living activity is collected. On the negative side, the use of some isotopes is restricted, and some, particularly oxygen-18, are extremely expensive. The techniques also require that a number of assumptions are made and, of course, the animals still have to be captured and handled. Although it is possible to look at some components of energy expenditure, the measures are still an average for the period between injections and sampling, so that small scale components, such as costs of swimming to feeding areas, cannot be identified separately.

Measurements of oxygen consumption and heat production provide the most direct assessment of energy expenditure that can be made, but require complex laboratory facilities or the construction of cumbersome field metabolic chambers. Captive facilities for seals, with their large size and specialized needs, are expensive to construct and maintain. The stress placed on the animals may produce results unrepresentative of free-living animals, and the range of behaviours which can be accommodated in a small chamber is limited.

Correlates of energy expenditure may provide only a crude estimate or index of energy expenditure, but they are often useful to study because they involve low-cost, practical field methods. If they are combined with laboratory or other field studies that provide validation of the correlate, then they can be the methods of choice, particularly in remote field conditions. They can have a specific advantage in providing small scale details of biological processes under study, such as the behavioural components of reproduction.

In summary, no one technique available for studying energy expenditure in pinnipeds is superior to another. Each has an appropriate use, and a combination of two or more techniques may produce a comprehensive assessment of energy expenditure. Where two different techniques can be used to answer the same question, an opportunity exists to validate one technique against another.

References

Agricultural Research Council. 1965. *The Nutrient Requirements of Farm Livestock*, nos. 1 and 2. Agricultural Research Council, London.

Agricultural Research Council. 1968. *The Nutrient Requirements of Ruminant Livestock*, pp. 24–117. Gersham, Farnham Royal.

Anderson, S.S. & Fedak, M.A. 1985. Grey seal males: energetic and behavioural links between size and sexual success. *Anim. Behav.* **33**, 829–38.

Anderson, S.S. & Fedak, M.A. 1987a. Grey seal energetics: females invest more in male offspring. *J. Zool. Lond.* **211**, 667–79.

312   *S.S. Anderson, D. Costa and M.A. Fedak*

Anderson, S.S. & Fedak, M.A. 1987b. The energetics of sexual success of grey
    seals and comparison with the costs of reproduction in other pinnipeds.
    In *Reproductive Energetics in Mammals*, ed. A.S.I. Loudon & P.A.
    Racey, pp. 319-42. *Symp. Zool. Soc. Lond.* **57**. Clarendon Press, Oxford.
Anderson, S.S. & Harwood, J. 1985. Time budgets and topography: how
    energy reserves and terrain determine the breeding behaviour of grey
    seals. *Anim. Behav.* **33**, 1343-8.
Ashwell-Erickson, S. & Elsner, R. 1982. The energy cost of free-existence for
    Bering Sea harbor and spotted seals. In *Eastern Bering Sea Shelf:
    Oceanography and Resources*, vol. 2, ed. D.W. Hood & J.A. Calder.
    University of Washington Press, Seattle.
Baldwin, R.L. & Bywater, A.C. 1984. Nutritional energetics of animals. *Ann.
    Rev. Nutr.* **4**, 101-14.
Brody. S. 1945. *Bioenergetics and Growth*. Hafner Press, New York.
Bryden, M.M., Smith, M.S.R., Tedman, R.A. & Featherston, D.W. 1984.
    Growth of the Weddell seal, *Leptonychotes weddelli*. *Aust. J. Zool.* **32**,
    33-41.
Costa, D.P. 1978. *Ecological energetics, water and electrolyte balance of the
    Californian sea otter*, Enhydra lutris. Ph.D. Dissertation University of
    California, Santa Cruz.
Costa, D.P. 1982. Energy, nitrogen, and electrolyte flux and sea water
    drinking in the sea otter, *Enhydra lutris*. *Physiol. Zool.* **55**, 35-44.
Costa, D.P. 1986. *Assessment of the Impact of the Californian Sea Lion and
    Elephant Seal on Commercial Fisheries R/F-92*. In Californian Sea Grant
    College Program Final Report, Institute of Marine Research, University
    of California, San Diego, CA.
Costa, D.P. 1987. Isotopic methods for quantifying material and energy
    intake of free-ranging marine mammals. In *Approaches to Marine
    Mammal Energetics*, ed. A.C. Huntley, D.P. Costa, G.A.J. Worthy &
    M.A. Castellini, pp. 42-65. Allen Press, Lawrence, KA.
Costa, D.P. 1988. Methods for studying energetics of freely diving animals.
    Can. J. Zool. **66**, 45-52.
Costa , D.P. & Gentry, R.L. 1986. Reproductive energetics of the northern
    fur seal. In *Fur Seals: Maternal Strategies at Land and Sea*, ed. R.L.
    Gentry & G.L. Kooyman, pp. 79-101. Princeton University Press,
    Princeton.
Costa, D.P. & Prince, P.A. 1987. Foraging energetics of Grey-headed
    Albatrosses *Diomedea chrysostoma* at Bird Island, South Georgia. *Ibis.*
    129, 149-58.
Costa, D.P. & Trillmich, F. 1988. Mass Changes and Metabolism during the
    perinatal fast, a comparison between Antarctic (*Arctocephalus gazella*)
    and Galapagos fur seals (*A. galapagoensis*). *Physiol. Zool.* **61**, 160-9.
Costa, D.P., Croxall, J.P. & Duck, C.D. (1989) Foraging energetics of
    Antarctic fur seals, *Arctocephalus gazella*, in relation to changes in prey
    availability. *Ecology.* **70**, 596-606.
Costa, D.P., Dann, P. & Disher, W. 1986. Energy requirements of free-
    ranging Little Penguin, *Eudyptula minor*. *Comp. Biochem. Physiol.* **85A**,
    135-8.
Costa, D.P., Le Boeuf, B.J., Ortiz, C.L. & Huntley, A.C. 1986. The
    energetics of lactation in the northern elephant seal. *J. Zool. Lond.* **209**,
    21-33.

Davis, R.W., Kooyman, G.L. & Croxall, J.P. 1983. Water flux and estimated metabolism of free-ranging Gentoo and Macaroni penguins at South Georgia. *Polar Biol.* **2**, 41-6.

Davis, R.W., Williams, T. & Kooyman, G.L. 1985. Swimming metabolism of yearling and adult harbor seals, *Phoca vitulina. Physiol. Zool.* **55**, 590-6.

Depocas, F., Hart, J.S. & Fisher, H.D. 1971. Sea water drinking and water flux in starved and in fed harbor seals, *Phoca vitulina. Can. J. Physiol. Pharmacol.* **49**, 53-62.

Fedak, M.A. 1986. Diving and exercise in seals: a benthic perspective. In *Diving in Animals and Man*, ed. A.O. Brubakk, J.W. Kanwisher & G. Sundnes, pp. 11-32. Roy. Norweg. Soc. Lett., Trondheim.

Fedak, M.A. & Anderson, S.S. 1982. The energetics of lactation: accurate measurements from a large wild mammal, the Grey seal (*Halichoerus grypus*). *J. Zool. Lond.* **198**, 473-9.

Fedak, M.A., Pullen, M.R. & Kanwisher, J. 1988. Circulatory responses of seals to periodic breathing: heart rate and breathing during exercise and diving in the laboratory and open sea. *Can. J. Zool.* **66**, 53-60.

Fedak, M.A., Rome, L. & Seeherman, H.J. 1981. One-step N2-dilution technique for calibrating open circuit VO2 measuring systems. *J. Appl. Physiol., Resp. Env. Exercise Physiol.* **51**, 772-6.

Feldkamp, S.D. 1988. Swimming in the Californian sea lion: morphometrics, drag and energetics. *J. Exper. Biol.* **131**, 117-35.

Flint, E.N. & Nagy, K.A. 1984. Flight energetics of free-living Sooty Terns. *Auk*, **101**, 288-94.

Foy, J.M. & Schnieden, H. 1960. Estimation of total body water (virtual tritium space) in the rat, cat, rabbit, guinea-pig and man, and of the biological half-life of tritium in man. *J. Physiol.* **154**, 169-76.

Gales, N.J. & Burton, H.R. 1987. Ultrasonic measurement of blubber thickness in southern elephant seal, *Mirounga leonina* (Linneus). *Aust. J. Zool.* **35**, 207-17.

Gentry, R.L. & Kooyman, G.L. (Eds). 1986. *Fur Seals: Maternal Strategies on Land and at Sea.* Princeton University Press, Princeton.

Higgins, L.V., Costa, D.P., Le Boeuf, B.J. & Huntley, A.C. 1988. Behavioural and physiological measurements of maternal investment in the Steller sea lion, *Eumetopias jubatus. Mar. Mamm. Sci.* **4**, 44-58.

Hill, R.D. 1986. Microcomputer monitor and blood sampler for free-diving Weddell seals. *J. Appl. Physiol.* **61**, 1570-6.

Keiver, K.M., Ronald, K. & Beamish, F.W.H. 1984. Metabolizable energy requirements for maintenance and faecal and urinary losses of juvenile harp seals (*Phoca groenlandica*). *Can. J. Zool.* **62**, 769-76.

Kleiber, M. 1975. *The Fire of Life.* Krieger Publishing, New York.

Kooyman, G.L., Davis, R.W., Croxall, J.P. & Costa, D.P. 1982. Diving depths and energy requirements of King Penguins. *Science* **217**, 726-7.

Ku, H.H. 1969. *Precision, measurement and calibration.* Special Publication, vol. 1. *Nat. Bur. Stand.*, Washington, D.C.

Lavigne, D.M., Barchard, W., Innes, S. & Øritsland, N.A. 1982. Pinniped Bioenergetics. In *Mammals in the Seas*. Vol. 4, pp. 191-235. FAO Fish. Ser. No. 5. FAO, Rome.

Laws, R.M. 1953. *The Elephant Seal* (Mirounga leonina *Linn.*). *I Growth and Age.* Falkland Islands Dependencies Survey Report No. 8, 1-62.

Leith, H. 1975. The measurement of caloric values. In Primary Productivity of the Biosphere, ed. H. Leith & R. Whittaker, pp. 119–29. Springer-Verlag, New York.

Lifson, N. & McClintock, R. 1966. Theory and use of the turnover rates of body water for measuring energy and material balance. *J. Theoret. Biol.* **12**, 46–74.

Lifson, N.A., Gordon, G.B. & McClintock, R. 1955. Measurement of total carbon dioxide production by means of D2180. *J. Appl. Physiol.* **7**, 704–10.

Lifson, N.A., Gordon, G.B., Visscher, M.B. & Nier, A.O.C. 1949. The fate of utilized molecular oxygen and the sources of the oxygen of respiratory carbon dioxide, studied with the aid of heavy oxygen. *J. Biol. Chem.* **180**, 803–11.

Loughlin, T.R. 1979. Radio telemetric determination of the 24-hour feeding activities of sea otters, *Enhydra lutris*. In *Handbook on Biotelemetry and Radio Tracking*, ed. C.J. Amlaner, D.W. McDonald, pp. 717–24. Pergamon Press, New York.

McCann, T.S., Fedak, M.A. & Harwood, J. 1989. Parental investment in southern elephant seals, *Mirounga leonina*. *Behav. Ecol. Sociobiol.* **25**, 81–7.

Miller, L.K. 1978. *Energetics of the Northern Fur Seal in Relation to Climate and Food Resources of the Bering Sea*. US Marine Mammal Commission, Washington, DC. Report MMC-75/08. [Available from US Department of Commerce, National Technical Information Service, Springfield, VA as PB 275 296].

Mook, W.G. & Grootes, P.M. 1973. The measuring procedure and corrections for the high precision mass spectrometric abundance ratios, especially referring to carbon, oxygen and nitrogen. *Int. J. Mass Spectr. and Ion Phys.* **12**, 273–98.

Nagy, K. 1980. CO_2 production in animals: an analysis of potential errors in the doubly labeled water technique. *Am. J. Physiol.* **238**, R466–73.

Nagy, K.A. & Costa D.P. 1980. Water flux in animals: analysis of potential errors in the tritiated water method. *Am. J. Physiol.* **238**, R454–65.

National Research Council. 1976. *Nutrient Requirements of Beef Cattle*. National Academy of Science, Washington, D.C.

National Research Council. 1981a. *Nutrient Requirements of Dairy Cattle*. National Academy of Science, Washington, D.C.

National Research Council. 1981b. *Nutritional energetics of Domestic Animals and Glossary of Terms*. National Academy of Science, Washington D.C.

Oftedal, O.T. & Iverson, S.J. 1987. Hydrogen isotope methodology for measurement of milk intake and energetics of growth in suckling pinnipeds. In *Approaches to Marine Mammal Energetics*. ed. A.C. Huntley, D.P. Costa, G.A.J. Worthy & M.A. Castellini, pp. 66–95. Allen Press, Lawrence KA.

Oftedal, O.T., Iverson, S.J. & Boness, D.J. 1987. Milk and energy intakes of suckling California sea lion (*Zalophus californianus*) pups in relation to sex, growth and predicted maintenance requirements. *Physiol. Zool.* **60**, 560–75.

Øritsland, N.A., Pasche, A.J., Markussen, N.H. & Ronald, K. 1985. Weight loss and catabolic adaptation to starvation in grey seal pups. *Comp. Biochem. Physiol.* **82**, 931–3.

Ortiz, C.L., Costa, D.P. & Le Boeuf, B.J. 1978. Water and energy flux in elephant seal pups fasting under natural conditions. *Physiol. Zool.* **238**, 166–78.

Ortiz, C.L., Le Boeuf, B.J. & Costa, D.P. 1984. Milk intake of elephant seal pups: an index of parental investment. *Amer. Nat.* **124**, 416-22.

Pace, N. & Rathbun, E.N. 1945. Studies on body composition. 11. The body water and chemically combined nitrogen content in relation to fat content. *J. Biol. Chem.* **158**, 685-91.

Paine, R.T. 1971. The measurement and application of the Calorie to ecological problems. *Ann. Rev. Ecol. Syst.* **2**, 145-64.

Powell, R.A. & Leonard, R.D. 1983. Sexual dimorphism and energy expenditure for reproduction in female fisher, *Martes pennanti. Oikos* **40**, 166-74.

Reilly, J.J. & Fedak, M.A. 1990. Measurement of body composition of living gray seals by hydrogen isotope dilution. *J. Appl. Physiol.* **69**, 885-91.

Ronald, K., Keiver, K.M., Beamish, F.W. & Frank, R. 1984. Energy requirements for maintenance and faecal and urinary losses of the grey seal (*Halichoerus grypus*). *Can. J. Zool.* **62**, 1101-5.

Schoeller, D.A. van Santen, E., Peterson, D.W., Dietz, W., Jaspan, J. & Klein, P.D. 1980. Total body water measurement in humans with 180 and 2D labeled water. *Amer. J. Clin. Nutr.* **33**, 2686-93.

Schoeller, D.A. & van Santen, E. 1982. Measurement of energy expenditure in humans by doubly labeled water method. *J. Appl. Physiol.* **53**, 955-9.

Schoeller, D.A. & Webb, P.A. 1984. Five day comparison of the doubly labeled water method with respiratory gas exchange. *Amer. J. Clin. Nutr.* **40**, 153-8.

Schoeller, D.A., Ravussin, E., Schutz, Y., Acheson, K.J., Baertschi, P. & Jequier, E. 1986. Energy expenditure by doubly labeled water: validation in humans and proposed calculation. *Amer. J. Physiol.* **250**, R823-30.

Shoemaker, V.H., Nagy, K.A. & Costa, W.R. 1976. Energy utilization and temperature regulation of jack rabbits in the Mojave Desert. *Physiol. Zool.* **49**, 364-75.

Smirnov, N. 1924. On the Eastern Harp Seal. *Tromsø Mus. Aarsh.* **47**, No. 2.

Smith, M.S.R. 1966. *Studies on the Weddell Seal in McMurdo Sound, Antarctica.* Ph.D. Dissertation, Univ. Canterbury, Christchurch, New Zealand.

Stewart, R.E.A. & Lavigne, D.M. 1984. Energy transfer and female condition in nursing harp seals, *Phoca groenlandica. Holarctic Ecol.* **7**, 182-94.

Stock, M. & Rothwell, N. 1982. *Obesity and Leaness: Basic Aspects.* John Libby and Sons, London.

Tarasoff, F.J. & Toews, D.P. 1972. The osmotic and ionic regulatory capacities of the kidney of the harbor seal, *Phoca vitulina. J. Comp. Physiol.* **81**, 121-32.

Terwilliger, D.T. 1977. An improved mass spectrometric method for determining the hydrogen abundance ratio. *Int. J. Mass Spectr. Ion. Phys.* **25**, 393-9.

Williams, T.M. & Kooyman, G.L. 1985. Swimming performance and hydrodynamic characteristics of harbor seals, *Phoca vitulina. Physiol. Zool.* **58**, 576-89.

Woods, P.E. 1982. Vertebrate digestive and assimilation efficiencies: taxonomic and trophic comparisons. *Biologist* **64**, 58-77.

Worthy G.A.J. 1982. *Energy sources of harp seals*, Phoca groenlandica, *during the post-weaning period.* Msc. thesis, Univ. Guelph, Ontario Canada., pp. 100.

15

Development of technology and research needs

R.M. LAWS

Introduction

In this chapter the suggested objectives of research and the types of research programmes needed to meet these objectives are outlined. They relate both to improved understanding of the basic biology of seals and to meeting the needs of two international conventions within the Antarctic Treaty System – the Convention for the Conservation of Antarctic Seals (CCAS) and the Convention for the Conservation of Antarctic Marine Living Resources (CCAMLR), which deals with all the component species of the Antarctic marine ecosystems (Appendices 16.5, 16.6). Typical research programmes are broadly framed so as to indicate where methodology and techniques particularly need to be improved. They are not static and will no doubt be subject to periodic review. The views of the Group of Specialists on Seals, and invited collaborators, on the main areas where developments of methods and techniques are needed are likewise as currently perceived. The emphasis may well change as existing technical problems are solved and new opportunities are grasped. It can be expected that new techniques will be developed and existing ones improved, particularly as computing, electronics, information technology and satellite instrumentation rapidly develop, making smaller but more effective equipment available to the Antarctic field scientist.

Objectives

The following objectives have emerged in discussion. The first is very broad and encompasses a wide diversity of topics in seal research.

1. To obtain as complete as possible an understanding of the basic biology and life history of all the species concerned. Literature

reviews of specific fields will indicate gaps in knowledge and methods and facilitate the formulation of research plans and procedures.

2. To identify unit breeding populations, their location, movements, degree of separation and possible mixing.

3. To identify appropriate management populations, their location, movements and degree of separation.

4. To obtain information on age, sex and seasonal variation in distribution patterns, and annual variation in age-specific survival and reproductive rates.

5. To obtain quantitative information on the principal prey species of Antarctic seals, by age, sex and season, and the numerical and functional relationships between the various seal species and other species – including whales, birds, fish and squid in the Antarctic marine food webs of which they are a part.

6. To obtain the knowledge and data necessary for informed decision making on suitable procedures for wise management and conservation, particularly in relation to the CCAS (Appendix 16.5) and CCAMLR (Appendix 16.6).

7. To develop the database and procedures necessary to assess and monitor the status of exploited stocks and, when appropriate, to estimate potential sustainable yields taking into consideration uncertainties concerning the identity and status of management stocks.

If an industry develops:

8. From the start of any industry, to compile accurate information on the hunting effort, by species, area and season; the number of animals taken, by species, area, age and sex; and the reproductive and physical condition of seals taken, from the start of any industry that may develop.

Objective 8 is presently met by the the requirements of CCAS (Appendix 16.5) which call for reporting of catches. Where regular non-commercial collections are made (e.g. for food for dog teams) every effort should be made to obtain data on seals killed. In relation to objective 5, it should be understood that seals are but one component of the Antarctic marine ecosystem and understanding the biology and ecology of seals, and determining the likely causes of observed changes in population size, growth rates, diet, and so on, requires reliable information on the structure, interrelationships and natural variability of the communities and ecosystems of which they are a part.

In order to achieve these objectives techniques will have to be developed and/or standardized. Before detailing these specific needs components of a desirable research programme are outlined.

Suggested research

Scope (seasonal and geographical)

In general, research should be carried out throughout the whole geographical range of species, throughout the year, and over a number of years. However, some questions could be addressed by research restricted both spatially and temporally. Some species and areas, because of their low densities or remoteness, present serious problems of access. Research in the pack ice, in the spring breeding season and in winter is particularly important, but is extremely difficult to mount owing to the logistic difficulties.

Basic biological studies on species

These include studies of behaviour, social structure, physiology, growth, reproduction, host–parasite relationships, and other fields relevant to the evolutionary and life histories of the various seal species. These studies will be particularly important for developing the information base necessary to recognize and interpret changes in population parameters that may occur in response to exploitation, or variation in food availability, or ecosystem structure caused by natural changes, fishery development or other human activities. Wherever possible studies should be made without killing seals, but some information can only be obtained from post-mortem examination of animals killed for research or in commercial operations.

For behaviour studies several categories of investigations are desirable. These include:

1. Determining the mating behaviour and factors influencing breeding success of crabeater, leopard and Ross seals.
2. Determining how fluctuations in food availability affect the care of pups and pup survival (for pack ice- as well as land-breeding forms); this is relevant to understanding the large fluctuations

in year class strength shown by some species (e.g. Weddell, crabeater and leopard seals (Laws, 1984; Testa *et al.*, 1991)).

3. Determining winter, as well as more easily determined summer, distribution and movement patterns and any variations or segregation by sex and age.
4. Determining whether vocalization patterns vary geographically and are useful indicators of stock separation.
5. Determining how disturbance from research and other activities may affect behaviour patterns and, perhaps, survival and productivity.

Abundance and distribution

Abundance

Observations should be made on land, or from the air, to establish absolute abundance of land-breeding fur seals, elephant seals and fast ice-breeding Weddell seals. This should preferably be undertaken in the pupping season, because counts of pups can be raised, to estimate total population, by applying a factor relating the size of this age class to the total population (chapter 2). Attempts should be made to mount surveys from ships (including ice-breakers) and aircraft to obtain reliable information on densities, relative abundance and if possible absolute abundance of species in the pack ice and, if possible in the open sea, at any time of year. Indices of abundance can be used to follow trends.

Current views on the abundance of the ice-breeding seals (crabeater, leopard, Ross and Weddell seals) are based on data obtained from past surveys in the pack ice zone, primarily as part of the US Antarctic Program from 1968 to 1983 and no significant surveys have been conducted since then. New analyses of the data (SCAR Group of Specialists on Seals, 1988) suggested a disturbing decline in population density values of crabeater seals, by far the most abundant species, between the late 1960s and 1983. This strongly reinforces the need for further work in the pack ice. Such investigations are impossible without a high level of logistic support, primarily the commitment to the pack ice zone of ice-strengthened or ice-breaking vessels (preferably carrying helicopters), and also if feasible, over-ice flights by fixed-wing aircraft. The Group of Specialists on Seals has emphasized repeatedly in its reports since 1983 that this is *the* outstanding problem in research on pack ice seals. Without such provision the Group is unable to make further progress in these investigations and SCAR will not be able to meet its obligations to provide advice to CCAS or to answer questions from CCAMLR. (Objectives: 2–5; methods: chapters 2, 3.)

Distribution

The mapping of seasonal pack ice distribution and type, preferably by satellite remote sensing, is particularly important in extrapolating sample seal densities to larger areas and in assessing the significance of residual pack ice regions in summer for the segregation of stocks. Seal biologists should press for this mapping to be done and can help by providing ground truth.

The use of mark–recapture data, studies of population genetics, morphometrics, and other indicators to identify and describe unit stocks is necessary. Such studies should preferably be restricted to breeding animals or pups at their breeding localities, in order to eliminate the confounding effects of movements between breeding localities. In the case of pack ice-breeding seals, abundances have been estimated during the summer (post-breeding) and it may be preferable to assess unit stocks of these species at the same time. (Objectives 2, 3, 5; methods: chapters 2, 5–10.)

Marking and telemetry

Existing marking programmes should be continued and new ones developed. Data on known-age individuals are also needed for confirmation of methods of age determination. More integration of these programmes internationally is necessary and greater effort should be put into reporting resightings or recoveries for use in demographic studies, including estimation of emigration and immigration rates. This is particularly relevant to land- or fast ice-breeding species (fur, elephant and Weddell seals) but for pack ice seals only if a substantial industry were to develop. For large scale movements and migrations of all species improved techniques are needed for the study of individuals, particularly the development of satellite tracking, radio-telemetry, sonic tagging and confirmation of indirect methods of age determination. (Objectives: 1, 2, 7; methods: chapters 4, 5, 11.)

Food resources and energetics

Investigations of the diet, feeding habits, ecological separation and the impact of seal stocks on the food resource by species, sex and age, should be continued. More detailed quantitative information is needed to refine existing knowledge and to establish possible temporal and spatial differences. There is a special need for information outside the breeding season and summer – that is for the winter months and from non-breeders.

The development of research programmes for estimating metabolic rates and energy flow in different species and populations is desirable. Just as biomass is a better measure than numbers for establishing the significance of a species in the ecosystem, so energetic terms are preferable to biomass for describing processes, especially in studies of interactions and competition with other consumers for a possibly limiting food resource. This will be particularly important in defining the direct response of seal stocks to exploitation and indirectly to exploitation of their food resource. (Objectives: 1, 3, 5–7; methods: chapters 5, 13, 14.)

Ecosystem monitoring

Article 2 of CCAMLR (Appendix 16.6) calls for an ecosystem approach to conservation, including allowance for the effects of harvesting a species at a lower trophic level (e.g. krill) on its consumers at a higher level (e.g. crabeater, leopard and fur seals). Changes with time in the patterns of behaviour, diet, growth and reproduction of individuals and populations of seals could throw light on changes in abundance and/or distribution of food resources. Monitoring methods need to be developed and CCAMLR has set up an Ecosystem Monitoring Program (CEMP) the objective of which is to 'detect and record significant changes in critical components of the ecosystem, to serve as a basis for the conservation of Antarctic marine living resources' (CCAMLR, 1991). This programme draws on advice from the SCAR Group of Specialists on Seals.

Antarctic fur seals

When concentrated in their breeding colonies, female fur seals feed almost exclusively on krill within a radius of about 100 miles. Changes in fur seal behaviour, physiology and life history characteristics may reflect changes in the abundance of their food supply during the breeding season. Methods of monitoring the changes in the biology of the predator need to be developed and particularly relevant aspects for monitoring, from year to year include: direct observations of the duration of the mother's foraging trips and pup attendance cycles; pup growth rates; fine structure of teeth (for indirect evidence of attendance cycles). Another character that may merit such attention is juvenile tooth size (for indirect evidence of growth rates).

Crabeater seals

Because of their great abundance, wide geographical range and dependence on krill as their staple food throughout the year, it is very important to attempt to monitor their populations. The problems of doing this effectively are formidable. Particularly relevant parameters for monitoring include cohort strength, body growth rates and body condition, age at sexual maturity and reproductive rate. However, these parameters all have long integration and lag times, generalizing the sum of environmental events over a lengthy period; the effect itself may only be detectable a number of years after the causative event occurred. (Objectives: 1, 6, 7; methods: chapters 8, 11, 12.)

Improvement, development and standardization of techniques

This chapter is concerned primarily with new methods and techniques that need to be devised, or existing procedures that need to be adapted and tested, so that they can be applied to improve knowledge of the biology of Antarctic seals. There is clearly much scope for improvement, but the Antarctic is not necessarily the best place to develop new field methods. However, once developed the applicability of such methods to Antarctic seals should be established. The following sections are all-embracing and many of the developments recommended should be related to the needs of specific research projects.

Field data recording

Until very recently field data were always recorded in notebooks or on proforma sheets. Now the use of small portable computers in the field allows the biologist to make preliminary analyses while the data are being collected. Errors commonly found with field recording of data are immediately recognized and the chance of understanding the cause of the error is much greater than it would be several months after the error was made. Also, preliminary analyses often allow the investigator to modify field procedures and experimental design, as these ongoing analyses may reveal shortcomings in the original design. Since there is a fair measure of standardization of computer formats, data sets from different investigators and studies are much more easily and quickly compared.

The availability of these portable computers that possess rather advanced analysis programs also now makes it possible to undertake many preliminary analyses that previously had to wait until biologists returned to their home laboratories. Such computers are particularly helpful for keeping track of tagged seals, or other data, that may be recorded year after year. For tag-resighting on a daily basis, new sightings can be entered into the computer and complete sighting histories that cover several years are then available. These procedures are particularly useful for studies of Weddell seals which usually return annually, with high probability, to the same area for pupping (Testa & Siniff, 1987). The keeping of tag records on a daily basis in the field also makes it possible to correct errors that would be impossible to correct once biologists have left the field. Further, once the field season is complete, these data are easily transferred to a larger computer for further analyses and permanent storage.

Field recording of behavioural data has also advanced to entry directly into computer format. Special 'keyboards' for coding behavioural observations now exist. Behavioural sequences are entered directly and the internal computer clock records the exact time each behaviour began and ended. Thus, the investigator is not constrained by the time required to write notes on a recording form or notebook. Again these data may be directly transferred to a larger machine once the field season has ended.

The major pitfall that must be avoided is the loss of information because of some failure (human or equipment), when entered data do not end up in the internal memory of the computer; if no written data exist then these data have been lost. There are several ways to avoid this problem. Data entry programs should contain codes that show the investigator when data have been entered into storage. Additionally, storage on an internal disc should be completed after each data set is entered. Behavioural observations may be entered without any written record being made and therefore are the type of data where care is essential to ensure proper storage. Other types of observations (e.g. tag-resightings) are best recorded in a field notebook and then entered each evening after that day's field work is completed. In this way a written record exists, so that re-entry is possible if a software or other computer failure occurs.

Estimation of population sizes

New counts of ice seals are urgently needed to provide empirical evidence to test hypotheses of population change since the original synthesis

(1960–83) (Erickson & Hanson, 1990). We also need surveys of fur seals at localities such as South Georgia and repeated surveys of elephant seals in view of the declining population trends that have been demonstrated at some breeding localities.

To improve methodology, further work on correction factors for counts of pack ice seals is necessary. This involves the study of behavioural patterns, particularly of diurnal, and longer, haul-out patterns – as well as how these patterns are affected by weather, time of year, ice type and other variables. Time budgets for surface and underwater activities at sea or inshore are also needed. This can be approached by direct, time-series counts of seals hauled out at a particular locality and season; also by telemetry, using time depth recorders and radio or satellite position fixing.

Stock identification

Satellite-linked tracking

Distribution and movement studies of individual animals can provide valuable indications of aggregations and unit populations, and thereby indicate the degree of geographical separation. (See section on radio-telemetry and satellite links).

Genetic-based studies

These hold promise for stock identification and with some exceptions adequate methods have already been developed and should be applied. Because of its remoteness the Antarctic is not the preferred location for conducting research to develop new techniques of immunological, biochemical and tissue-typing methods, which are more easily developed on more accessible species. However, as new techniques are developed they should be introduced to Antarctic research.

Body measurements

If separate stocks exist they may be characterized also by differing body size. A comparison should be made between the standard lengths of seals handled and released and those collected for specimens, to evaluate the potential for bias in the measurements currently being taken on live animals. Predictive formulae should be further developed for each species to relate girth and length measurements to body weight at different times of year.

Specimen collection

Pressure to demonstrate that the methods used in research (or in commercial operations) are humane will continue to increase. Further attention should be given to investigating methods of humane and 'non-destructive killing' by drugs, currently barbiturates (e.g. phenobarbitone). Studies needed for the collection of biometric data have been given in section on estimation of population size. There is also a need to develop and standardize preserving techniques and to improve preservatives for long-term storage. Developments in histology and histochemistry need to be kept under review, but the Antarctic is not the place where new techniques should be developed, unless they are specific to needs in the Antarctica.

With the increasing power and availability of computers (see section on field data recording) there may be a need for more specific relational databasing facilities, which would require study.

Immobilization and handling

Drugs

There is a tendency towards conservatism in the use of immobilizing drugs, and to continue with the use of a drug/species combination, once it has been shown to be successful. However, there is room for improvement and the aim should be to identify a drug or drugs which are both as safe and as effective as possible for use in the field. The efficacy of such drugs may vary by age, sex, time of year and weather conditions, as well as by species. Trials of existing and new immobilizing drugs over the range of species are called for. Research will be constrained by, for example drug availability and licensing requirements. Trials should be followed through to full recovery; this has frequently not been the case in past experimental work, with the consequence that valuable information has been lost.

Physical restraint

The 'bagging' or netting technique has been applied with success to restrain crabeater seals, leopard and Ross seals and a variety of capture and restraint methods have been developed for fur seals. P.R. Condy (pers. comm.), in one three-week spell caught and tagged about 800 crabeater seals, about 20 leopard seals and about 50 Ross seals, simply by having three men throw a nylon cargo net over the seal and hanging on for about a minute, after which the seal quietened down and submitted to capture. Surprisingly the leopard seals showed little if any fight, while crabeaters struggled the most. As soon as the seal had quietened, a dark blue cloth was put gently over its head (not wrapped around) and in most cases there was no further problem. Crabeater and leopard seals were only tagged but a blood sample was taken from most Ross seals, which required no drugs or additional physical restraint.

Similar and other methods should be developed for other species, because the disturbance to the animal may be less in the long run than using drugs.

Marking and telemetry

Development of improved marking techniques

Research is still required to develop improved tags for seals. The value of a specific type of tag varies according to the species of seal involved, as does the value of a single tag type on the same species in different habitats (i.e. ice vs. land). It is essential in many studies to have a truly persistent mark. For seals tags may always be inadequate in this respect; no tags developed to date are permanent and in general few last more than a few years.

Specific directed research is therefore needed to improve tag design, including the search for better materials, identification of the best design and material, attachment sites and application for individual species. Improving the efficiency of marking requires research into the causes and rates of tag loss. Toe- or web-clip marks are used in fur seal research to denote cohorts so that even if a tag is lost and the individual's identity with it, that animal can still be used as part of cohort population analysis. However, it requires an experienced observer to identify even the best clipped seals after a few years.

Other research needed on marking techniques includes the identification of improved dyes and paints, use of tattoo marking (probably combined with another more visible method) development of a simple and effective field freeze branding device, the provision of an effective lightweight forge for hot-branding, and assessment and possible development of the explosive branding technique. Natural patterns (e.g. scars or pelage markings) are very important for studies on some species (e.g. fur seals) and may have potential for other species; computer matching of characters has been used on northern species and should be investigated for Antarctic seals.

Radio-telemetry and satellite links

Radio- or satellite-telemetry offers a very important technique to determine short-term and year round movement patterns and may help to demonstrate stock discreteness for marine mammals. Fixing the position of instrumented animals with satellites is a rapidly developing field and further work on this is strongly encouraged. Positional data are particularly useful for determining feeding and haul-out patterns, and for assessing other behaviour patterns. For seals that can be reliably recovered, to obtain data on position only, it may be feasible simply to deploy a receiver (with a data storage unit) and use this to store successive position fixes from overpassing satellites, or based on daily light levels (chapter 5). It avoids the need to transmit (and hence reduces instrument package size). As soon as data on behaviour or physiology are required however, transmission becomes desirable, because of the volume of data, and it is for this purpose that satellite telemetry is particularly useful.

The urgent need to extend the working lifespan of instruments could be met by improving battery life, storing data and transmitting on interrogation or at infrequent intervals. This will be particularly important as satellite-linked instruments evolve. Also the use of microprocessor controlled recorders opens up new opportunities for telemetry, particularly of physiological data.

These difficulties could possibly be resolved – at a cost – by placing listening and relaying stations at key field locations. This approach would allow a smaller transmitter to be used, but suffers from the limitation that data would be obtained only when the animals are in the vicinity of the repeater station. However, developments in micro-electronics are leading to reductions in the size of instrument packages. This has several advantages.

First, it allows more room for batteries, offering either a greater signal strength or a longer operational lifespan for the instrument.

Secondly, it allows for greater complexity. More processing power means that the instrument can perform some of the data reduction and filtering, thereby extending the operational lifespan. The added 'intelligence' can also save power because the instrument can selectively switch components off or on as they are required, so that for example, the transmitter need only operate when the instrument is at the surface.

Thirdly, there is more space to accommodate memory. Memory chips are becoming smaller and this may be where the immediate future gains lie. The amount of information which can be stored in memory chips with a volume corresponding to the batteries required to give uplift to the satellite is enormous. This suggests concentration on developing recoverable devices, except for some applications such as positional information on crabeater seals.

Fourthly, a further consequence is that there is more room and processing capacity for attaching additional transducers. This will permit physiological data to be obtained in conjunction with data about diving and the environment of the seal. (Such developments should also have value for physical oceanographers interested in the properties of water masses by locality and depth.)

The great leap forward in the monitoring of pelagic species will be the production of a device which will automatically detach from a seal, to float on the surface where it can either transmit its position for subsequent collection by a ship, or transmit the contents of its memory to a ship or shore station.

The urgent need to extend the working lifespan of instruments could be met by improving battery life, storing data, and transmitting on interrogation or at infrequent intervals. This will be particularly important as satellite-linked instruments evolve. Also the use of microprocessor controlled recorders opens up new opportunities for telemetry, particularly of physiological data, such as heart rate.

Behaviour

The kinds of studies needed to advance knowledge of seal behaviour have been outlined earlier. New methods and techniques are being developed elsewhere, such as the use of portable sound recording apparatus, electronic notebooks and field computers, and should be applied where appropriate in the Antarctic (see section on field data recording).

Age determination

Improvement of age determination methods

Reliable age determination is the key to understanding many aspects of population ecology. A major comparative study of age-determination as an end in itself is required for the Antarctic seals, and in fact for the Pinnipedia as a whole. In such studies all the best methods developed to date should be applied to all tissues with annual growth layers in selected individuals of each species to assess what methods are best for that species. (The value of known-age specimens obtained from marking programmes is enormous and every effort should be made to acquire these and to make detailed observations and collections of material from them). Similarly, the reliability and accuracy of each method should be compared between species. A key question to be resolved is the age at which the first cementum layer is deposited. Considerable emphasis should be placed on careful experimentation with different preparations modified from both accepted and untested methods, in order to improve upon those techniques presently in use. Since most research workers have been interested in age-determination merely as a tool, they have tended to develop their respective techniques only to the point where they were functional. It will be helpful, for example, to further investigate cementum layers in post-canine and incisor teeth of elephant seals for ageing, because some female canine teeth may give misleadingly low ages based on dentine counts, and post-canine and incisor teeth can be extracted from living animals. An extensive and intensive, quantitative, experimental study of age determination will significantly improve the validity and reliability of techniques currently in use. In addition there is academic interest in the physiological and biochemical processes and causative factors in dentine and cementum deposition.

Micro-layer patterns in teeth

The annual growth layers in the teeth of most species usually show varying numbers of distinct micro-layers. Further study of the thickness, number and patterns of deposition of these micro-layers in individually tagged seals with known reproductive histories, moulting patterns and periods of feeding, nursing or suckling, could provide a basis for obtaining such information from the teeth of seals with unknown histories.

Method of tooth preservation and treatment

A series of teeth should be collected from each of a number of individual seals. They should be treated by boiling or in some other way and without such treatment, to establish – in quantitative terms – whether or not this damages the tissues and adversely affects readability, stainability or etching qualities.

Tooth etching techniques

The application of etching techniques for age determination should be evaluated for all Antarctic seal species. In this context, careful experimentation should be undertaken to establish the optimum acid concentrations, etching duration and most suitable peel techniques for each species.

Methods for aging live animals

Experimentation is required to develop techniques which can be used on the foreflipper nails of crabeater, leopard and elephant seals to facilitate age determination in the field similar to the method developed for Weddell seals. It seems likely that incisor teeth can only be removed by surgical means in most species, but may be more easily extractable in Ross and fur seals; investigations should be undertaken on the ease of extraction of post-canine teeth of all species, since this has recently been developed as a collecting method for elephant seal and fur seal teeth (chapter 11).

Moult patterns

Because of the probable implication of moulting physiology in formation of the tooth layer pattern, detailed, day to day documentation of the timing and duration of moult pattern and its relation to tooth growth is required for both pups and adults in most species of Antarctic seals.

Reproduction

Endocrinology

Techniques for establishing the hormonal status of individuals, so far mainly developed for domesticated and zoo animals, need to be adapted and applied to determine whether it is possible to establish the reproductive status of living immobilized seals in the field – whether individuals are immature, mature, in rut, in oestrus or pregnant. For example, colorimetric pregancy tests for humans could be applied to urine, serum or blood samples from seals. More specialized methods for determining hormone levels in blood and fluids of mammals could be applied to Antarctic seals, once the relevant normal variation in hormone levels is determined. A conventional method is radioimmunoassay (Hodges, 1985), but as a practical method of hormone measurement the ELISA assay (enzyme-linked immunosorbent assay) has important advantages. It has been used to diagnose pregnancy or monitor reproductive cycles in a number of species, including the Antarctic fur seal (Boyd, 1991; J.A. Hector, pers. comm.). Pregnancy can be diagnosed from urine or blood samples.

DNA fingerprinting techniques

Recent developments in DNA technology (Amos & Dover, 1990; Kirby, 1990) have revealed very powerful methods for studying relatedness, with a resolution capable of accurately assigning a seal pup to one or both parents. This information is required to establish paternity and hence reproductive success, particularly in polygynous species. The method is described in chapters 9 and 12, but there is a need for further studies of its application to Antarctic seals.

Diet

Collection and analysis of stomach samples

There is a need for information outside the breeding season and summer – that is, for the winter months and from non-breeders.

1. Collection of stomach contents from killed seals provides the best information, but lavage techniques do not involve sacrificing the animal; they have proved useful and could be refined. In the Antarctic the provision of water at warmer temperatures may be an advantage, although no real problems have emerged so far and glaciologists have solved greater problems in developing the hot-water ice drill.
2. Immunological techniques for identifying food material are in their infancy, but could be useful when stomach contents cannot be determined by gross examination. Type collections should be made to raise antigens for proteins against specific prey. We need quantitative methods, to give estimates of the amount of a certain prey present in the sample. The ELISA method (see also chapter 12) could be used in this way and studies should be undertaken.
3. There is a need to develop further the use of eyes, otoliths and beaks of krill, fish and squid, to estimate the original mass and length of ingested food items. Statoliths of squid could be useful for estimating their age (Clarke *et al.*, 1980) but this method needs further work on its application to seal stomach samples.

Experimental feeding studies

Experiments should be conducted on captive animals (especially crabeater seals) to establish the biases which may apply. These would include: studies of passage rates, digestibility and digestion rates of food material. A critical study of otoliths, by fish species and families, would yield valuable results and indicate which otolith measurements are most useful for estimating the original size of the prey; the same applies to squid beaks. For faecal samples the calculation of correction factors to allow for the loss of hard parts of prey animals (Murie, 1989), needs to be developed for Antarctic seal species.

Bioenergetics studies

In general the Antarctic is not a preferred environment for developing new techniques. It is probably best to apply existing methods, and new methods as they are worked up, to Antarctic species and situations. DNA fingerprinting (see also chapters 9 and 12) is vital for studies of reproductive success in polygynous species. Other methods dealing with radioisotopes, heart rate telemetry, meters to assess the energy costs of swimming, the costs of lactation, etc., need to be further developed.

The hydroxyproline method of estimating instantaneous growth rates has been applied to humans and other mammals (e.g. McCullagh, 1969). It would seem to have potential value for Antarctic seals and merits research on its applicability to them.

The ultrasound technique for determining blubber thickness, perhaps a key to developing realistic condition indices, is mentioned in chapter 14; further studies are needed, specifically on Antarctic seals.

Remote sensing

Identification from aerial photographs or satellite imagery

The identification of ice-breeding seals from aerial photographs or satellite imagery is often a problem. It would be worth while looking for cues that could be used for such identification.

Estimation of pack ice area, distribution and types

Estimation of the total area of pack ice present at the time of a census voyage is essential to extrapolate for total population size in the area. Satellite imagery is necessary for this, pertinent aspects being: (a) stratification of the imagery according to ice type and concentration (requiring high-resolution imagery) and (b) real-time availability of the satellite imagery is limited because of the cost, and many workers may not have access to either the imagery, or to the derived data concerning pack ice area and distribution.

Studies needed if an industry should ever develop

Harvest strategies, harvest monitoring and experimentation

Although there is currently no large-scale exploitation of Antarctic seals, and it seems unlikely that an industry will develop, an effort should be made to evaluate in advance the consequences of different harvest strategies and to identify the information which would be required to verify conclusions and to detect unforeseen effects should an industry develop. The analysis should assess how different age-specific harvest strategies (e.g. pup harvest vs. adult males) will affect population size, age/sex structure, reproductive capacity and sustainable yield levels. This depends on the relationships between density and age-specific survival rates and between density and age-specific reproductive rates. Computer simulation studies should be carried out to determine critical variables and data needs. However, data for validation of models are difficult to obtain, especially in the absence of an industry, when developed methods of analysing catch per unit of effort cannot be applied, and will necessarily involve long-term studies.

If an industry develops, information from independent scientific research can be supplemented by data from sealing operations. Accurate catch and effort information should be collected, reported, archived and analyzed to detect and monitor changes in such measures as catch per unit of effort, and the sex, age and size distributions of the catch. In addition, the harvest effort should be structured and complementary research programmes and perturbation experiments should be carried out to verify sustainable yield estimates. A programme of pelagic research captures from icebreakers in the pack ice would be desirable to complement and extend the results from commercial operations. (Objectives: 3, 5, 7; methods: chapters 3, 4, 7–9, 11–13.)

Basic studies on seals killed commercially

If an industry should ever develop it would be essential to obtain the maximum information on animals killed during commercial operations, to monitor the catches and to improve rational conservation and management measures. The following basic investigations are desirable:

Body measurements

Details of morphometrics, especially efficient methods of taking the standard body measurements are given in chapter 8. Methods of measuring and estimating body weights can be improved, and measurements of body length and proportions could be made by photogrammetric methods (Haley, Deutsch & Le Boeuf, 1991).

Age determination

Age determination is necessary to help establish growth rates, population structure and mortality rates, as well as sex- and age-specific reproductive parameters. Because of the numerous sex and age groups relatively large samples will ideally be required. For example, sample sizes of at least 800 (and preferably 1500) individuals have been found necessary in certain population ecology studies of exploited species in the Northern Hemisphere (T. Øritsiand, pers. comm.). However, other age-related studies can be carried out on much smaller samples.

Reproductive state and reproductive history

Age-specific and time-specific pregnancy rates are required. This would include estimation of ages at sexual maturity to establish possible trends in the mean age at sexual maturity.

Reporting, archiving and analysis

It is necessary to establish how these data should be reported, archived and analyzed so as to obtain maximum value from them.

Acknowledgements

In preparing this chapter I have been helped by the contributors to the individual chapters, but I wish to acknowledge particularly the contributions by J.L. Bengtson, W.N. Bonner, I.L. Boyd, P.R. Condy, J.P. Croxall, R.J. Hofman, T.S. McCann, P.D. Shaughnessy, A.A. Sinha and D.B. Siniff.

References

Amos, B. & Dover, G. 1990. DNA fingerprinting and the uniqueness of whales. *Mammal Rev.*, **20**, 23–30.
Boyd, I.L. 1991. Changes in plasma progesterone and prolactin concentrations during the annual cycle and the role of prolactin in the maintenance of lactation and luteal development in the Antarctic fur seal (*Arctocephalus gazella*). *J. Reprod. Fert.* **91**, 637–47.
CCAMLR 1991. *Standard methods for monitoring studies.* Scientific Committee for the Conservation of Antarctic Marine Living Resources, Hobart, Tasmania.

Clarke, M.R., Fitch, J.E., Kristensen, T., Kubodera, T. & Maddock, L. 1980. Statoliths of one fossil and four living squids (Gonatidae: Cephalopoda). *J. Mar. Biol. Assoc., U.K.* **60**, 329–47.

Erickson, A.W. & Hanson, M.B. 1990. Continental estimates and population trends of Antarctic ice seals. In *Antarctic Ecosystems. Ecological Change and Conservation*, ed. K.R. Kerry & G. Hempel. pp. 253–64. Springer-Verlag. Berlin.

Haley, M.P., Deutsch, C.J. & Le Boeuf, B.J. 1991. *Mar. Mamm. Sci.*, **7**, 157–64.

Hodges, J.K. 1985. The endocrine control of reproduction. *Symp. Zool. Soc. Lond.*, no. 54: 149–68.

Kirby, L.T. 1990. *DNA Fingerprinting: An Introduction*. Macmillan, London.

Laws, R.M. 1984. Seals. In *Antarctic Ecology*, vol. 2, ed. R.M. Laws, pp. 621–715. Academic Press, London.

McCullagh, K.G. 1969. The growth and nutrition of the African elephant. I. Seasonal variations in the rate of growth and the urinary excretion of hydroxyproline. *E. Afr. Wildl. J.* **7**, 85–90.

Murie, D.J. 1989. Experimental approaches to stomach content analyses of piscivorous marine mammals. In *Marine Mammal Energetics*, ed. A.C. Huntley, D.P. Costa, G.A.J. Worthy, and M.A. Castellini, pp. 147–63. Society for Marine Mammalogy, Special Publication No. 1.

SCAR Group of Specialists on Seals. 1988. Meeting of the SCAR Group of Specialists on Seals, Hobart, Tasmania, Australia, 23–25 August 1988. *BIOMASS Report Series*, No. 59, 1–61.

Testa, J.W., Oehlert, G., Ainley, D., Bengtson, J.L., Siniff, D.B., Laws, R.M. & Rounsevell, D. 1991. Temporal variability in Antarctic marine ecosystems: periodic fluctuations in the phocid seals. *Can. J. Fish. Aquatic Sci.* **48**, 631–9.

Testa, J.W. & Siniff, D.B. 1987. Population dynamics of Weddell seals (*Leptonychotes weddelli*) in McMurdo Sound, Antarctica. *Ecol. Monogr.* **57**, 149–65.

16

Appendices

16.1 Origins of scientific names of Antarctic seals

Elephant seal

Mirounga: from *miouroung*, the Australian aboriginal name for the species.

leonina: from Latin *leoninus*, lion-like, probably referring to both size and roar.

Weddell seal

Leptonychotes: from Greek *leptos*, small, slender, and *onux*, claw plus suffix *-otes*, denoting possession. Refers to small claws on hind flippers.

weddellii: named after James Weddell, who commanded the British sealing expedition, 1822–24, which penetrated to the head of the Weddell Sea. Description of the species based on a drawing by Weddell and skeletons collected on this voyage in the South Orkney Islands.

Ross seal

Ommatophoca: from Greek *omma*, eye, plus *phoca*; refers to enormous orbits.

rossii: named after Sir James Clark Ross, Commander of the British Expedition to the Antarctic in 1839–43. Description of the species first based on two skeletons collected on this voyage in the Ross Sea.

Crabeater seal

Lobodon: from Greek *lobos*, lobe, plus *odons*, tooth; refers to the conspicuously lobed postcanine teeth.

carcinophagus: from Greek *karkinos*, crab, plus *phagein*, to eat; a mistaken reference to its diet.

Leopard seal

Hydrurga: from Greek *hudor*, water, and (possibly) suffix *ourgos*, a worker in.

leptonyx: from Greek, as in *Leptonychotes*.

Fur seals

Arctocephalus: from Greek *arktos*, a bear, and *kephale*, head; bear-like appearance of head.

gazella: first described from a specimen collected at Iles Kerguelen by the German vessel S.M.S. *Gazelle* in 1874.

tropicalis: a reference to the locality as the first specimen was said, mistakenly, to have been collected on the north coast of Australia.

16.2 Vernacular names of Antarctic seals in different languages

Species (English)	Spanish	Portugese	Norwegian	German	French
Lobodon carcinophagus (crabeater)	Foca cangrejere	Foca caranguejeira	Crabeater sel	Krabbenfresser	Phoque crabier
Leptonychotes weddellii (Weddell)	Foca de Weddell	Foca de Weddell	Weddell sel	Weddellrobbe	Phoque de Weddell
Ommatophoca rossii (Ross)	Foca de Ross	Foca de Ross	Ross sel	Rossrobbe	Phoque de Ross
Hydrurga leptonyx (leopard)	Leoperdo Marino	Leopardo marinho	Sjøleopard	Seeleopard	Leopard de mer
Mirounga leonina (southern elephant)	Elefante Marino (Austrel)	Elefante marinho	Antarktisk sjøelefant	See-Elefant	Eléphant de mer
Arctocephalus gazella (Antarctic fur)	Lobo Fino Antartico Foca peletera	Lobo marinho antártico	Antarktisk pels sel	Antarktische pelzrobbe	Otarie de Kerguelen
Arctocephalus tropicalis (Sub-Antarctic fur)	Lobo Fino Austrel (Sub Antartico)	Lobo marinho subantártico	Sub-Antarktisk pels sel	Subantarktische pelzrobbe	Otarie de subantarctique

338

Japanese	Chinese	Italian	Russian
Kanikul-Azarashi		Foca crabeater	
カニクイアザラシ	鋸歯海豹		ТЮЛЕНЬ-КРАБОЕД
Wederu-Azarashi		Foca di Weddell	
ウエッデルアザラシ	威德爾海豹		ТЮЛЕНЬ УЭДДЕЛА
Rosu-Azarashi		Foca di Ross	
ロスアザラシ	羅斯海豹		ТЮЛЕНЬ РОССА
Hyo-Azarashi		Leopardo di mare	
ヒヨウアザラシ	豹形海豹		МОРСКОЙ ЛЕОПАРД
Zou-Azarashi		Elefante di mare	
ゾウアザラシ	象海豹		ЮЖНЫЙ МОРСКОЙ СЛОН
Minami-Ottosei	南極	Lupo di mare antartico	
ミナミオットセイ	毛皮海獅		КЕРГЕЛЕНСКИЙ МОРСКОЙ КОТИК
Minami-Ottosei	亞南極	Lupo di mare subantartico	
ミナミオットセイ	毛皮海豹		

16.3 Scientific names of mammal species referred to in this book

Antarctic fur seal	*Arctocephalus gazella*
Australian sea lion	*Neophoca cinerea*
Bearded seal	*Erignathus barbatus*
California sea lion	*Zalophus californianus*
Cape fur seal	*Arctocephalus pusillus*
Common seal	*Phoca vitulina*
Crabeater seal	*Lobodon carcinophagus*
Dusky dolphin	*Lagenorhynchus obscurus*
Elephant seal, southern	*Mirounga leonina*
Elephant seal, northern	*Mirounga angustirostris*
Fisher	*Martes pennanti*
Grey seal	*Halichoerus grypus*
Harbour seal	*Phoca vitulina*
Harp seal	*Phoca groenlandica*
Hooded seal	*Cystophora cristata*
Hooker's sea lion	*Phocarctos hookeri*
Humpback whale	*Megaptera novaeangliae*
Killer whale	*Orcinus orca*
Leopard seal	*Hydrurga leptonyx*
Monk seal, Pacific	*Monachus schauinslandi*
New Zealand fur seal	*Arctocephalus forsteri*
Northern fur seal	*Callorhinus ursinus*
Ringed seal	*Pusa hispida*
Ross seal	*Ommatophoca rossii*
Sea otter	*Enhydra lutris*
South American fur seal	*Arctocephalus australis*
Southern right whale	*Eubalaena australis*
Southern sea lion	*Otaria byronia* (= *flavescens*)
Steller's sea lion	*Eumetopias jubatus*
Sub-Antarctic fur seal	*Arctocephalus tropicalis*
Walrus	*Odobenus rosmarus*
Waterbuck	*Kobus defassa*
Weddell seal	*Leptonychotes weddellii*

16.4 Agreed Measures for the Conservation of Antarctic Fauna and Flora

The Representatives, taking into consideration Article IX of the Antarctic Treaty, and recalling Recommendation I-VIII of the First Consultative Meeting and Recommendation II-II of the Second Consultative Meeting, recommend to their Governments that they approve as soon as possible and implement without delay the annexed 'Agreed Measures for the Conservation of Antarctic Fauna and Flora'.

Preamble

The Governments participating in the Third Consultative Meeting under Article IX of the Antarctic Treaty,

Desiring to implement the principles and purposes of the Antarctic Treaty;

Recognising the scientific importance of the study of Antarctic fauna and flora, their adaptation to their rigorous environment, and their inter-relationship with that environment;

Considering the unique nature of these fauna and flora, their circumpolar range, and particularly their defencelessness and susceptibility to extermination;

Desiring by further international collaboration within the framwork of the Antarctic Treaty to promote and achieve the objectives of protection, scientific study, and rational use of these fauna and flora; and

Having particular regard to the conservation principles developed by the Scientific Committee on Antarctic Research (SCAR) of the International Council of Scientific Unions;

Hereby consider the Treaty Area as a Special Conservation Area and have agreed on the following measures:

Article I

1. These Agreed Measures shall apply to the same area to which the Antarctic Treaty is applicable (hereinafter referred to as the Treaty Area) namely the area south of 60° South Latitude, including all ice shelves.

   However, nothing in these Agreed Measures shall prejudice or in any way affect the rights, or the exercise of the rights, of any State under international law with regard to the high seas within the Treaty Area, or restrict the implementation of the provisions of the Antarctic Treaty with respect to inspection.

2. The Annexes to these Agreed Measures shall form an integral part thereof, and all references to the Agreed Measures shall be considered to include the Annexes.

Article II

For the purposes of these Agreed Measures:

   (a) 'Native mammal' means any member, at any stage of its life cycle, of any species belonging to the Class Mammalia indigenous to the Antarctic or occurring there through natural agencies of dispersal, excepting whales;

   (b) 'native bird' means any member, at any stage of its life cycle (including eggs), of any species of the Class Aves indigenous to the Antarctic or occurring there through natural agencies of dispersal;

   (c) 'native plant' means any kind of vegetation at any stage of its life cycle (including seeds), indigenous to the Antarctic or occurring there through natural agencies of dispersal;

   (d) 'appropriate authority' means any person authorised by a Participating Government to issue permits under these Agreed Measures;

   (e) 'permit' means a formal permission in writing issued by an appropriate authority;

   (f) 'participating government' means any Government for which these Agreed Measures have become effective in accordance with Article XIII of these Agreed Measures.

Article III

Each Participating Government shall take appropriate action to carry out these Agreed Measures.

Article IV

The Participating Governments shall prepare and circulate to members of expeditions and stations information to ensure understanding and observance of the provisions of these Agreed Measures, setting forth in particular prohibited activities, and providing lists of specially protected species and specially protected areas.

Article V

The provisions of these Agreed Measures shall not apply in cases of extreme emergency involving possible loss of human life or involving the safety of ships or aircraft.

Article VI

1. Each Participating Government shall prohibit within the Treaty Area the killing, wounding, capturing or molesting of any native mammal or native bird, or any attempt at any such act, except in accordance with a permit.
2. Such permits shall be drawn in terms as specific as possible and issued only for the following purposes:
   (a) to provide indispensable food for men or dogs in the Treaty Area in limited quantities, and in conformity with the purposes and principles of these Agreed Measures;
   (b) to provide specimens for scientific study or scientific information;
   (c) to provide specimens for museums, zoological gardens, or other educational or cultural institutions or uses.
3. Permits for Specially Protected Areas shall be issued only in accordance with the Provisions of Article VIII.
4. Participating Governments shall limit the issue of such permits so as to ensure as far as possible that:
   (a) no more native mammals or birds are killed or taken in any year than can normally be replaced by natural reproduction in the following breeding season;
   (b) the variety of species and the balance of the natural ecological systems existing within the Treaty Area are maintained.
5. The species of native mammals and birds listed in Annex A of these Measures shall be designated 'Specially Protected Species', and shall be accorded special protection by Participating Governments.
6. A Participating Government shall not authorise an appropriate authority to issue a permit with respect to a Specially Protected Species except in accordance with paragraph 7 of this Article.

7. A permit may be issued under this Article with respect to a Specially Protected Species, provided that:
   (a) it is issued for a compelling scientific purpose, and;
   (b) the actions permitted thereunder will not jeopardise the existing natural ecological system or the survival of that species.

Article VII

1. Each Participating Government shall take appropriate measures to minimize harmful interference within the Treaty Area with the normal living conditions of any native mammal or bird, or any attempt at such harmful interference, except as permitted under Article VI.
2. The following acts and activities shall be considered as harmful interference:
   (a) allowing dogs to run free;
   (b) flying helicopters or other aircraft in a manner which would unnecessarily disturb bird and seal concentrations, or landing close to such concentrations (e.g. within 200 metres);
   (c) driving vehicles unnecessarily close to concentrations of birds and seals (e.g. within 200 metres);
   (d) use of explosives close to concentrations of birds and seals;
   (e) discharge of firearms close to bird and seal concentrations (e.g. within 300 metres);
   (f) any disturbance of bird and seal colonies during the breeding period by persistent attention from persons on foot.
   However, the above activities, with the exception of those mentioned in a) and e) may be permitted to the minimum extent necessary for the establishment, supply and operation of stations.
3. Each Participating Government shall take all reasonable steps towards the alleviation of pollution of the waters adjacent to the coast and ice shelves.

Article VIII

1. The areas of outstanding scientific interest listed in Annex B shall be designated 'Specially Protected Areas' and shall be accorded special protection by the Participating Governments in order to preserve their unique natural ecological system.
2. In addition to the prohibitions and measures of protection dealt with in other Articles of these Agreed Measures, the Participating Governments shall in Specially Protected Areas further prohibit:

    (a) the collection of any native plant, except in accordance with a permit;

    (b) the driving of any vehicle.

3. A permit issued under Article VI shall not have effect within a Specially Protected Area except in accordance with paragraph 4 of the present Article.

4. A permit shall have effect within a Specially Protected Area provided that:

    (a) it was issued for a compelling scientific purpose which cannot be served elsewhere; and

    (b) the actions permitted thereunder will not jeopardise the natural ecological system existing in that Area.

Article IX

1. Each Participating Government shall prohibit the bringing into the Treaty Area of any species of animal or plant not indigenous to that Area, except in accordance with a permit.

2. Permits under paragraph 1 of this Article shall be drawn in terms as specific as possible and shall be issued to allow the importation only of the animals and plants listed in Annex C. When any such animal or plant might cause harmful interference with the natural system if left unsupervised within the Treaty Area, such permits shall require that it be kept under controlled conditions and, after it has served its purpose, it shall be removed from the Treaty Area or destroyed.

3. Nothing in paragraphs 1 and 2 of this Article shall apply to the importation of food into the Treaty Area so long as animals and plants used for this purpose are kept under controlled conditions.

4. Each Participating Government undertakes to ensure that all reasonable precautions shall be taken to prevent the accidental introduction of parasites and diseases into the Treaty Area. In particular, the precautions listed in Annex D shall be taken.

Article X

Each Participating Government undertakes to exert appropriate efforts, consistent with the Charter of the United Nations, to the end that no one engages in any activity in the Treaty Area contrary to the principles or purposes of these Agreed Measures.

Article XI

Each Participating Government whose expeditions use ships sailing under flags of nationalities other than its own shall, as far as feasible, arrange with the owners of such ships that the crews of these ships observe these Agreed Measures.

Article XII

1. The Participating Governments may make such arrangements as may be necessary for the discussion of such matters as:
   (a) the collection and exchange of records (including records of permits) and statistics concerning the numbers of each species of native mammal and bird killed or captured annually in the Treaty Area;
   (b) the obtaining and exchange of information as to the status of native mammals and birds in the Treaty Area, and the extent to which any species needs protection;
   (c) the number of native mammals or birds which should be permitted to be harvested for food, scientific study, or other uses in the various regions;
   (d) the establishment of a common form in which this information shall be submitted by Participating Governments in accordance with paragraph 2 of this Article.
2. Each Participating Government shall inform the other Governments in writing before the end of November of each year of the steps taken and information collected in the preceding period of July 1st to June 30th relating to the implementation of these Agreed Measures. Governments exchanging information under paragraph 5 of Article VII of the Antarctic Treaty may at the same time transmit the information relating to the implementation of these Agreed Measures.

Article XIII

1. After the receipt by the Government designated in Recommendation I-XIV (5) of notification of approval by all Governments whose representatives are entitled to participate in meetings provided for under Article IX of the Antarctic Treaty, these Agreed Measures shall become effective for those Governments.
2. Thereafter any other Contracting Party to the Antarctic Treaty may, in consonance with the purposes of Recommendation III-VII, accept these Agreed Measures by notifying the designated Government of its intention to apply the Agreed Measures and to be bound by them.

The Agreed Measures shall become effective with regard to such Governments on the date of receipt of such notification.

3. The designated Government shall inform the Governments referred to in paragraph 1 of this Article of each notification of approval, the effective date of these Agreed Measures and of each notification of acceptance. The designated Government shall also inform any Government which has accepted these Agreed Measures of each subsequent notification of acceptance.

Article XIV

1. These Agreed Measures may be amended at any time by unanimous agreement of the Governments whose Representatives are entitled to participate in meetings under Article IX of the Antarctic Treaty.
2. The Annexes, in particular, may be amended as necessary through diplomatic channels.
3. An amendment proposed through diplomatic channels shall be submitted in writing to the designated Government which shall communicate it to the Governments referred to in paragraph 1 of the present Article for approval; at the same time, it shall be communicated to the other Participating Governments.
4. Any amendment shall become effective on the date on which notifications of approval have been received by the designated Government from all of the Governments referred to in paragraph 1. of this article.
5. The designated Government shall notify those same Governments of the date of receipt of each approval communicated to it and the date on which the amendment will become effective for them.
6. Such amendment shall become effective on that same date for all other Participating Governments, except those which before the expiry of two months after that date notify the designated Government that they do not accept it.

Annexes to these agreed measures

Annex A
Specially protected species

Annex B
Specially protected areas

Annex C

Importation of animals and plants

The following animals and plants may be imported into the Treaty Area in accordance with permits issued under Article IX (2) of these Agreed Measures:

(a) sledge dogs;

(b) domestic animals and plants;

(c) laboratory animals and plants.

Annex D

Precautions to prevent accidental introduction of parasites and diseases into the Treaty Area

The following precautions shall be taken:

1. *Dogs:* All dogs imported into the Treaty Area shall be inoculated against the following diseases:

   (a) distemper;

   (b) contagious canine hepatitis;

   (c) rabies;

   (d) leptospirosis (*L. canicola* and *L. icterohaemor-rhagicae*).

   Each dog shall be inoculated at least two months before the time of its arrival in the Treaty Area.

2. *Poultry:* Notwithstanding the provisions of Article IX (3) of these Agreed Measures, no living poultry shall be brought into the Treaty Area after July 1st 1966.

Note added in proof:

The 'Agreed Measures' have been modified from time to time by Recommendations of the Antarctic Treaty Consultative Meetings, and are incorporated in a modified form in the Protocol on Environmental Protection to the Antarctic Treaty, signed in Madrid in June 1991, but not yet in force. Annex II to the Protocol deals with 'Conservation of Antarctic Fauna and Flora'; Annex V deals with 'Area Protection and Management'.

16.5 Convention for the Conservation of Antarctic Seals

The Contracting Parties,

Recalling the Agreed Measures for the Conservation of Antarctic Fauna and Flora, adopted under the Antarctic Treaty signed at Washington on 1 December 1959;

Recognizing the general concern about the vulnerability of Antarctic seals to commercial exploitation and the consequent need for effective conservation measures;

Recognizing that the stocks of Antarctic seals are an important living resource in the marine environment which requires an international agreement for its effective conservation;

Recognizing that this resource should not be depleted by over-exploitation, and hence that any harvesting, should be regulated so as not to exceed the levels of the optimum sustainable yield;

Recognizing that in order to improve scientific knowledge and so place exploitation on a rational basis, every effort should be made both to encourage biological and other research on Antarctic seal populations and to gain information from such research and from the statistics of future sealing operations, so that further suitable regulations may be formulated;

Noting that the Scientific Committee on Antarctic Research of the International Council of Scientific Unions (SCAR) is willing to carry out the tasks requested of it in this Convention;

Desiring to promote and achieve the objectives of protection, scientific study and rational use of Antarctic seals, and to maintain a satisfactory balance within the ecological system,

Have agreed as follows:

Article 1

Scope

1. This Convention applies to the seas south of 60° South Latitude, in respect of which the Contracting Parties affirm the provisions of Article IV of the Antarctic Treaty.
2. This Convention may be applicable to any or all of the following species:
   Southern elephant seal *Mirounga leonina*,
   Leopard seal *Hydrurga leptonyx*,
   Weddell seal *Leptonychotes weddelli*,
   Crabeater seal *Lobodon carcinophagus*,
   Ross seal *Ommatophoca rossi*,
   Southern fur seals *Arctocephalus* sp.
3. The Annex to this Convention forms an integral part thereof.

Article 2

Implementation

1. The Contracting Parties agree that the species of seals enumerated in Article 1 shall not be killed or captured within the Convention area by their nationals or vessels under their respective flags except in accordance with the provisions of this Convention.
2. Each Contracting Party shall adopt for its nationals and for vessels under its flag such laws, regulations and other measures, including a permit system as appropriate, as may be necessary to implement this Convention.

Article 3

Annexed measures

1. This Convention includes an Annex specifying measures which the Contracting Parties hereby adopt. Contracting Parties may from time to time in the future adopt other measures with respect to the conservation, scientific study and rational and humane use of seal resources, prescribing *inter alia:*
   (a) permissible catch;

(b) protected and unprotected species;

(c) open and closed seasons;

(d) open and closed areas, including the designation of reserves;

(e) the designation of special areas where there shall be no disturbance of seals;

(f) limits relating to sex, size, or age for each species;

(g) restrictions relating to time of day and duration, limitations of effort and methods of sealing;

(h) types and specifications of gear and apparatus and appliances which may be used;

(i) catch returns and other statistical and biological records;

(j) procedures for facilitating the review and assessment of scientific information;

(k) other regulatory measures including an effective system of inspection.

2. The measures adopted under paragraph 1. of this Article shall be based upon the best scientific and technical evidence available.

3. The Annex may from time to time be amended in accordance with procedures provided for in Article 9.

Article 4

Special permits

1. Notwithstanding the provisions of this Convention, any Contracting Party may issue permits to kill or capture seals in limited quantities and in conformity with the objectives and principles of this Convention for the purposes:

(a) to provide indispensable food for men or dogs;

(b) to provide for scientific research; or

(c) to provide specimens for museums, educational or cultural institutions.

2. Each Contracting Party shall, as soon as possible, inform the other Contracting Parties and SCAR of the purpose and content of all permits issued under paragraph 1. of this Article and subsequently of the numbers of seals killed or captured under these permits.

Article 5

Exchange of information and scientific advice

1. Each Contracting Party shall provide to the other Contracting Parties and to SCAR the information specified in the Annex within the period indicated therein.
2. Each Contracting, Party shall also provide to the other Contracting Parties and to SCAR before 31 October each year information on any steps it has taken in accordance with Article 2 of this Convention during the preceding period 1 July to 30 June.
3. Contracting Parties which have no information to report under the two preceding paragraphs shall indicate this formally before 31 October each year.
4. SCAR is invited:
   (a) to assess information received pursuant to this Article; encourage exchange of scientific data and information among the Contracting Parties; recommend programmes for scientific research; recommend statistical and biological data to be collected by sealing expeditions within the Convention area; and suggest amendments to the Annex; and
   (b) to report on the basis of the statistical, biological and other evidence available when the harvest of any species of seal in the Convention area is having a significantly harmful effect on the total stocks of such species or on the ecological system in any particular locality.
5. SCAR is invited to notify the Depositary which shall report to the Contracting Parties when SCAR estimates in any sealing season that the catch limits for any species are likely to be exceeded and, in that case, to provide an estimate of the date upon which the permissible catch limits will be reached. Each Contracting Party shall then take appropriate measures to prevent its nationals and vessels under its flag from killing or capturing seals of that species after the estimated date until the Contracting Parties decide otherwise.
6. SCAR may if necessary seek the technical assistance of the Food and Organization of the United Nations in making its assessments.
7. Notwithstanding the provisions of paragraph 1. of Article 1 the Contracting Parties shall, in accordance with their internal law, report to each other and to SCAR, for consideration, statistics relating to

the Antarctic seals listed in paragraph 2. of Article 1 which have been killed or captured by their nationals and vessels under their respective flags in the area of floating sea ice north of 60° South Latitude.

Article 6

Consultations between contracting parties

1. At any time after commercial sealing has begun a Contracting Party may propose through the Depositary that a meeting of Contracting Parties be convened with a view to:
   (a) establishing by a two-thirds majority of the Contracting Parties, including the concurring votes of all States signatory to this Convention present at the meeting, an effective system of control, including inspection, over the implementation of the provisions of this Convention:
   (b) establishing a commission to perform such functions under this Convention as the Contracting Parties may deem necessary; or
   (c) considering other proposals, including:
      (i) the provision of independent scientific advice;
      (ii) the establishment, by a two-thirds majority, of a scientific advisory committee which may be assigned some or all of the functions requested of SCAR under this Convention, if commercial sealing reaches significant proportions;
      (iii) the carrying out of scientific programmes with the participation of the Contracting Parties; and
      (iv) the provision of further regulatory measures, including moratoria.
2. If one-third of the Contracting Parties indicate agreement the Depositary shall convene such a meeting, as soon as possible.
3. A meeting shall be held at the request of any Contracting Party, if SCAR reports that the harvest of any species of Antartic seal in the area to which this Convention applies is having a significantly harmful effect on the total stocks or the ecological system in any particular locality.

Article 7

Review of operations

The Contracting Parties shall meet within five years after the entry into force of this Convention and at least every five years thereafter to review the operation of the Convention.

Article 8

Amendments to the Convention

1. This Convention may be amended at any time. The text of any amendment proposed by a Contracting Party shall be submitted to the Depositary, which shall transmit it to all the Contracting Parties.
2. If one-third of all Contracting Parties request a meeting to discuss the proposed amendment the Depositary shall call such a meeting.
3. An amendment shall enter into force when the Depositary has received instruments of ratification or acceptance thereof from all the Contracting Parties.

Article 9

Amendments to the Annex

1. Any Contracting Party may propose amendments to the Annex to this Convention. The text of any such proposed amendment shall be submitted to the Depositary which shall transmit it to all Contracting Parties.
2. Each such proposed amendment shall become effective for all Contracting Parties six months after the date appearing on the notification from the Depositary to the Contracting Parties, if within 120 days of the notification date, no objection has been received and two-thirds of the Contracting Parties have notified the Depositary in writing of their approval.
3. If an objection is received from any Contracting Party within 120 days of the notification date, the matter shall be considered by the Contracting Parties at their next meeting. If unanimity on the matter is not reached at the Contracting, Parties shall notify the Depositary within 120 days from the date of closure of the meeting of their approval

or rejection of the original amendment or of any new amendment proposed by the meeting. If, by the end of this period, two-thirds of the Contracting Parties have approved such amendment, it shall become effective six months from the date of the closure of the meeting for those Contracting Parties which have by then notified their approval.

4. Any Contracting Party which has objected to a proposed amendment may at any time withdraw that objection, and the proposed amendment shall become effective with respect to such Party immediately if the amendment is already in effect, or at such time as it becomes effective under the terms of this Article.

5. The Depositary shall notify each Contracting Party immediately upon receipt of each approval or objection, of each withdrawal of objection, and entry into force of any amendment.

6. Any State which becomes a party to this Convention after an amendment to the Annex has entered into force shall be bound by the Annex as so amended. Any State which becomes a Party to this Convention during the period when a proposed amendment is pending may approve or object to such an amendment within the time limits applicable to other Contracting Parties.

Article 10

Signature

This Convention shall be open for signature at London from 1 June to 31 December 1972 by States participating in the Conference on the Conservation of Antarctic Seals held at London from 3 to 11 February 1972.

Article 11

Ratification

This Convention is subject to ratification or acceptance. Instruments of ratification or acceptance shall be deposited with the Government of the United Kingdom of Great Britain and Northern Ireland, hereby designated as the Depositary.

Article 12

Accession

This Convention shall be open for accession by any State which may be invited to accede to this Convention with the consent of all the Contracting Parties.

Article 13

Entry into Force

1. This Convention shall enter into force on the thirtieth day following the date of deposit of the seventh instrument of ratification or acceptance.
2. Thereafter this Convention shall enter into force for each ratifying, or acceding State on the thirtieth day after deposit by such State of instrument of ratification, acceptance or accession.

Article 14

Withdrawal

Any Contracting Party may withdraw from this Convention on 30 June of any year by giving notice on or before 1 January of the same year to the Depositary, which upon receipt of such a notice shall at once communicate it to the other Contracting Parties. Any other Contracting Party may, in like manner, within one month of the receipt of a copy of such a notice from the Depositary, give notice of withdrawal, so that the Convention shall cease to be in force on 30 June of the same year with respect to the Contracting Party giving such notice.

Article 15

Notifications by the depositary

The Depositary shall notify all signatory and acceding States of the following:

    (a) signatures of this Convention, the deposit of instruments of ratification, acceptance or accession and notices of withdrawal;

(b) the date of entry into force of this Convention and of any amendments to it or its Annex.

Article 16

Certified copies and registration

1. This Convention, done in the English, French, Russian and Spanish languages, each version being equally authentic, shall be deposited in the archives of the Government of the United Kingdom of Great Britain and Northern Ireland, which shall transmit duly certified copies thereof to all signatory and acceding States.
2. This Convention shall be registered by the Depositary pursuant to Article 102 of the Charter of the United Nations.

IN WITNESS WHEREOF, the undersigned, duly authorized, have signed this Convention.

DONE at London, this 1st day of June 1972.

Annex

1. *Permissible Catch*

   The Contracting Parties shall in any one year, which shall run from 1 July to 30 June inclusive, restrict the total number of seals of each species killed or captured to the numbers specified below. These numbers are subject to review in the light of scientific assessments:
   (a) in the case of Crabeater seals *Lododon carcinophagus*, 175 000;
   (b) in the case of Leopard seals *Hydrurga leptonyx*, 12 000;
   (c) in the case of Weddell seals *Leptonychotes weddelli*, 5000.

2. *Protected Species*

   (a) It is forbidden to kill or capture Ross seals *Ommatophoca rossi*, Southern elephant seals *Mirounga leonina*, or fur seals of the genus *Arctocephalus*.
   (b) In order to protect the adult breeding stock during the period when it is most concentrated and vulnerable, it is forbidden to kill or capture any Weddell seal *Leptonychotes weddelli* one year old or older between 1 September and 31 January inclusive.

3. *Closed Season and Sealing Season*

   The period between 1 March and 31 August inclusive is a Closed Season, during which the killing or capturing of seals is forbidden.

The period 1 September to the last day in February constitutes a Sealing Season.

4. *Sealing Zones*

Each of the sealing zones listed in this paragraph shall be closed in numerical sequence to all sealing operations for the seal species listed in paragraph 1 of this Annex for the period 1 September to the last day of February inclusive. Such closures shall begin with the same zone as is closed under paragraph 2 of Annex B to Annex 1 of the Report of the Fifth Antarctic Treaty Consultative Meeting at the moment the Convention enters into force. Upon the expiration of each closed period, the affected zone shall reopen:

Zone 1–between 60° and 120° West Longitude

Zone 2–between 0° and 60° West Longitude, together with that part of the Weddell Sea lying westward of 60° West Longitude

Zone 3–between 0° and 70° East Longitude

Zone 4–between 70° and 130° East Longitude

Zone 5–between 130° East Longitude and 170° West Longitude

Zone 6–between 120° and 170° West Longitude.

5. *Seal Reserves*

It is forbidden to kill or capture seals in the following reserves, which are seal breeding areas or the site of long-term scientific research:

(a) The area around the South Orkney Islands between 60° 20′ and 60° 56′ South Latitude and 44° 05′ and 46° 25′ West Longitude.

(b) The area of the southwestern Ross Sea south of 76° South Latitude and west of 170° East Longitude.

(c) The area of Edisto Inlet south and west of a line drawn between Cape Hallett at 72° 19′ South Latitude, 170° 18′ East Longitude, and Helm Point, at 72° 11′ South Latitude, 170° 00′ East Longitude.

6. *Exchange of Information*

(a) Contracting Parties shall provide before 31 October each year to other Contracting Parties and to SCAR a summary of statistical information on all seals killed or captured by their nationals and vessels under their respective flags in the Convention area, in respect of the preceding period 1 July to 30 June. This information shall include by zones and months:

(i) The gross and nett tonnage, brake horse-power, number of crew, and number of days' operation of vessels under the flag of the Contracting Party;

(ii) The number of adult individuals and pups of each species taken.

When specially requested, this information shall be provided in respect of each ship, together with its daily position at noon each operating day and the catch on that day.

(b) When an industry has started, reports of the number of seals of each species killed or captured in each zone shall be made to SCAR in the form and at the intervals (not shorter than one week) requested by that body.

(c) Contracting Parties shall provide to SCAR biological information, in particular:

  (i) Sex

  (ii) Reproductive condition

  (iii) Age

SCAR may request additional information or material with the approval of the Contracting Parties.

(d) Contracting Parties shall provide to other Contracting Parties and to SCAR at least 30 days in advance of departure from their home ports, information on proposed sealing expeditions.

7. *Sealing Methods*

(a) SCAR is invited to report on methods of sealing and to make recommendations with a view to ensuring that the killing or capturing of seals is quick, painless and efficient. Contracting Parties, as appropriate, shall adopt rules for their nationals and vessels under their respective flags engaged in the killing and capturing of seals, giving due consideration to the views of SCAR.

(b) In the light of the available scientific and technical data, Contracting Parties agree to take appropriate steps to ensure that their nationals and vessels under their respective flags refrain from killing or capturing seals in the water, except in limited quantities to provide for scientific research in conformity with the objectives and principles of this Convention. Such research shall include studies as to the effectiveness of methods of sealing from the viewpoint of the management and humane and rational utilization of the Antarctic seal resources for conservation purposes. The undertaking and the results of any such scientific research programme shall be communicated to SCAR and the Depositary which shall transmit them to the Contracting Parties.

Note added in proof:

In September 1988 a meeting was held in London to review the operation of the Convention for the Conservation of Antarctic Seals (CCAS). The principal decisions taken at the meeting were:

(a) the annual reporting dates were changed;
(b) the amount of detail to be reported for each seal killed and captured was increased;
(c) the period of advance notice and information on sealing operations was increased;
(d) a ban on killing Weddell seal pups was put in place.

16.6 Convention on the Conservation of Antarctic Marine Living Resources

Preamble

The Contracting Parties,

Recognizing the importance of safeguarding the environment and protecting the integrity of the ecosystem of the seas surrounding Antarctica;

Noting the concentration of marine living resources found in Antarctic waters and the increased interest in the possibilities offered by the utilization of these resources as a source of protein;

Conscious of the urgency of ensuring the conservation of Antarctic marine living resources;

Considering that it is essential to increase knowledge of the Antarctic marine ecosystem and its components so as to be able to base decisions on harvesting on sound scientific information;

Believing that the conservation of Antarctic marine living resources calls for international co-operation with due regard for the provisions of the Antarctic Treaty and with the active involvement of all States engaged in research or harvesting activities in Antarctic waters;

Recognizing the prime responsibilities of the Antarctic Treaty Consultative Parties for the protection and preservation of the Antarctic environment and, in particular, their responsibilities under Article IX, paragraph l(f) of the Antarctic Treaty in respect of the preservation and conservation of living resources in Antarctica;

Recalling the action already taken by the Antarctic Treaty Consultative Parties including in particular the Agreed Measures for the Conservation of Antarctic Fauna and Flora, as well as the provisions of the Convention for the Conservation of Antarctic Seals;

Bearing in mind the concern regarding the conservation of Antarctic marine living resources expressed by the Consultative Parties at the Ninth Consultative Meeting of the Antarctic Treaty and the importance of the provisions of Recommendation IX-2 which led to the establishment of the present Convention;

Believing that it is in the interest of all mankind to preserve the waters surrounding the Antarctic continent for peaceful purposes only and to prevent their becoming the scene or object of international discord;

Recognizing in the light of the foregoing, that it is desirable to establish suitable machinery for recommending, promoting, deciding upon and co-ordinating the measures and scientific studies needed to ensure the conservation of Antarctic marine living organisms;

Have agreed as follows:

Article I

Scope and definitions

1. This Convention applies to the Antarctic marine living resources of the area south of 60° South latitude and to the Antarctic marine living resources of the area between that latitude and the Antarctic Convergence which form part of the Antarctic marine ecosystem.
2. Antarctic marine living resources means the populations of fin fish, molluscs, crustaceans and all other species of living organisms, including birds, found south of the Antarctic Convergence.
3. The Antarctic marine ecosystem means the complex of relationships of Antarctic marine living resources with each other and with their physical environment.
4. The Antarctic Convergence shall be deemed to be a line joining the following points along parallels of latitude and meridians of longitude:
   50°S, 0°; 50°S, 30°E; 45°S, 30°E; 45°S, 80°E; 55°S, 80°E; 55°S, 150°E; 60°S, 150°E; 60°S, 50°W; 50°S, 50°W; 50°S, 0°.

Article II

Objective

1. The objective of this Convention is the conservation of Antarctic marine living resources.
2. For the purposes of this Convention, the term 'conservation' includes rational use.

3. Any harvesting and associated activities in the area to which this Convention applies shall be conducted in accordance with the provisions of this Convention and with the following principles of conservation:
   (a) prevention of decrease in the size of any harvested population to levels below those which ensure its stable recruitment. For this purpose its size should not be allowed to fall below a level close to that which ensures the greatest net annual increment;
   (b) maintenance of the ecological relationships between harvested, dependent and related populations of Antarctic marine living resources and the restoration of depleted populations to the levels defined in sub-paragraph (a) above; and
   (c) prevention of changes or minimization of the risk of changes in the marine ecosystem which are not potentially reversible over two or three decades, taking into account the state of available knowledge of the direct and indirect impact of harvesting, the effect of the introduction of alien species, the effects of associated activities on the marine ecosystem and of the effects of environmental changes, with the aim of making possible the sustained conservation of Antarctic marine living resources.

Article III

Antarctic Treaty

The Contracting Parties, whether or not they are Parties to the Antarctic Treaty, agree that they will not engage in any activities in the Antarctic Treaty area contrary to the principles and purposes of that Treaty and that, in their relations with each other, they are bound by the obligations contained in Articles I and V of the Antarctic Treaty.

Article IV

Territorial sovereignty and coastal state jurisdiction

1. With respect to the Antarctic Treaty area, all Contracting Parties, whether or not they are Parties to the Antarctic Treaty, are bound by Articles IV and VI of the Antarctic Treaty in their relations with each other.

2. Nothing in this Convention and no acts or activities taking place while the present Convention is in force shall:
   (a) constitute a basis for asserting, supporting or denying a claim to territorial sovereignty in the Antarctic Treaty area or create any rights of sovereignty in the Antarctic Treaty area;
   (b) be interpreted as a renunciation or diminution by any Contracting Party of, or as prejudicing, any right or claim or basis of claim to exercise coastal state jurisdiction under international law within the area to which this Convention applies;
   (c) be interpreted as prejudicing the position of any Contracting Party as regards its recognition or non-recognition of any such right, claim or basis of claim;
   (d) affect the provision of Article IV, paragraph 2, of the Antarctic Treaty that no new claim, or enlargement of an existing claim, to territorial sovereignty in Antarctica shall be asserted while the Antarctic Treaty is in force.

Article V

Agreed Measures for the Conservation of Antarctic Fauna and Flora, etc

1. The Contracting Parties which are not Parties to the Antarctic Treaty acknowledge the special obligations and responsibilities of the Antarctic Treaty Consultative Parties for the protection and preservation of the environment of the Antarctic Treaty area.
2. The Contracting Parties which are not Parties to the Antarctic Treaty agree that, in their activities in the Antarctic Treaty area, they will observe as and when appropriate the Agreed Measures for the Conservation of Antarctic Fauna and Flora and such other measures as have been recommended by the Antarctic Treaty Consultative Parties in fulfilment of their responsibility for the protection of the Antarctic environment from all forms of harmful human interference.
3. For the purposes of this Convention, 'Antarctic Treaty Consultative Parties' means the Contracting Parties to the Antarctic Treaty whose Representatives participate in meetings under Article IX of the Antarctic Treaty.

Article VI

Relationship to existing conventions relating to the conservation of whales and seals

Nothing in this Convention shall derogate from the rights and obligations of Contracting Parties under the International Convention for the Regulation of Whaling and the Convention for the Conservation of Antarctic Seals.

Article VII

Commission for the conservation of Antarctic Marine Living Resources: Membership

1. The Contracting Parties hereby establish and agree to maintain the Commission for the Conservation of Antarctic Marine Living Resources (hereinafter referred to as 'the Commission').
2. Membership in the Commission shall be as follows:
   (a) each Contracting Party which participated in the meeting at which this Convention was adopted shall be a Member of the Commission;
   (b) each State Party which has acceded to this Convention pursuant to Article XXIX shall be entitled to be a Member of the Commission during such time as that acceding party is engaged in research or harvesting activities in relation to the marine living resources to which this Convention applies;
   (c) each regional economic integration organization which has acceded to this Convention pursuant to Article XXIX shall be entitled to be a Member of the Commission during such time as its States members are so entitled;
   (d) a Contracting Party seeking to participate in the work of the Commission pursuant to subparagraphs (b) and (c) above shall notify the Depositary of the basis upon which it seeks to become a Member of the Commission and of its willingness to accept conservation measures in force. The Depositary shall communicate to each Member of the Commission such notification and accompanying information. Within two months of receipt of such communication from the Depositary, any Member of the Commission may request that a special meeting of the Commission be held to

consider the matter. Upon receipt of such request, the Depositary shall call such a meeting. If there is no request for a meeting, the Contracting Party submitting the notification shall be deemed to have satisfied the requirements for Commission Membership.

3. Each Member of the Commission shall be represented by one representative who may be accompanied by alternate representatives and advisers.

Article VIII

Commission: legal personality, privileges and immunities

The Commission shall have legal personality and shall enjoy in the territory of each of the States Parties such legal capacity as may be necessary to perform its function and achieve the purposes of this Convention. The privileges and immunities to be enjoyed by the Commission and its staff in the territory of a State Party shall be determined by agreement between the Commission and the State Party concerned.

Article IX

Commission: functions, conservation measures, implementation, objection procedure

1. The function of the Commission shall be to give effect to the objective and principles set out in Article II of this Convention. To this end, it shall:
   (a) facilitate research into and comprehensive studies of Antarctic marine living resources and of the Antarctic marine ecosystem;
   (b) compile data on the status of and changes in population of Antarctic marine living resources and on factors affecting the distribution, abundance and productivity of harvested species and dependent or related species or populations;
   (c) ensure the acquisition of catch and effort statistics on harvested populations;
   (d) analyse, disseminate and publish the information referred to in sub-paragraphs (b) and (c) above and the reports of the Scientific Committee;
   (e) identify conservation needs and analyse the effectiveness of conservation measures;

(f)   formulate, adopt and revise conservation measures on the basis of the best scientific evidence available, subject to the provisions of paragraph 5 of this Article;

(g)   implement the system of observation and inspection established under Article XXIV of this Convention;

(h)   carry out such other activities as are necessary to fulfil the objective of this Convention.

2.   The conservation measures referred to in paragraph 1 (f) above include the following:

(a)   the designation of the quantity of any species which may be harvested in the area to which this Convention applies;

(b)   the designation of regions and sub-regions based on the distribution of populations of Antarctic marine living resources;

(c)   the designation of the quantity which may be harvested from the populations of regions and sub-regions;

(d)   the designation of protected species;

(e)   the designation of the size, age and, as appropriate, sex of species which may be harvested;

(f)   the designation of open and closed seasons for harvesting;

(g)   the designation of the opening and closing of areas, regions or subregions for purposes of scientific study or conservation, including special areas for protection and scientific study;

(h)   regulation of the effort employed and methods of harvesting, including fishing gear, with a view, *inter alia*, to avoiding undue concentration of harvesting in any region or sub-region;

(i)   the taking of such other conservation measures as the Commission considers necessary for the fulfilment of the objective of this Convention, including measures concerning the effects of harvesting and associated activities on components of the marine ecosystem other than the harvested populations.

3.   The Commission shall publish and maintain a record of all conservation measures in force.

4.   In exercising its functions under paragraph 1 above, the Commission shall take full account of the recommendations and advice of the Scientific Committee.

5.   The Commission shall take full account of any relevant measures or regulations established or recommended by the Consultative Meetings pursuant to Article IX of the Antarctic Treaty or by existing fisheries commissions responsible for species which may enter the area to which this Convention applies, in order that there shall be no inconsistency

between the rights and obligations of a Contracting Party under such regulations or measures and conservation measures which may be adopted by the Commission.

6. Conservation measures adopted by the Commission in accordance with this Convention shall be implemented by Members of the Commission in the following manner;

   (a) the Commission shall notify conservation measures to all Members of the Commission;

   (b) conservation measures shall become binding upon all Members of the Commission 180 days after such notification, except as provided in sub-paragraphs (c) and (d) below;

   (c) if a Member of the Commission, within ninety days following the notification specified in sub-paragraph (a), notifies the Commission that it is unable to accept the conservation measure, in whole or in part, the measure shall not, to the extent stated, be binding upon that Member of the Commission;

   (d) in the event that any Member of the Commission invokes the procedure set forth in sub-paragraph (c) above, the Commission shall meet at the request of any Member of the Commission to review the conservation measure. At the time of such meeting and within thirty days following the meeting, any Member of the Commission shall have the right to declare that it is no longer able to accept the conservation measure, in which case the Member shall no longer be bound by such measure.

Article X

Commission: monitoring function

1. The Commission shall draw the attention of any State which is not a Party to this Convention to any activity undertaken by its nationals or vessels which, in the opinion of the Commission, affects the implementation of the objective of this Convention.

2. The Commission shall draw the attention of all Contracting Parties to any activity which, in the opinion of the Commission, affects the implementation by a Contracting Party of the objective of this Convention or the compliance by that Contracting Party with its obligations under this Convention.

Article XI

Commission: relations with adjacent areas

The Commission shall seek to co-operate with Contracting Parties which may exercise jurisdiction in marine areas adjacent to the area to which this Convention applies in respect of the conservation of any stock or stocks of associated species which occur both within those areas and the area to which this Convention applies, with a view to harmonizing the conservation measures adopted in respect of such stocks.

Article XII

Commission: making of decisions

1. Decisions of the Commission on matters of substance shall be taken by consensus. The question of whether a matter is one of substance shall be treated as a matter of substance.
2. Decisions on matters other than those referred to in paragraph 1 above shall be taken by a simple majority of the Members of the Commission present and voting.
3. In Commission consideration of any item requiring a decision, it shall be made clear whether a regional economic integration organization will participate in the taking of the decision and, if so, whether any of its member States will also participate. The number of Contracting Parties so participating shall not exceed the number of member States of the regional economic integration organization which are Members of the Commission.
4. In the taking of decisions pursuant to this Article, a regional economic integration organization shall have only one vote.

Article XIII

Commission: headquarters, meetings, officers, subsidiary bodies

1. The headquarters of the Commission shall be established at Hobart, Tasmania, Australia.
2. The Commission shall hold a regular annual meeting. Other meetings shall also be held at the request of one-third of its members and

as otherwise provided in this Convention. The first meeting of the Commission shall be held within three months of the entry into force of this Convention, provided that among the Contracting Parties there are at least two States conducting harvesting activities within the area to which this Convention applies. The first meeting shall, in any event, be held within one year of the entry into force of this Convention. The Depositary shall consult with the signatory States regarding the first Commission meeting, taking into account that a broad representation of such States is necessary for the effective operation of the Commission.

3. The Depositary shall convene the first meeting of the Commission at the headquarters of the Commission. Thereafter, meetings of the Commission shall be held at its headquarters, unless it decides otherwise.

4. The Commission shall elect from among its members a Chairman and Vice-Chairman, each of whom shall serve for a term of two years and shall be eligible for re-election for one additional term. The first Chairman shall, however, be elected for an initial term of three years. The Chairman and Vice-Chairman shall not be representatives of the same Contracting Party.

5. The Commission shall adopt and amend as necessary the rules of procedure for the conduct of its meetings, except with respect to the matters dealt with in Article XII of this Convention.

6. The Commission may establish such subsidiary bodies as are necessary for the performance of its functions.

Article XIV

Scientific committee: membership, meetings, other experts

1. The Contracting Parties hereby establish the Scientific Committee for the Conservation of Marine Living Resources (hereinafter referred to as 'the Scientific Committee') which shall be a consultative body to the Commission. The Scientific Committee shall normally meet at the headquarters of the Commission unless the Scientific Committee decides otherwise.

2. Each Member of the Commission shall be a member of the Scientific Committee and shall appoint a representative with suitable scientific qualifications who may be accompanied by other experts and advisers.

3. The Scientific Committee may seek the advice of other scientists and experts as may be required on an ad hoc basis.

Article XV

Scientific committee: functions

1. The Scientific Committee shall provide a forum for consultation and co-operation concerning the collection, study and exchange of information with respect to the marine living resources to which this Convention applies. It shall encourage and promote co-operation in the field of scientific research in order to extend knowledge of the marine living resources of the Antarctic marine ecosystem.
2. The Scientific Committee shall conduct such activities as the Commission may direct in pursuance of the objective of this Convention and shall:
   (a) establish criteria and methods to be used for determinations concerning the conservation measures referred to in Article IX of this Convention;
   (b) regularly assess the status and trends of the populations of Antarctic marine living resources;
   (c) analyse data concerning the direct and indirect effects of harvesting on the populations of Antarctic marine living resources;
   (d) assess the effects of proposed changes in the methods or levels of harvesting and proposed conservation measures;
   (e) transmit assessments, analyses, reports and recommendations to the Commission as requested or on its own initiative regarding measures and research to implement the objective of this Convention;
   (f) formulate proposals for the conduct of international and national programs of research into Antarctic marine living resources.
3. In carrying out its functions, the Scientific Committee shall have regard to the work of other relevant technical and scientific organizations and to the scientific activities conducted within the framework of the Antarctic Treaty.

Article XVI

Scientific committee: first meeting, procedure, subsidiary bodies

1. The first meeting of the scientific committee shall be held within three months of the first meeting of the Commission. The Scientific Committee shall meet thereafter as often as may be necessary to fulfil its functions.
2. The Scientific Committee shall adopt and amend as necessary its rules of procedure. The rules and any amendments thereto shall be approved by the Commission. The rules shall include procedures for the presentation of minority reports.
3. The Scientific Committee may establish, with the approval of the Commission, such subsidiary bodies as are necessary for the performance of its functions.

Article XVII

Commission and scientific committee: secretariat

1. The Commission shall appoint an Executive Secretary to serve the Commission and Scientific Committee according to such procedures and on such terms and conditions as the Commission may determine. His term of office shall be for four years and he shall be eligible for re-appointment.
2. The Commission shall authorize such staff establishment for the Secretariat as may be necessary and the Executive Secretary shall appoint, direct and supervise such staff according to such rules and procedures and on such terms and conditions as the Commission may determine.
3. The Executive Secretary and Secretariat shall perform the functions entrusted to them by the Commission.

Article XVIII

Languages

The official languages of the Commission and of the Scientific Committee shall be English, French, Russian and Spanish.

Article XIX

Budget and financial obligations

1. At each annual meeting, the Commission shall adopt by consensus its budget and the budget of the Scientific Committee.
2. A draft budget for the Commission and the Scientific Committee and any subsidiary bodies shall be prepared by the Executive Secretary and submitted to the Members of the Commission at least sixty days before the annual meeting of the Commission.
3. Each Member of the Commission shall contribute to the budget. Until the expiration of five years after the entry into force of this Convention, the contribution of each Member of the Commission shall be equal. Thereafter the contribution shall be determined in accordance with two criteria: the amount harvested and an equal sharing among all Members of the Commission. The Commission shall determine by consensus the proportion in which these two criteria shall apply.
4. The financial activities of the Commission and Scientific Committee shall be conducted in accordance with financial regulations adopted by the Commission and shall be subject to an annual audit by external auditors selected by the Commission.
5. Each Member of the Commission shall meet its own expenses arising from attendance at meetings of the Commission and of the Scientific Committee.
6. A Member of the Commission that fails to pay its contributions for two consecutive years shall not, during the period of its default, have the right to participate in the taking of decisions in the Commission.

Article XX

Information: collection and provision

1. Members of the Commission shall, to the greatest extent possible, provide annually to the Commission and to the Scientific Committee such statistical, biological and other data and information as the Commission and Scientific Committee may require in the exercise of their functions.
2. The Members of the Commission shall provide, in the manner and at such intervals as may be prescribed, information about their harvesting activities, including fishing areas and vessels, so as to enable reliable catch and effort statistics to be compiled.

3. The Members of the Commission shall provide to the Commission at such intervals as may be prescribed information on steps taken to implement the conservation measures adopted by the Commission.
4. The Members of the Commission agree that in any of their harvesting activities, advantage shall be taken of opportunities to collect data needed to assess the impact of harvesting.

Article XXI

Domestic measures to ensure compliance

1. Each Contracting Party shall take appropriate measures within its competence to ensure compliance with the provisions of this Convention and with conservation measures adopted by the Commission to which the Party is bound in accordance with Article IX of this Convention.
2. Each Contracting Party shall transmit to the Commission information on measures taken pursuant to paragraph 1 above, including the imposition of sanctions for any violation.

Article XXII

Activities contrary to objective of Convention

1. Each Contracting Party undertakes to exert appropriate efforts, consistent with the Charter the United Nations, to the end that no one engages in any activity contrary to the objective of this Convention.
2. Each Contracting Party shall notify the Commission of any such activity which comes to its attention.

Article XXIII

Relations with other international organizations

1. The Commission and the Scientific Committee shall co-operate with the Antarctic Treaty Consultative Parties on matters falling within the competence of the latter.
2. The Commission and the Scientific Committee shall co-operate, as appropriate, with the Food and Agriculture Organisation of the United Nations and with other Specialised Agencies.

3. The Commission and the Scientific Committee shall seek to develop co-operative working relationships, as appropriate, with inter-governmental and non-governmental organizations which could contribute to their work, including the Scientific Committee on Antarctic Research, the Scientific Committee on Oceanic Research and the International Whaling Commission.

4. The Commission may enter into agreements with the organizations referred to in this Article and with other organizations as may be appropriate. The Commission and the Scientific Committee may invite such organizations to send observers to their meetings and to meetings of their subsidiary bodies.

Article XXIV

Observation and inspection

1. In order to promote the objective and ensure observance of the provisions of this Convention, the Contracting Parties agree that a system of observation and inspection shall be established.

2. The system of observation and inspection shall be elaborated by the Commission on the basis of the following principles:

   (a) Contracting Parties shall co-operate with each other to ensure the effective implementation system of observation and inspection, taking account of the existing international practice. This system shall include, *inter alia*, procedures for boarding and inspection by observers and inspectors designated by the Members of the Commission and procedures for flag state prosecution and sanctions on the basis of evidence resulting from such boarding and inspections. A report of such prosecutions and sanctions imposed shall be included in the information referred to in Article XXI of this Convention;

   (b) In order to verify compliance with measures adopted under this Convention, observation and inspection shall be carried out on board vessels engaged in scientific research or harvesting of marine living resources in the area to which this Convention applies, through observers and inspectors designated by the Members of the Commission and operating under terms and conditions to be established by the Commission;

   (c) designated observers and inspectors shall remain subject to the jurisdiction of the Contracting Party of which they are nationals.

They shall report to the Member of the Commission by which they have been designated which in turn shall report to the Commission.

3. Pending the establishment of the system of observation and inspection, the Members of the Commission shall seek to establish interim arrangements to designate observers and inspectors and such designated observers and inspectors shall be entitled to carry out inspections in accordance with the principles set out in paragraph 2. above.

Article XXV

Dispute settlement

1. If any dispute arises between two or more of the Contracting Parties concerning the interpretation or application of this Convention, those Contracting Parties shall consult among themselves with a view to having the dispute resolved by negotiation, inquiry, mediation, conciliation, arbitration, judicial settlement or other peaceful means of their own choice.

2. Any dispute of this character not so resolved shall, with the consent in each case of all Parties to the dispute, be referred for settlement to the International Court of Justice or to arbitration; but failure to reach agreement on reference to the International Court or to arbitration shall not absolve Parties to the dispute from the responsibility of continuing to seek to resolve it by any of the various peaceful means referred to in paragraph 1 above.

3. In cases where the dispute is referred to arbitration, the arbitral tribunal shall be constituted as provided in the Annex to this Convention.

Article XXVI

Signature

1. This Convention shall be open for signature at Canberra from 1 August to 31 December 1980 by the States participating in the Conference on the Conservation of Antarctic Marine Living Resources held at Canberra from 7 to 20 May 1980.

2. The States which so sign will be the original signatory States of the Convention.

Article XXVII

Ratification, acceptance or approval

1. This Convention is subject to ratification, acceptance or approval by signatory States.
2. Instruments of ratification, acceptance or approval shall be deposited with the Government of Australia, hereby designated as the Depositary.

Article XXVIII

Entry into force

1. This Convention shall enter into force on the thirtieth day following the date of deposit of the eighth instrument of ratification, acceptance or approval by States referred to in paragraph 1 of Article XXVI of this Convention.
2. With respect to each State or regional economic integration organization which subsequent to the date of entry into force of this Convention deposits an instrument of ratification, acceptance, approval or accession, the Convention shall enter into force on the thirtieth day following such deposit.

Article XXIX

Accession

1. This Convention shall be open for accession by any State interested in research or harvesting activities in relation to the marine living resources to which this Convention applies.
2. This Convention shall be open for accession by regional economic integration organizations constituted by sovereign States which include among their members one or more States Members of the Commission and to which the States Members of the organization have transferred, in whole or in part, competences with regard to the matters covered by this Convention. The accession of such regional economic integration organizations shall be the subject of consultations among Members of the Commission.

Article XXX

Amendment

1. This Convention may be amended at any time.
2. If one-third of the Members of the Commission request a meeting to discuss a proposed amendment the Depositary shall call such a meeting.
3. An amendment shall enter into force when the Depositary has received instruments of ratification, acceptance or approval thereof from all the Members of the Commission.
4. Such amendment shall thereafter enter into force as to any other Contracting Party when notice of ratification, acceptance or approval by it has been received by the Depositary. Any such Contracting Party from which no such notice has been received within a period of one year from the date of entry into force of the amendment in accordance with paragraph 3 above shall be deemed to have withdrawn from this Convention.

Article XXXI

Withdrawal

1. Any Contracting Party may withdraw from this Convention on 30 June of any year, by giving written notice not later than 1 January of the same year to the Depositary, which, upon receipt of such a notice, shall communicate it forthwith to the other Contracting Parties.
2. Any other Contracting Party may, within sixty days of the receipt of a copy of such a notice from the Depositary, give written notice of withdrawal to the Depositary in which case the Convention shall cease to be in force on 30 June of the same year with respect to the Contracting Party giving such notice.
3. Withdrawal from this Convention by any Member of the Commission shall not affect its financial obligations under this Convention.

Article XXXII

Function of depositary

The Depositary shall notify all Contracting Parties of the following:

(a) signatures of this Convention and the deposit of instruments of ratification, acceptance, approval or accession;
(b) the date of entry into force of this Convention and of any amendment thereto.

Article XXXIII

Texts

1. This Convention, of which the English, French, Russian and Spanish texts are equally authentic, shall be deposited with the Government of Australia which shall transmit duly certified copies thereof to all signatory and acceding Parties.
2. This Convention shall be registered by the Depositary pursuant to Article 102 of the Charter of the United Nations.

Drawn up at Canberra this twentieth day of May 1980.

IN WITNESS WHEREOF the undersigned, being duly authorized, have signed this Convention.

Annex for an arbitral tribunal

The arbitral tribunal referred to in paragraph 3 of Article XXV shall be composed of three arbitrators who shall be appointed as follows:

The Party commencing proceedings shall communicate the name of an arbitrator to the other Party which, in turn, within a period of forty days following such notification, shall communicate the name of the second arbitrator. The Parties shall, within a period of sixty days following the appointment of the second arbitrator, appoint the third arbitrator, who shall not be a national of either Party and shall not be of the same nationality as either of the first two arbitrators. The third arbitrator shall preside over the tribunal.

If the second arbitrator has not been appointed within the prescribed period, or if the Parties have not reached agreement within the prescribed period on the appointment of the third arbitrator, that arbitrator shall be appointed, at the request of either Party, by the Secretary-General of the Permanent Court of Arbitration, from among persons of international standing not having the nationality of a State which is a Party to this Convention.

The arbitral tribunal shall decide where its headquarters will be located and shall adopt its own rules of procedure.

The award of the arbitral tribunal shall be made by a majority of its members, who may not abstain from voting.

Any Contracting Party which is not a Party to the dispute may intervene in the proceedings with the consent of the arbitral tribunal.

The award of the arbitral tribunal shall be final and binding on all Parties to the dispute and on any Party which intervenes in the proceedings and shall be complied with without delay. The arbitral tribunal shall interpret the award at the request of one of the Parties to the dispute or of any intervening Party.

Unless the arbitral tribunal determines otherwise because of the particular circumstances of the case, the expenses of the tribunal, including the remuneration of its members, shall be borne by the Parties to the dispute in equal shares.

16.7 SCAR Code of Conduct for use of Animals for Scientific Purposes in Antarctica

Preamble

Recognizing that Man has a moral obligation to respect all animals and to have due consideration for their capacity for suffering and memory;

Accepting nevertheless that Man in his quest for knowledge has a need to use animals where there is a reasonable expectation that the result will provide a significant advance in knowledge or be of overall benefit for animals;

Resolved to limit the use of animals for experimental and other scientific purposes, with the aim of replacing such use wherever practical, in particular by seeking alternative measures and encouraging the use of these alternative measures;

Desiring to adopt common provisions in order to protect animals used in those procedures which may possibly cause pain, suffering, distress or lasting harm and to ensure that where unavoidable they shall be kept to a minimum;

SCAR has adopted a Code of Conduct which is based on the international guiding principles for biomedical research involving animals as developed by the Council for International Organization of Medical Sciences.

Code of conduct

1. The advancement of biological knowledge and the development of improved means to the protection of the health and well-being both of man and of the animals require recourse to experimentation on intact live mammals and birds of a wide variety of species.
2. Methods such as mathematical models, computer simulation and *in vitro* biological systems should be used wherever appropriate.

3. Animal experiments should be undertaken only after due consideration of their relevance for human or animal health and the advancement of biological knowledge.

4. The animals selected for an experiment should be of an appropriate species and quality, and the minimum number required to obtain scientifically valid results.

5. Investigators and other personnel should never fail to treat animals as sentient, and should regard their proper care and use and the avoidance or minimization of discomfort, distress, or pain as ethical imperatives.

6. Investigators should assume that procedures that would cause pain in human beings cause pain in other mammals and in birds.

7. Procedures with animals that may cause more than momentary or minimal pain or distress should be performed with appropriate sedation, analgesia, or anaesthesia in accordance with accepted veterinary practice. Surgical or other painful procedures should not be performed on unanaesthetized animals paralyzed by chemical agents.

8. Where waivers are required in relation to the provisions of Article 7, the decisions should not rest solely with the investigators directly concerned but should be made, with due regard to the provisions of articles 4, 5 and 6, by a suitably constituted review body. Such waivers should not be made solely for the purposes of teaching or demonstration.

9. At the end, or, when appropriate, during an experiment animals that would otherwise suffer severe or chronic pain, distress, discomfort, or disablement that cannot be relieved should be painlessly killed.

10. The best possible living conditions and supervision should be maintained for animals kept for biomedical purposes.

11. It is the responsibility of the director of an institute or department using animals to ensure that investigators and personnel have appropriate qualifications or experience for conducting procedures on animals. Adequate opportunities shall be provided for inservice training, including the proper and humane concern for the animals under their care.

Index